Collins

KT-152-278

A2 Biology
for AQA

FOR
REFERENCE ONLY

Keith Hirst & Mary Jones

Series Editor: Lesley Higginbottom

NORWICH CITY COLLEGE

Stock No.	238254	
Class	570 HIR	3wk
Cat.	SSAD	Proc	NATL

238 254

William Collins's dream of knowledge for all began with the publication of his first book in 1819. A self-educated mill worker, he not only enriched millions of lives, but also founded a flourishing publishing house. Today, staying true to this spirit, Collins books are packed with inspiration, innovation and practical expertise. They place you at the centre of a world of possibility and give you exactly what you need to explore it.

Collins. Freedom to teach.

Published by Collins
An imprint of HarperCollinsPublishers
77-85 Fulham Palace Road
Hammersmith
London
W6 8JB

Browse the complete Collins catalogue at
www.collinseducation.com

© HarperCollinsPublishers Limited 2008

10 9 8 7 6 5 4 3 2 1

ISBN-13 978-0-00-726822-1

Keith Hirst and Mary Jones assert their moral rights to be identified as the authors of this work.

All rights reserved. No part of this publication may be reproduced, stored in a retrieval system or transmitted in any form or by any means – electronic, mechanical, photocopying, recording or otherwise – without the prior written consent of the Publisher or a licence permitting restricted copying in the United Kingdom issued by the Copyright Licensing Agency Ltd, 90 Tottenham Court Road, London W1T 4LP.

British Library Cataloguing in Publication Data. A Catalogue record for this publication is available from the British Library.

Commissioned by Penny Fowler
Project management by Laura Deacon
Edited by Rosie Parrish and Mitch Fitton
Proof read by Helen Barham and Gudrun Kaiser
Indexing by Laurence Errington
Design by Bookcraft Ltd
Cover design by Angela English
Production by Leonie Kellman
Printed and bound in Hong Kong by Printing Express

Mixed Sources
Product group from well-managed forests and other controlled sources
www.fsc.org Cert no. SW-COC-1806
© 1996 Forest Stewardship Council

FSC is a non-profit international organisation established to promote the responsible management of the world's forests. Products carrying the FSC label are independently certified to assure consumers that they come from forests that are managed to meet the social, economic and ecological needs of present and future generations.

Find out more about HarperCollins and the environment at
www.harpercollins.co.uk/green

Contents

Acknowledgments

Text and diagrams reproduced by kind permission of:

Biological Sciences Review, British Agricultural Association, Cambridge University Press, *Conservation Review*, Granada, Macmillan magazines, Mosby, *NCBA newsletter*, *New Scientist*, Pearson, *RSPB Conservation Review*, *Science*, Simon and Schuster, Wiltshire Trust for Nature Conservation, www.bourn-hall-clinic.co.uk

Every effort has been made to contact the holders of copyright material, but if any have been inadvertently overlooked the publishers will be pleased to make the necessary arrangements at the first opportunity.

The publishers would like to thank the following for permission to reproduce photographs
(T = Top, B = Bottom, C = Centre, L = Left, R= Right):

Alamy 68L, 68R, 80, 114B
Ardea/Weisser 9
Biophoto Associates 246B
bone clones 140BL, 140TR
British Columbia Ministry of Forests 100B
Bruce Colman Collection 246T
Cardiff.ac.uk 240
Corbis 252, 253
Cygnus Inc. 222
Dartmoor National Park Authority 116R
DEFRA 100T
Dreamstime/David Davis 206
Environmental Picture Library 260, 285
Field Studies Council 16BL, 16TR
Flickr Creative Commons aeu04117 158R; Allan Harris 105; Anirudh Koul 141; Charles-Edouard Coste 104; Chronic urticaria 207T; David Cartier 23BL; jez̀ 226T; jonboy mitchell 122C; jurvetson 234; Karmalize 230L; Lawrie Phipps 6TL; M Poudyal 94TL; Mark Robinson 102L; Marlene 23BR; Miss Baker 87; Nicholas_T 112L; pfly 40; r neches 158L; Smudge 9000 194; SuperFantastic 145; takomabibelot 95; we must reinvent love 20

Frank Lane 8TR, 232R; P Moulu 231T; R Wilmshurst 286;
GGS-greenhouse 88L
Heather Angel 8BL
Holt Studios International/N Cattlin 250
John Walmsley 11
Justin McClellan 118
Markku Savela 102BR
NASA 171
NHPA D Woodfall 14; Ernie James 187; R Erwin 232L
OSF 159
Picasa/Harrayhen 102TR
Portsdown Hill 120R
PPL Therapeutics 272
Professor Don Grierson 266
Res-Q-Air 239
Roslin Institute Edinburgh 249
SPL 85; 117B; 124; 125T; 133L; 161; 165; 174; 182; 195; 196; 224; 229; 230R; 264; 274; 277; 279
Turly at fnarr.net 158C
UKTV Dave 156
USA Photo Library/Bruce Coleman 71
Wikimedia Commons 6BL; 21; 24; 25; 38; 55; 56; 66; 81; 88BL; 88TR; 92; 94CL; 112R; 114T; 115; 116TL; 117T; 120L; 122B; 125B; 130C; 130T; 135; 140L; 146; 148; 154; 172L; 207B; 226B; 231B; 233
Wordpress 133R

Cover photograph © Steve Dibblee/istockphoto.com

This book contains references to fictitious characters in fictitious case studies. For educational purposes only, photographs have been used to accompany these case studies. The juxtaposition of photographs and case studies is not intended to identify the individual in the photograph with the character in the case study. The publishers cannot accept any responsibility for any consequences resulting from this use of photographs and case studies except as expressly provided by law.

To the student

This book aims to make your study of advanced science successful and interesting. Science is constantly evolving and, wherever possible, modern issues and problems have been used to make your study stimulating and to encourage you to continue studying science after you complete your current course.

Using the book

Don't try to achieve too much in one reading session. Science is complex and some demanding ideas need to be supported with a lot of facts. Trying to take in too much at one time can make you lose sight of the most important ideas – all you see is a mass of information.

Each chapter starts by showing how the science you will learn is applied somewhere in the world. At other points in the chapter you will find the *How Science Works* boxes. These will help you to pose scientific questions and analyse, interpret and evaluate evidence and data. Using these boxes and the *How Science Works* assignments at the end of each chapter will help you to tackle questions in your final examination with a *How Science Works* element.

The numbered questions in the main text allow you to check that you have understood what is being explained. These are all short and straightforward in style – there are no trick questions. Don't be tempted to pass over these questions, they will give you new insights into the work. Answers are given in the back of the book.

Stretch and Challenge really test your knowledge of biology allowing you to go beyond the specification and achieve the maximum grade.

This book covers the content needed for AQA Biology at A2-level. The Key Facts for each section summarise the information you will need in your examination. However, the examination will test your ability to apply these facts rather than simply to remember them. The *How Science Works* boxes will help you to develop this skill.

Words written in bold type appear in the glossary at the end of the book. If you don't know the meaning of one of these words, check it out immediately – don't persevere, hoping all will become clear.

Past paper questions are included at the end of each chapter. These will help you to test yourself against the sorts of questions that will come up in your examination. You can find the answers to these questions on the website **www.collinseducation.com/advancedscienceaqa.**

The website also provides sample student answers to these questions – a stronger and a weaker answer for each question – to help you to improve your own answers. On this website you will also find mathematical and examination technique guidance to help you to prepare for your examinations, and PSA (Practical Skills Assessment) and ISA (Investigative Skills Assignment) guidance to help you achieve your best in your practical work.

1 Populations

Badgers and hedgehogs are two of Britain's most loved animals. Until recently, most people would have been familiar with hedgehogs – they have been regular visitors and residents in many gardens, where they are usually welcomed because of their enjoyment of slugs and snails. Badgers are less likely to be seen, generally being more nervous of people than hedgehogs, and tending to live outside urban areas and well away from villages.

But recently, things have been changing. The sight of a hedgehog in a garden – either in the town or in the country – is becoming increasingly rare. Even the number of hedgehogs killed on the roads has decreased, probably a sign that there are fewer of them around rather than that they have developed better road sense. Although there has not been a definitive count of the population of hedgehogs in Britain, some researchers think that it has decreased from about two million in the mid 1990s to less than half of that now.

Over a similar time period, the population of badgers has been steadily increasing. At least part of the reason for this is the increased protection for badgers that has been provided by the Badgers Act of 1973 and the Protection of Badgers Act of 1992. It is now illegal to disturb badgers, let alone kill them.

Badgers are predators of hedgehogs – indeed, probably the only significant one, as no other animal is prepared to deal with the prickles. Badgers simply tip over the hedgehog, force it to uncurl using their hugely strong forepaws and claws, and scoop out the hedgehog from its prickly covering. People have frequently reported finding the remains of scooped-out hedgehogs in their gardens shortly after badgers have appeared in the area. A research paper into the relationship between hedgehog populations and badger populations, published in 2006, showed that hedgehogs were far less common where badgers were abundant (for example, in pasture fields in rural habitats) than where badgers were rare. There appears to be a strong negative spatial relationship between badgers and hedgehogs. Lots of badgers mean few hedgehogs.

So, do we have to make a choice? Save the hedgehogs or save the badgers? The situation is probably more complex than it seems. Many people do not accept that badgers are the main cause of hedgehog decline. For example, they cite the warmer winters we have experienced recently, which make it difficult for hedgehogs to hibernate in their normal way. Some people have strong protective feelings about furry animals such as badgers, and cannot accept that perhaps we do need to control their numbers to maintain some kind of balance between different species. We need to find out more about what is affecting hedgehog numbers. But if we are going to act to save them, we need to do it quickly. Some ecologists are predicting that, if things continue as they are, there will be no hedgehogs in Britain by 2025.

A badger.

A hedgehog.

1.1 Populations and ecosystems

In this chapter, we will be looking at some of the many factors that influence population size. This is of importance to us for numerous reasons – not only so that we can understand what is happening to species that share our world with us, but also in order to take control of our own population, which expanded hugely during the 20th century and continues to do so (although at a decreasing rate) today.

There are several terms which we will be using throughout this chapter, and it is important that you know exactly what they mean, especially as many of them have other, broader meanings in everyday language.

Ecosystem – a group of interrelated organisms and their physical environment in a particular area, such as a pond, desert or forest. The living and non-living components of an ecosystem interact with each other.

Habitat – the part of an ecosystem in which particular organisms live. The mud on the bottom of a pond and the surface layer of water are both habitats.

Population – all the organisms of one species living in a particular habitat at the same time, such as all the pond snails or water lilies in a pond. Sometimes the term is used to mean all the members of a species in a much larger area, such as the human population of the world or the rabbit population of Australia.

Community – all the populations of all the species that live together in a particular ecosystem or habitat. The pond community comprises all the plants, snails, insects, frogs and fish living there, as well as the microscopic algae, bacteria and so on.

Abiotic factors – the non-living, physical conditions in an ecosystem, such as temperature, light, soil conditions (sometimes called **edaphic factors**) and pH.

Biotic factors – the effects of the activities of living organisms on other organisms. Food availability, predation and competition are examples of biotic factors.

1 Which term correctly describes each of the following?

a all the oak trees in a wood

b the surface of the bark of an oak tree

c the Siberian tundra

d the salt concentration of a rock pool

e the organisms living in a rock pool

f the effect of snails grazing on pondweed

The pond community shown in Fig. 1 contains many different species of plants and animals, many more than can be shown in the illustration. Each species is a specialist in some way. There are examples of every trophic (feeding) level in the food chain, some producers, some primary consumers, and so on.

The total size and richness of the community depends on the resources available to the producers. A pond that has few mineral nutrients dissolved in the water will support a limited amount of plant life, and hence relatively few consumers.

Different species are not evenly distributed throughout the pond. The greater the variety in the physical environment, such as shallow

Fig. 1   A pond ecosystem

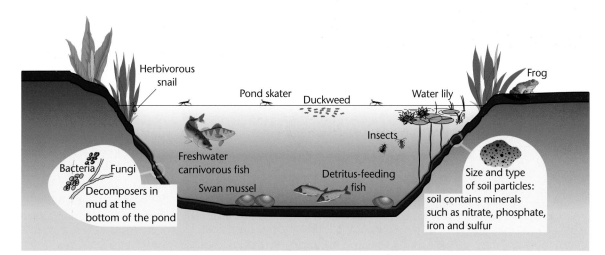

edges, deep water, shaded banks and so on, the greater the diversity is likely to be. Those plants that must have their roots in the soil can grow only near the margins and in shallow water. Others, such as the duckweed, float and can therefore extend across the surface of the pond. Swan mussels lie on the bottom and filter food particles from the water that they suck through their bodies, whereas pond skaters are fast-moving predators that can literally walk and run on water.

The concept of the ecological niche

In the community, each species is adapted to a set of conditions, and it is only successful where the abiotic and biotic factors in an ecosystem are suited to its way of life.

The way of life of a particular species in a habitat is called its **ecological niche**. The organism can only live where the abiotic and biotic factors fulfil its requirements. As an example, we can look at the ecological niche of the swan mussel, which lives predominantly on the bed of shallow ponds.

The body structure of the swan mussel obviously rules out a similar lifestyle to the pond skater; it could not easily chase insects across the surface of the water. Instead, it uses its siphon and cilia to draw a current of water through its body and filter out microscopic organisms for food. This may seem very lazy, but is an energy-efficient way of obtaining food, as long as the pond water contains a good supply of microscopic food particles. This is more likely in shallow ponds than in deep

A freshwater swan mussel; the frilly edge of the siphons can be seen projecting between the valves (shell) on the right of the photograph.

ponds, since microscopic plant plankton need light. Very deep water could also contain less oxygen and more silt, which could clog the delicate filter system of the mussel. All these factors contribute to the specific requirements of the mussel and determine its niche in the pond.

Niches within a habitat

In a particular habitat every species has its own niche. The niche is governed by the adaptation of the species to both biotic and abiotic factors.

Organisms of different species, with different niches, are able to coexist in a community because they do not compete for exactly the same resources. Even species that live in the same habitat and that appear to be very similar often have different ways of life. For example, several snail species live in pond habitats, but different species are part of different food chains, and the smaller and lighter ones can live on the more delicate plants shunned by their heavier neighbours. Similarly, different species of the hummingbird have very slightly different-shaped beaks that enable them to feed from one type of flower but not another. If they had identical niches, then competition would be likely to result in all but one of the species becoming extinct in that habitat. We will look at competition in more detail later in this chapter.

A pond skater, *Gerris lacustris*, on the surface of water.

A Rufous hummingbird drinks nectar from the scarlet gilla, pollinating it at the same time.

marsupial species because of their more efficient method of reproduction. Similarly, when the grey squirrel was introduced to Britain from North America, it was able to exploit woodland food sources such as nuts and acorns more effectively than the native red squirrel and has largely supplanted it.

Competing for the same niche

When two species occupy exactly the same niche, one usually out-competes the other, perhaps by being better able to use the food supply, or being better at protecting itself from predators. In the end the less successful competitor disappears from the habitat.

Interestingly, similar niches in different ecosystems in other parts of the world may be occupied by different species. Australia used to have numerous marsupial species that grazed on the vegetation, but when rabbits were introduced they out-competed some of the

2 The European mole is adapted to burrow through soil and feed on organisms such as earthworms. It has powerful clawed front legs, a pointed snout and eyes buried in its fur. In Australia there is a marsupial mole which occupies a very similar niche and which has similar behaviour and adaptations. However, the two animals have totally different ancestors.

Use your knowledge of evolution to explain how both the common and marsupial moles could have evolved from quite different ancestors to have similar features, such as powerful clawed front legs and a pointed snout. (You can answer this question using what you learned about natural selection in your GCSE and AS course, but if you want to look ahead, there is more about evolution in Chapter 8.)

key facts

● A community is a group of organisms of different species, all living in the same place at the same time. A population is all the members of one species, living together at the same time. The community and its non-living environment make up an ecosystem, in which the living and non-living components interact.

● Each species has its own ecological niche – its role in the community. The niche is determined by the adaptations that the species has evolved, so that it is able to be successful in a particular range of abiotic and biotic conditions.

● Abiotic factors are those resulting from non-living parts of an ecosystem. Biotic factors result from living organisms.

● No two species can occupy exactly the same niche in a community, although niches do often overlap. However, in different ecosystems there may be different species that have evolved similar adaptations that make them successful in similar niches to one another.

1.2 Sampling populations

When an ecologist begins to study an ecosystem, there are almost always three questions that need to be answered. These are:

- What organisms live there?
- Where do these organisms live?
- How many of them are there?

In order to answer these questions, numerical data must be collected from the ecosystem. The data that you collect, and the way in which you collect them, will be determined by the precise questions that you want to answer.

You will probably already have studied an ecosystem yourself, and will have attempted to answer some or all of the questions above. If you had an infinite amount of time, and an infinite number of willing assistants to help you collect the data, it might just be possible – in a very small ecosystem – to count and record every individual that lived there. In reality, however, this is simply not possible. Instead, we use various **sampling techniques**, attempting to do so in such a way that the results we obtain are genuinely representative of the habitat.

Whichever of the following techniques you are able to use in your own fieldwork, you must never forget that your studies do not take precedence over the lives of the animals and plants you are investigating. Take great care not to disturb the habitat any more than necessary. For example, if you are working on a seashore, turning over rocks to see what lives beneath them, always replace the rock in the same position and orientation in which you found it. If you are working in a meadow, avoid trampling all over the plants – tread carefully, and retrace your steps rather than making new paths in all directions. If you are trapping beetles, check traps regularly and release your captives as soon as you can.

Random and systematic sampling

There are many ways of collecting information about the organisms living in a particular habitat, some of which are described below. But whatever sampling technique you decide on, you will need to decide whether to use random sampling or systematic sampling.

Random sampling means that you do not make a conscious decision about exactly where the samples are taken. Random sampling ensures that each part of, say, a meadow has an equal chance of being sampled. This is often the best thing to do where the distribution of species within the area you are interested in is fairly uniform. For example, you may want to know what species of plants are present in a meadow and the relative areas of ground covered by each one. If it looks as though the meadow has fairly similar vegetation growing all over it, then a random sample should give you data that are representative of the whole meadow. This would also be the best thing to do if there were obvious patches of different vegetation – perhaps clumps of nettles – that were distributed randomly throughout the meadow.

One method of obtaining a random sample is to use a set of random numbers, either taken from a book or generated by a computer, to tell you where to put the quadrat. The numbers are used as coordinates on a pair of imaginary graph axes along two edges of your sampling area, as shown in Fig. 2a. You place your quadrat at the intersection of these coordinates.

Systematic sampling means that you decide where to take your samples, and take them at regular spatial intervals within the area you are interested in. You might decide, for example, to make a map of the meadow and to draw a set of grid lines on it, intersecting each other at right angles at regular intervals, as shown in Fig. 2b. You then count and record what is growing at each intersection. This could be useful if you were planning to come back and sample the same area again to find out if any changes had taken place, as you would then be sure that your second sample would be taken in the same places as the first one.

Another reason for deciding on systematic rather than random sampling is if you can see a gradation of some kind in the habitat, and you want to know more about it. For example, it may be apparent that the species growing in the meadow gradually change as you move from a dry part into a wetter part, or from the meadow into the edge of adjoining woodland. In these

Fig. 2 Random and systematic sampling grids

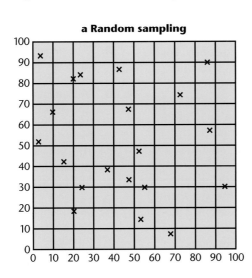

a Random sampling

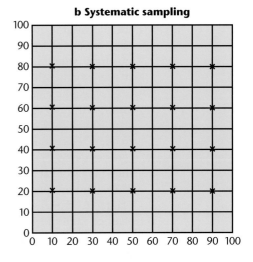

b Systematic sampling

Using a wire quadrat to collect data.

instances, you might choose to sample along a line that runs through these changing areas, allowing you to collect data about any changes in the species present, and their numbers, as the habitat changes. Such a line is called a **transect**, and is described more fully on pages 13–15.

Frame quadrats

A **quadrat** is a defined area within which you collect data. The quadrat is placed on the ground, and you then identify and record the organisms inside it. Quadrats are the usual way of collecting data about which plants are growing in a habitat. They can also be used for **sessile** animals – that is, animals that are immobile for long periods of time, such as limpets and sea anemones. You have probably used some type of **frame quadrat**, a square frame that you place on the ground, then identify and count the organisms inside it (Fig. 3).

Fig. 3 Frame quadrats

Quadrats can be made from wood, wire or plastic.

50 cm

50 cm

A quadrat with sides of 0.5 m has an area of 0.25 m². Wire fixed at 10 cm intervals gives 25 smaller units, each of 0.01 m², to make counting easier.

Larger quadrats, for example, for sampling in woodland, can be laid out with string and pegs.

All the species within a quadrat can be counted, or the abundance of each species estimated.

Exactly what you record in your quadrats, and how you record it, depends both on the kind of organisms that are there, and what you intend to do with the data. A very simple, yet useful, count is of the **species frequency**. For this, you simply record in how many of your quadrats you found a particular species, no matter whether there was just one specimen or 20. For example, if you placed your quadrat on a lawn 50 times and found one or more dandelions in 12 of these quadrats, then the frequency of dandelions is $12/50 \times 100 = 24\%$.

Another set of data to record is the number of individuals of each species inside the quadrat. This could be appropriate if, for example, you wanted to know what animals were living on a seashore and the sizes of their populations.

It can sometimes be possible to record data about plants in this way, for example, on an area of disturbed soil in a garden that has weeds growing on it. However, in many cases it is simply not possible to tell where one individual plant ends and another starts. In this instance, it is better to estimate the percentage of the area inside the quadrat that is occupied by each species. This is known as **percentage cover**. To help you to judge this, it is useful to divide a quadrat into several smaller ones (Fig. 4).

Fig. 4   Percentage cover

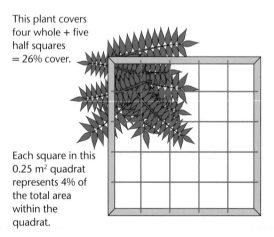

This plant covers four whole + five half squares = 26% cover.

Each square in this 0.25 m² quadrat represents 4% of the total area within the quadrat.

To measure percentage cover:

- Lay a frame quadrat over the selected area.

- Count the number of whole and half squares occupied by the species.

- Calculate this as a percentage of the whole area of the quadrat.

- Repeat for all the other species in the quadrat.

3

a   The dimensions of a quadrat are 25 cm × 25 cm. Calculate its area in m².

b   A student uses this quadrat to survey the plants in 20 positions in a habitat. What is the total area surveyed?

c   The total area of the habitat being studied is 220 m². What percentage of the habitat has the student surveyed?

4   A student is studying the distribution of clover in a field, using a 50 × 50 cm quadrat.

a   He placed the quadrat 30 times, and found clover growing in 25 quadrats. What is the frequency of clover in this habitat? Round your answer to the nearest whole number.

b   Fig. 5 represents one quadrat, showing the parts where clover is growing. Estimate the percentage cover of clover in this quadrat.

Fig. 5   Estimating the percentage cover of clover in a quadrat

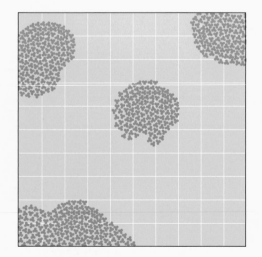

c   The student notices that the grass growing near the patches of clover is a darker green than in other areas. Clover contains nitrogen-fixing bacteria in its roots. Suggest a reason for the student's observation.

A third method is to use some kind of **abundance scale**. The ACFOR scale is a frequently used example in which you record each species as being abundant, common, frequent, occasional or rare. You can make an abundance scale semi-quantitative by relating it to percentage cover, for example, by making A equal to 80%–100% cover, C equal to 60%–80% cover and so on down the scale.

> **5** Suggest the advantages and disadvantages of using the ACFOR scale, rather than estimating percentage cover.

Point quadrats

Estimating percentage cover, even in a quadrat divided into many smaller ones, is not easy to do accurately; and the larger the quadrat, the more difficult it gets. One solution is to make the quadrat so small that its area is a single point. Such a quadrat is known as a **point quadrat**. Fig. 6 shows a frame that can be used for sampling with point quadrats.

Fig. 6   Point quadrats

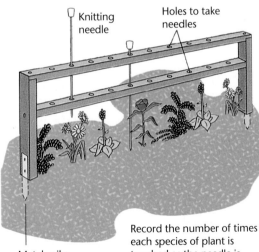

Knitting needle

Holes to take needles

Metal spike pushed into ground

Record the number of times each species of plant is touched as the needle is lowered to the ground through the hole.

Point quadrats are an excellent way of determining the percentage cover of all the different plant species in an area of relatively short vegetation. You can use random numbers to determine where to place the point quadrat frame. Then you drop the first needle through

its hole, and count what it touches on its way to the ground – that is, the species that are present in the tiny quadrat represented by the end of the needle. If the vegetation is quite thick, with a mix of tall and short plants, the point may touch more than one species. You repeat this with all the other needles in the frame, and then repeat the whole exercise over and over again in the habitat. You can then work out the percentage of times you scored a 'hit' for any particular species, and this gives you the percentage cover of that species in that habitat. As leaves can lie above one another and you can hit more than one plant with each point, the percentage cover will probably add up to more than 100%.

Point quadrats can be a much more objective and repeatable way of collecting data from which to calculate percentage cover than frame quadrats, especially in relatively short vegetation. However, they are not so useful where the vegetation is long and thick, because the needle may touch so many leaves on its way to the ground that you have difficulty in determining what it did hit and what it did not.

Line and belt transects

If you have decided to use a transect to collect data, then you need to consider whether it is better to use a **line transect** or a **belt transect** (Fig. 7). In either case, you begin by

Fig. 7   Line and belt transects

a Line transect
For a line transect, record the species of plant touching the tape at regular intervals along it.

b Belt transect
For a belt transect, place quadrats all along the tape (belt transect) or at regular intervals along it (interrupted belt transect) and record the percentage cover or abundance within each quadrat.

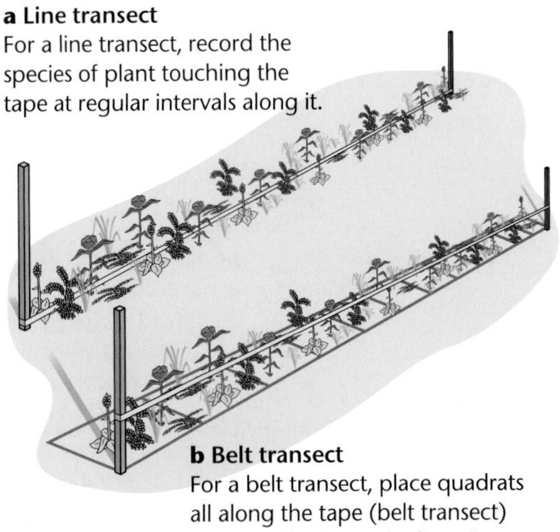

running a tape or piece of string along the line you are going to sample. If you are using a line transect, you then count and record individual organisms that are touching the tape. If the vegetation is very sparse, you could record every single individual. More usually, however, this is impractical, and you will need to choose a suitable interval along the tape – say, every 10 cm – at which to do this.

For a belt transect, you place frame quadrats so that one edge lies against the tape, and record the organisms within each quadrat. Once again, in some cases you might decide to record all along the tape, not missing out anything. Alternatively, if this would be too time-consuming, you can place the quadrats at equal intervals along the tape, perhaps every two metres. This is known as **an interrupted belt transect**.

6 Different species of seaweed live at different heights above the low tide line. Use Fig. 8 to find the range of heights at which the different seaweeds live.

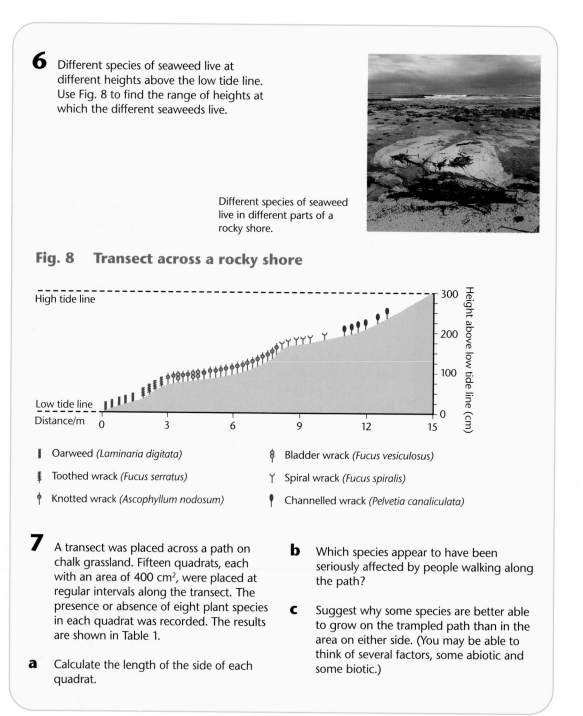

Different species of seaweed live in different parts of a rocky shore.

Fig. 8   Transect across a rocky shore

Oarweed (*Laminaria digitata*)	Bladder wrack (*Fucus vesiculosus*)
Toothed wrack (*Fucus serratus*)	Spiral wrack (*Fucus spiralis*)
Knotted wrack (*Ascophyllum nodosum*)	Channelled wrack (*Pelvetia canaliculata*)

7 A transect was placed across a path on chalk grassland. Fifteen quadrats, each with an area of 400 cm², were placed at regular intervals along the transect. The presence or absence of eight plant species in each quadrat was recorded. The results are shown in Table 1.

a Calculate the length of the side of each quadrat.

b Which species appear to have been seriously affected by people walking along the path?

c Suggest why some species are better able to grow on the trampled path than in the area on either side. (You may be able to think of several factors, some abiotic and some biotic.)

Table 1   Transect results

Species	Rough grass					Trampled path					Rough grass				
	Quadrat														
	1	2	3	4	5	6	7	8	9	10	11	12	13	14	15
Daisy					×	×		×	×	×		×			
Hoary plantain						×	×	×	×	×	×				
Ribwort plantain	×	×	×	×	×							×		×	×
Rock rose		×	×	×										×	×
Salad burnet	×	×	×	×	×			×		×	×	×		×	×
Meadow grass						×	×	×	×	×	×				
Sheep's fescue grass		×	×	×	×	×	×	×	×	×	×	×	×	×	×
Tor grass	×	×	×	×	×							×	×	×	×

× indicates that the species was present.

Judging the results of quadrat studies

Interpreting means

After using quadrats to survey an area, you may want to find the mean numbers of different species. For example, if you are investigating the earthworm populations in two different areas of grassland, you might extract the worms from the soil beneath 10 quadrats in each area, and then calculate the mean number of worms per quadrat. Suppose that the numbers of worms in the 10 quadrats in one area were: 12, 14, 11, 15, 13, 12, 16, 11, 14 and 12. The mean number is 13, and you could be reasonably sure that if you count a lot more quadrats in the same area the mean would be roughly similar.

Suppose that in the other area the results were: 2, 1, 0, 112, 5, 2, 3, 0, 4 and 1. The mean is again 13. If you count another 10 quadrats in this area it is quite unlikely that you would get the same again. To suggest that the populations in the two areas were similar would obviously be silly. It is much more likely that there happened to be something odd about the quadrat with 112 worms – maybe it was the site of a large recently deposited cowpat.

You therefore need to be careful when comparing means, but the differences will not always be so obvious as in this case. More commonly, the individual pieces of data that are collected will be fairly evenly distributed about the mean, as shown in the graphs in Fig. 9.

Fig. 9   Normal distributions with different ranges

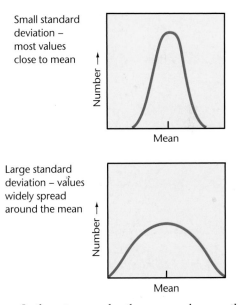

Small standard deviation – most values close to mean

Large standard deviation – values widely spread around the mean

In these two graphs, the mean values are the same, but one has a much greater **range** than the other. As well as calculating a mean, statisticians use a formula to work out how big a range the data has. This formula gives the **standard deviation**. The larger the standard deviation, the larger the spread of the data. If the standard deviations of two sets of data are large, comparing them is likely to be unreliable because the values may overlap. Chapter 8 of your AS book explains in more detail how standard deviations are calculated and used.

Netting

Quadrats are a very good way of sampling anything that stays in one place. They can be used for collecting data about plants and sessile animals. Many seashore animals are sessile, and quadrats are an excellent way of investigating the distribution and abundance of seaweeds and animals on a rocky shore. However, the shore is a relatively unusual situation, and in most cases you will need to find different methods for collecting data about animals. In a meadow, for example, you would be most unlikely to find a short-tailed field vole sitting in the middle of your quadrat, even though these animals might be very common.

In grassland, you can collect samples of insects and other invertebrates using **sweep netting**. A standard technique should be used, always using the same size of net and moving it through the vegetation in the same way.

Nets are also useful for sampling aquatic organisms. In a stream or river, you can use a technique called **kick sampling**. You hold the net with its mouth facing into the water flow, as you stand upstream of it and disturb the mud or stones on the bottom with one foot. Again, you should use a standard technique in each area that you sample, moving your foot in the same way and for the same length of time. In still water, you could use a technique similar to sweep netting, pulling your net through the water in a systematic way.

Using the kick sampling technique to sample aquatic organisms.

Trapping

Nets do not catch everything that lives in a habitat. Even sweep netting is unlikely to catch a vole in a meadow. In terrestrial habitats, some kind of trap is often used to collect larger and more mobile animals. Four different kinds of traps are shown in Fig. 10.

Longworth traps are used to catch small mammals such as mice and voles. The animal is not harmed by the trap, but it is very important to check traps at least once a day, so that any trapped animals can be released as soon as possible. Individual animals that have been trapped are usually marked in some way, often by clipping away a small area of hair. In this way, you can tell whether a particular animal has been caught before.

A **pitfall trap** is simply a container that is sunk into the soil, into which invertebrates fall as they move over the ground. Slow-moving invertebrates can be caught in a **cover trap**, and some flying insects can be trapped in a **water trap**. Night-flying moths are attracted to a bright light such as a mercury vapour lamp, which can be arranged so that the moths fall into a container from which they can be retrieved the next morning.

Using sweep netting to collect samples of insects and other invertebrates.

Fig. 10   Traps

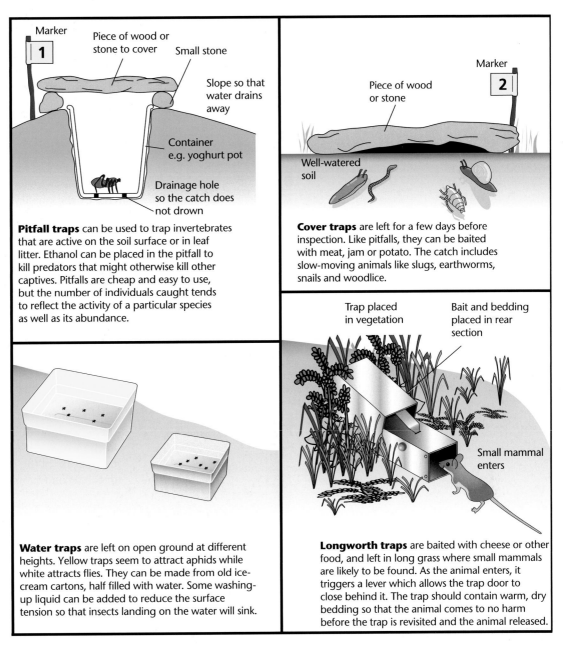

Marker 1

Piece of wood or stone to cover

Small stone

Slope so that water drains away

Container e.g. yoghurt pot

Drainage hole so the catch does not drown

Pitfall traps can be used to trap invertebrates that are active on the soil surface or in leaf litter. Ethanol can be placed in the pitfall to kill predators that might otherwise kill other captives. Pitfalls are cheap and easy to use, but the number of individuals caught tends to reflect the activity of a particular species as well as its abundance.

Piece of wood or stone

Marker 2

Well-watered soil

Cover traps are left for a few days before inspection. Like pitfalls, they can be baited with meat, jam or potato. The catch includes slow-moving animals like slugs, earthworms, snails and woodlice.

Water traps are left on open ground at different heights. Yellow traps seem to attract aphids while white attracts flies. They can be made from old ice-cream cartons, half filled with water. Some washing-up liquid can be added to reduce the surface tension so that insects landing on the water will sink.

Trap placed in vegetation

Bait and bedding placed in rear section

Small mammal enters

Longworth traps are baited with cheese or other food, and left in long grass where small mammals are likely to be found. As the animal enters, it triggers a lever which allows the trap door to close behind it. The trap should contain warm, dry bedding so that the animal comes to no harm before the trap is revisited and the animal released.

Source: Adapted from Wiltshire Trust for Nature Conservation

Data from traps are not usually reliable enough to determine accurately how many different species of animals live in a habitat, or to make anything better than a rough estimate of the sizes of their populations. For example, some mammals may be much more wary of Longworth traps than others. Some individuals become 'trap-happy', returning night after night to find the food that they have learnt is inside a trap. Another potential problem is that, if you set a pitfall trap in the evening and collect its contents next morning, you may find one rather fat predatory beetle, and nothing else. Short of doing an analysis of the beetle's stomach contents, you will never know what else had fallen into the trap. However, you can put ethanol into the bottom of the trap, so that everything that falls in is killed immediately.

The mark–release–recapture technique

The **mark–release–recapture** technique is used for estimating the size of a population of mobile animals (Fig. 11).

Fig. 11 The mark–release–recapture technique

Catch a large sample of the animals you want to study and count them.

⬇

Mark each animal in a way that will not harm it or attract predators.

⬇

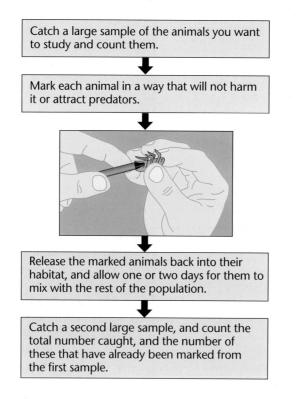

⬇

Release the marked animals back into their habitat, and allow one or two days for them to mix with the rest of the population.

⬇

Catch a second large sample, and count the total number caught, and the number of these that have already been marked from the first sample.

First, a large number of the animals are caught. The method you use for this depends on the species you are investigating. For aquatic insects such as water boatmen, you could use nets. For small mammals in a meadow, you could use Longworth traps. For woodlice, you could just search under stones and pieces of decaying wood.

The animals that you have caught are counted and then marked. The method of marking also depends on the species you are working with. Water boatmen or woodlice could be marked with a spot of red paint; small mammals by clipping their fur. Whatever method you choose, you need to try to ensure that it will not increase the likelihood of the marked animals being eaten by predators.

The marked animals are then released back into their original habitat. You need to give them enough time to mix thoroughly with the

rest of the population, and then you catch another large sample using the same method. Count the total number of marked and unmarked animals in the second sample.

You can now calculate the size of the population using a formula called the **Lincoln Index**:

$$\text{total number of animals in population} = \frac{\text{number in sample 1} \times \text{number in sample 2}}{\text{number of marked animals in sample 2}}$$

An easy way of remembering this is that you multiply the two biggest numbers together, and divide by the smallest one.

This method only gives you relatively reliable results if:

- the original number of animals caught and marked is large
- there is no significant immigration into, or emigration out of, the population between the collection of the first sample and the collection of the second sample
- the marked animals are no more or less likely to die than the unmarked ones
- the marked animals do mix fully and randomly into the population after they have been released
- the population does not change significantly in size as a result of births or deaths between the capture of the first sample and the capture of the second sample.

It is probably almost impossible to be certain that your data meet all of these criteria, but nevertheless this method can give a useful approximation of population sizes for many small, mobile animals.

8 In an attempt to measure the size of a population of woodlice under a large piece of dead wood, a student captured and marked 54 animals. She released them and waited 48 hours before capturing another sample of 63 animals. Of these, 18 were marked. Use the Lincoln Index to estimate the number of woodlice in the population.

The chi-squared test

Scientific investigations proceed by a process of formulating and testing ideas (hypotheses). You can never prove anything in science, but you *can* gather support for your ideas. This is where the chi-squared test comes in – it helps scientists to decide how much weight to give to experimental results. (Chi is the Greek letter χ, pronounced 'ky', as in Kylie, and is the mathematical symbol in the formula.)

Suppose that you are trying to find whether using fertiliser makes any difference to the species that grow in a field. You count the number of each of four species in 20 quadrats from fertilised and unfertilised areas and obtain the results shown in Table 2.

Table 2   Species distribution in fertilised and unfertilised fields

Species	Total number in 20 quadrats	
	Fertilised area	Unfertilised area
A	0	182
B	2	51
C	88	124
D	91	102

From the results shown in Table 2 you would feel pretty certain that fertilising the field affected species A and B. But the results are much less clear for species C and D. You need to know whether there really is a significant difference between the numbers in the two areas, or whether the apparent difference might be due purely to chance. One way of deciding might be to keep on counting, but a quicker method is to use a statistical test, the **chi-squared test**.

The chi-squared test can only be used when you are counting things, rather than measuring features such as height or mass. It is based on working out how likely it is that the results might have occurred by chance. If you flip a coin 20 times, you could expect that it would fall as 'heads' 10 times and as 'tails' 10 times. However, you might also get, say, 12 heads and eight tails, or even 15 and five. The further the result is from the expected 10 heads and 10 tails, the rarer it will be. The test formula calculates how likely the *actual* results are compared with the *expected* results. In the example above, if the species were randomly distributed between the two areas, you would expect the numbers in each area to be the same.

For the purposes of the test, we assume that any difference between the actual results and the expected results is just due to chance. This is called the **null hypothesis**. The results of the calculation will tell us how rare it would be for the actual results to be due to chance. If the test results show that such results would occur by chance only five times out of 100 or less, we would consider it unlikely that the results were just chance and we would reject the null hypothesis. This means we are suggesting that we think our research has probably found a significant difference. Details of how the test is done and how to interpret the results are explained in Chapter 16.

key facts

● An area can be sampled to find out which species live there, and to estimate the size of their populations. Quadrats are used to delineate an area within which data will be collected.

● Quadrats may be placed randomly, using random numbers as coordinates. The percentage of quadrats in which a species occurs gives the species frequency. The mean percentage area that a species covers in all the quadrats gives the mean percentage cover. An abundance scale can be used where it is difficult or too time-consuming to determine percentage cover.

● Quadrats may be of different sizes, including point quadrats, which have a tiny area.

● A transect is a line along which sampling takes place. Quadrats may be placed all along the line, or at intervals. This is useful to get an indication of species distribution across a range of conditions, for example, at different heights on a rocky shore.

● The mark–release–recapture technique can be used to estimate the size of a population of mobile animals. Animals can be caught using a suitable technique, then marked and released. The proportion of marked animals in a second sample can be used to calculate the population, using the Lincoln index.

1.3 Population size

What determines which species live in a particular habitat, and the sizes of their populations? It may simply be that a species has never arrived at that habitat. For example, although all polar bears live in the Arctic, and all penguins live south of the equator, it is likely that polar bears could thrive in the Antarctic, and penguins could live successfully in high northern latitudes. The reason that they do not is that they have never dispersed into those regions.

Often, however, the reasons for the absence of a species from a particular habitat are that certain features of that habitat make it unsuitable for the species to live there. Features such as availability of food and range of temperatures are known as **ecological factors**, and they have a great influence on the distribution and abundance of organisms. Within any ecosystem, a very large number of ecological factors can be identified. As we have seen, we can classify these into abiotic and biotic factors.

Abiotic factors

Abiotic factors result from non-living parts of the ecosystem. They include temperature, light intensity, availability and salinity of water, and availability of gases such as carbon dioxide and oxygen. Most features of the soil, known as **edaphic factors**, are abiotic factors and include the mineral content of the soil, its pH and its water-holding and drainage capacities.

Many organisms are very sensitive to the pH of their environment. Not many plant species can survive the acid soils found on peat moorlands like these. Cotton grass and sphagnum moss are typical of the few that are adapted to a pH below 5. High rainfall causes basic ions, such as calcium, to be washed out of the soil. Waterlogging reduces the rate of decay, and organic acids are released from partially decayed peat. These factors both contribute an excess of hydrogen ions to the soil, resulting in the low pH.

9 Low pH can affect the activities of cells in two ways. The hydrogen ions may affect enzymes directly, and they may cause an increase in the concentration of heavy metal ions. Use your knowledge of enzymes to explain why each of these factors affects cell activity.

10 The photograph shows a rock pool on a rocky shore. Abiotic factors in the rock pool vary considerably at different times. For part of the day it is covered by sea water. When the tide goes out it is exposed to air, and some of the water evaporates. Fig. 12 shows variations in the temperature of the water in the pool on a summer's day.

a Explain how a rise in temperature could affect the oxygen concentration in the water of the rock pool. How would this affect the animal life?

b Explain how the presence of seaweeds in a pool can affect the concentration of dissolved oxygen in the water while the tide is out, during the daytime, and at night.

c The concentration of salt in the water (salinity) is an abiotic factor that may vary considerably in a rock pool. Suggest what might cause the salt concentration in the pool to rise above the normal salinity of sea water, and what might cause it to fall below. How could changes in concentration affect living organisms in the pool?

A rock pool in Santa Cruz.

Fig. 12   Variations in rock pool temperature

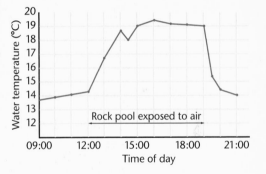

Within any particular habitat, abiotic factors can vary hugely from place to place. In a hedgerow, for example, light intensity is relatively high at the top of the hedge, much lower at the base, and practically zero beneath a stone lying among the hedgerow shrubs. Variations in temperature are highest at the top of the hedge, and much lower under the stone. Humidity, too, will vary most at the top of the hedge, but remain much more constant – and will often be higher – beneath the stone.

The different areas within a habitat each have their own **microclimate**. Differences between microclimates within a habitat influence the distribution of the species within the habitat. For example, invertebrates such as woodlice, which are not very efficient at reducing water loss from their bodies, would not survive for long if they perched on a leaf at the top of a hedge, in full sunlight, fully exposed to drying winds. The microclimate beneath the stone, however, is ideal for them, and this is where you are most likely to find them.

11 Suggest which abiotic factors are likely to be most important in limiting the growth of plants in each of the following habitats:

a   a desert

b   the summit of a high mountain in Scotland

c   a mountain stream

d   a sandy shore

e   peat moorland.

Biotic factors

Biotic factors result from living components of the ecosystem. They include availability of food, competition, predation, parasitism and disease.

For consumers, one of the most important resources is food supply. The population size of primary consumers (herbivores) may be strongly affected by the availability of plants to eat. And this, of course, is strongly influenced by abiotic factors such as water supply, light intensity, temperature and edaphic factors such as the mineral content of the soil. This is true for most biotic factors – they are themselves directly or indirectly affected by abiotic factors.

If any one of the resources that an organism requires is in short supply, then organisms requiring that resource have to compete with one another in order to obtain it. **Competition** is frequently a major factor that affects the distribution and the population size of organisms.

Imagine a grassy meadow. The plants in the meadow all need light. The taller a plant is, the more light it is able to obtain. Shorter plants are shaded by taller ones, and receive less light. If the soil is rich in nutrients, then the taller plants such as some of the more vigorous grasses, thistles and docks will grow so large and shade the smaller plants so much that these are unable to thrive. Small, less vigorous plants such as orchids or cowslips may disappear completely from the meadow. The result of this competition

has been to affect the distribution of the smaller plants. In this example, the resource in short supply is light. The plants competing for it belong to different species, so this is an example of **interspecific competition**.

Competition also occurs between individuals of the same species, known as **intraspecific** competition. Fig. 13 shows the results of sowing different numbers of seeds into small pots of soil. You can see that the more seeds that are sown in the pot and therefore the larger the numbers of plants that grow in it, the fewer seeds these plants manage to produce. The plants are probably competing with one another for water and mineral ions from the soil and also for light. The effect of this competition has been to reduce the reproductive capacity of each individual plant.

Fig. 13   An example of intraspecific competition

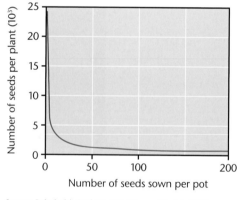

Source: Palmbald, *Ecology*, Vol. 49, pp. 26–34, 1968

Intraspecific competition occurs whenever a population rises to a level that is near to the maximum that can be supported sustainably in a particular habitat. This maximum level is known as the **carrying capacity** for that species in that habitat. Interspecific competition occurs whenever two different species living in the same habitat require the same resource, and this resource is in short supply. The more similar the niches of the two species, the more likely it is that interspecific competition will occur between them. If niches of two species are identical or extremely similar, then it is probably impossible for the two species to coexist. This is known as the 'competitive exclusion principle'.

The severity of **predation** on a population is, like interspecific competition, an example of a biotic factor acting between species. For example, hedgehogs are predators of slugs. If there are large numbers of hedgehogs in a garden, this could help to keep the population size of the slugs smaller than it would otherwise be. On the other hand, the size of a predator population may be influenced by the size of the prey population. Records for the number of skins of Canadian lynx (predator) and snowshoe hare (prey) sold to the Hudson Bay Company go back to 1845. They show regular fluctuations with changes in the lynx population lagging behind changes in the hare population (Fig. 14).

Parasitism and **disease** are very similar to predation in the ways in which they affect population sizes. Imagine a rabbit population living on a grassy hillside, where the rabbits spend much of their time resting and hiding from predators inside burrows. As their population increases, the rabbits become more and more crowded in their burrows. It becomes much easier for disease-causing organisms such as the virus that causes myxomatosis, or for parasites, to spread from one animal to another. Moreover, the individual rabbits may be less able to fight disease or parasites, because competition for food means that some of them may be undernourished. Therefore, as the rabbit population size increases, so the incidence and severity of disease and infection with parasites also increases. The rabbit population falls. When it reaches a lower level, disease and parasitism become less common and less severe, allowing the population to rise again.

Fig. 14 Population fluctuations in Canadian lynx and snowshoe hare

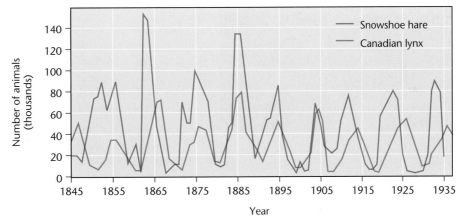

Source: King, *Ecology*, Thomas Nelson, 1980

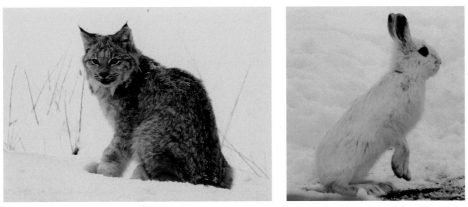

A Canadian lynx.

A Snowshoe hare.

hsw

how science works

Red grouse populations

The population size of red grouse tends to swing up and down, oscillating widely over a period of about four or five years. Several theories have been put forward to try to explain what is causing the fluctuations. Whatever it is must be a density-dependent factor – that is, a factor that acts more strongly when the population is high then when it is low. As the grouse population increases, this factor increases too and causes the population to decrease. As the population decreases, the factor decreases, allowing the population to increase again.

Suggestions for this density-dependent factor have included food supply and predation. However, neither of these seems to be the answer. Grouse eat only 2%–3% of the available heather, so supply of heather is very unlikely

Red grouse, *Lagopus lagopus*, live on heather moorland. They spend much of their time on the ground, where they feed on young heather shoots.

to be affecting them. And predators such as foxes and birds of prey do not seem to take enough grouse to have such large effects. Two more likely possibilities are intraspecific competition for space, and interactions between the red grouse and a parasite.

Competition for space arises during the autumn, when each breeding male takes control of an area of moorland. Territorial behaviour in autumn determines the number of breeding birds the following spring: the more aggressive the male grouse, the bigger its territory, the greater the emigration rate and the lower the number of breeding pairs. It is possible that aggression of the males is related to population density – the bigger the population, the more aggressive the males become, thus reducing the population size next year.

The second theory is to do with a parasite of red grouse, a nematode worm called *Trichostrongylus tenuis*. This parasite reduces growth of grouse populations by decreasing breeding rates. Both the rate and intensity of infection by the parasite increase as the population density of the birds increases. Evidence for this comes from studies showing that:

- Both grouse and *T. tenuis* have similar cyclical population densities.
- Higher levels of parasite infection match greater losses from the grouse population.
- Some grouse populations do not have any parasites, and their numbers do not oscillate.

During the 1980s and 1990s, six different grouse populations, living on six different areas of moorland in northern England, were investigated. First, using long-term data about grouse populations, predictions were made about when the next population 'crashes' would occur: 1989 and 1993. In 1989, the grouse in four of the six populations were caught and treated with a worm-killing drug. Two of the treated populations were then treated again in 1993. Fig. 15 shows the results for three of the six populations.

Fig. 15   Results for three of six grouse populations

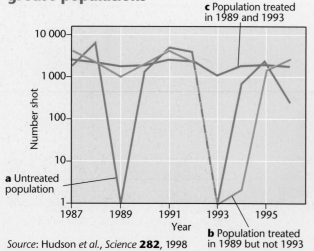

c Population treated in 1989 and 1993

a Untreated population

b Population treated in 1989 but not 1993

Source: Hudson *et al.*, *Science* **282**, 1998

12 Explain how these results support the hypothesis that the parasite is responsible for the cyclical fluctuations in red grouse populations.

13 What evidence is there from these results that the nematode parasite may not be the only cause of these fluctuations?

Population growth

Fig. 16 shows the results of an experiment in which a culture of single-celled algae was grown in a laboratory beaker. At first the population grows slowly, but then, after a few days, the rate of growth increases rapidly. During this phase, the rate of growth of the population is approximately exponential – that is, the population repeatedly doubles over a certain time period.

Fig. 16   Growth of an algal population

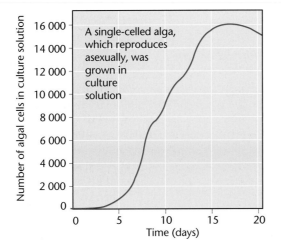

A single-celled alga, which reproduces asexually, was grown in culture solution

14 This question is about the graph in Fig. 16.

Suppose that 10 algal cells were originally put into the beaker and that they all survived after every division.

a How many cells would there be in each of the first 10 generations? Show your results in a simple table.

b Draw a graph showing the growth of this population. Label the x-axis 'Number of generations' and the y-axis 'Number of cells in population'. Describe the shape of the curve.

c How does the theoretical population growth in your graph compare with the actual growth shown in Fig. 16?

d Which abiotic factors probably limited the actual growth?

However, in a small beaker, this rise in numbers is soon restricted by a shortage of resources. Once the surface is covered by algae, there is less light lower down for other cells to photosynthesise. Also, the algae are likely to have absorbed most of the mineral ions that they need for the synthesis of proteins and chlorophyll. In the beaker, population growth stops and starts to fall. This happens when more cells are dying than are being replaced by new ones.

In real situations, changes in population size are much more complex, because a species would hardly ever live in total isolation from other species. Figs 17, 18 and 19 overleaf show the results of an experiment originally done by a Russian ecologist to investigate competition between two species. The species chosen were both single-celled protoctists: *Paramecium aurelia* and *Paramecium caudatum*. Both feed on bacteria and yeast cells.

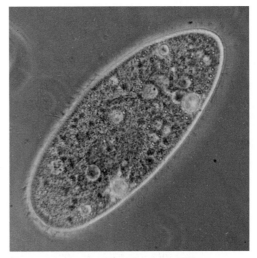

Paramecium aurelia (magnification ×300).

Fig. 17 Growth of a *Paramecium aurelia* population

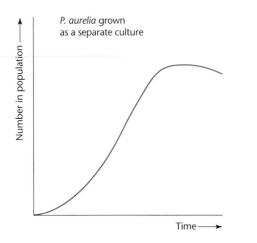

Fig. 18 Growth of a *Paramecium caudatum* population

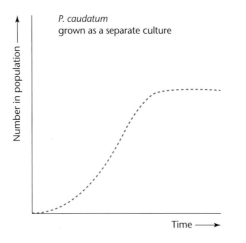

Fig. 19 Growth of both *Paramecium* species when cultured together

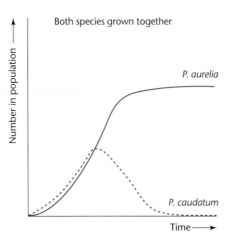

As Fig. 19 shows, when both species were grown together in the same culture, both species grew equally well at first. Then, once the food became scarcer *P. caudatum* was less successful at obtaining food. Its numbers declined while those of its competitor species, *P. aurelia*, increased. Interspecific competition is strongly affecting the population size of *P. caudatum*, causing it to be smaller than it would otherwise be.

Reaching a balance

Given time, most ecosystems reach a balanced situation in which the populations of most species remain fairly constant. However, the interactions between organisms in an ecosystem are highly complex, and quite small changes can alter the balance fairly rapidly.

Obviously, animals that feed on plants damage the plants, but most plants have a remarkable capacity to withstand damage without individuals being killed off. An oak tree in summer may have thousands of insects of many different species feeding on its leaves. Yet it can generate enough new growth to maintain itself for hundreds of years.

The most successful herbaceous plants, the grasses, are adapted to tolerate almost continuous grazing, because the growing point of the stem is tucked away at the base of the leaves, very close to the ground. Grasses will only suffer serious losses if the number of herbivores escalate excessively, as occurred with rabbits in Australia.

Normally, excessive growth of the herbivore populations is kept under control by predators. If the number of herbivores start to increase, there will be more food for predators, so their numbers will also increase. The increase in predation then brings down the population of herbivores again. If the herbivore population falls too low, the predators fail to find enough prey, so the predator population falls again. This can sometimes result in a regular fluctuation in predator and prey numbers as shown for the Canadian lynx and snowshoe hare in Fig. 14. Such a situation is relatively rare, however, as most predators feed on more than one prey species, so if the population of one falls it can turn to another, more abundant species for a period of time.

15 Grazing rabbits are in constant danger from predators. One way in which they are protected is by living in underground burrows. When feeding above ground, their acute hearing and 'all-round' eyesight enables them to sense danger and escape quickly. However, narrow-bodied predators such as the weasel and polecat can follow them into their burrows.

Table 3 shows the results of a study of rabbit numbers in a particular colony.

a Suggest why the number of young surviving in the colony at the end of the first season is so much lower than the number of young that were counted during the study.

b What would you expect to happen to the rabbit population in this colony if the same level of loss carried on? Explain your answer.

c Suggest how the colony's numbers might be restored by natural means.

Table 3   A study of a colony of rabbits

Number of adult rabbits at start of the study	70
Number of breeding females	36
Total number of young emerging to feed during the time period of the study	280
Number of young in the colony at the end of the first season	28
Number of adults in colony at the end of the first season	11

key facts

● Factors that influence the distribution and abundance (population size) of organisms can be categorised as abiotic and biotic factors.

● Abiotic factors result from the non-living part of the environment, such as temperature or water supply. Biotic factors result from the living parts of the ecosystem, such as competition and predation.

● Abiotic factors can vary greatly in a habitat, forming different microclimates that are important in determining the distribution of a species within the habitat.

● Biotic factors include food supply, predation, parasitism, disease, intraspecific competition and interspecific competition.

● In some cases, one particular factor has a very strong influence on population size. For example, if a predator depends on one species of prey for food, and if that predator kills mostly that prey, then the populations of one may be largely determined by the other. This can result in repeated oscillations of the population sizes of both species.

1.4 Human populations

Fig. 20 shows what has been happening to the human population in the last 4 000 years. It is rather frightening. We have seen that populations do not go on growing forever – sooner or later, some abiotic or biotic factor acts to slow or prevent further increase. In some cases, the population falls rapidly once some upper limit is reached.

Is this going to happen to us? Will we reach a crisis point, at which some factor causes a

Fig. 20   World population growth

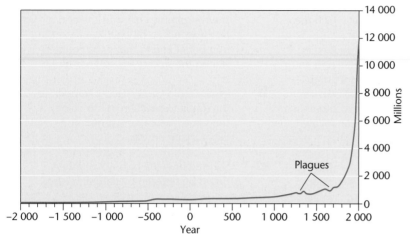

Fig. 21   Three different predictions of what may happen to the human population by 2050

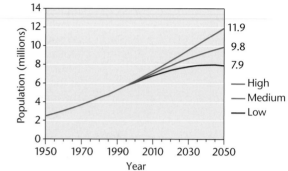

catastrophic fall in our population – for example, water shortage, food shortage, the spread of a new disease through our crowded cities and transport systems? Or will we manage to take control of the situation, slowing and eventually stopping the increase and settling into a stable situation in which there is enough of everything for all of us?

No-one knows the answer to this. We are not even sure what is going to happen in the next few years, let alone the next few hundred years. Fig. 21 shows three different predictions, made by the Population Division of the United Nations, of what may happen by 2050. The

signs are that the overall rate of our population growth is levelling out, so hopefully we will follow something between the 'Medium' and 'Low' lines, not the 'High' line.

Population growth varies widely in different parts of the world (Fig. 22). In general, the rate of growth in most developed countries, including Europe and North America, is slow; in some cases, it is even negative (that is, numbers of people are falling). Growth rates are fastest in developing countries. Later, we will look at some possible reasons for this pattern, and why we can expect the rates of population growth in developing countries to fall in the future, if living conditions improve.

Fig. 22   Populations of major world regions from 1750

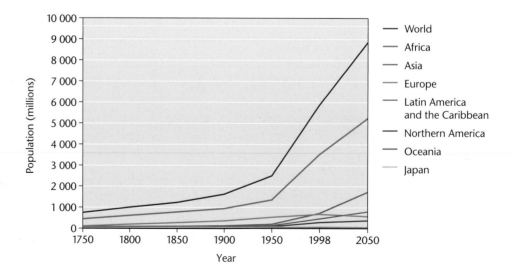

16 Table 4 contains data about the populations of five areas of the world since 1750 and predicted to 2050. The numbers are population in millions.

a Display these data as a line graph. Draw a separate curve for each area, on the same pair of axes.

b In which country did the greatest **percentage** growth take place between 1750 and 1998?

c What do these results suggest about population growth rates in developed countries, compared with those in developing countries?

Table 4   Populations in five areas of the world

Part of world	Year						
	1750	**1800**	**1850**	**1900**	**1950**	**1998**	**2050**
Europe	163	203	276	408	547	729	628
South America	16	24	38	74	167	504	809
North America	2	7	26	82	172	305	392
Oceania	2	2	2	6	13	30	46
Japan	26	26	27	45	83	126	100

The numbers are the populations in millions.

17 Table 5 shows the birth and death rates for five countries between 1990 and 2007. The birth and death rates are given as the number of people born or dying in one year, per 1 000 of the population.

a In which country or countries did the population grow in 2007?

b In which country or countries did the population decrease in 2007?

c Describe the pattern of population change in Japan between 1990 and 2007.

d Panama and Tunisia are developing countries, in which health is generally not as good as in developed countries. Yet the table shows their death rates per 1 000 to be lower than in the three developed countries. Suggest why this is so. (Think about what the birth rate data tell you about the ages of the people in their populations.)

e Apart from birth rates and death rates, what other factors could affect the size of the population in the United Kingdom in future years?

Table 5   Annual birth and death rates in five countries

Country	Birth rate per 1 000				Death rate per 1 000			
	1990	**2005**	**2006**	**2007**	**1990**	**2005**	**2006**	**2007**
Austria	11.6	8.8	8.7	8.7	10.6	9.7	9.8	9.8
Japan	9.9	9.5	9.4	9.2	6.7	9.0	9.2	9.4
Panama	23.9	22.0	21.7	21.5	not known	5.3	5.4	5.4
Tunisia	25.8	15.5	15.5	15.5	not known	5.1	5.1	5.2
United Kingdom	13.9	10.8	10.7	10.7	11.2	10.2	10.1	10.1

The population of the United Kingdom

Fig. 23 shows the changes in the UK population between 1991–92 and 2005–06.

Fig. 23 Changes in the UK population since 1991–92

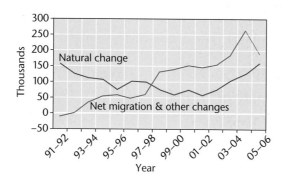

The line labelled 'natural change' shows the increase, in thousands, in the population as a result of differences between birth rate and death rate. The line labelled 'net migration' shows the increase, in thousands, in the population as a result of people coming to live in the UK, or leaving it to live elsewhere. To find the total increase at any point, you need to add together the increases due to both of these factors.

It is interesting to see how much more important net migration has become since 1997–98, compared with natural changes. Of course, one impacts on the other. Once new migrants have settled in the UK, then their births and deaths become part of the 'natural change' figures. This is probably the main reason why the 'natural change' curve began rising in 2001–02, after a period in which it fell.

Population pyramids

Fig. 24 is a different way of presenting information about what is happening to a population. It is called a **population pyramid**. It gives a 'snapshot' of the numbers of people in different age groups at any one point in time. The pyramid in Fig. 24 shows the age distribution of the UK population in 2006, when the total population was estimated to be 60.6 million people.

This population pyramid shows that in 2006 the age group with the largest numbers of people was between about 35 and 45. There was another peak amongst 60-year olds – these are the so-called 'baby boomers', who were born soon after the Second World War ended in 1945. Each year, those peaks will move one step

further up the pyramid, so we can use the shape of the pyramid in any one year to predict what shape it will be in the future.

18 Table 6 shows the numbers of people in different age groups living in the United Kingdom in 1999.

Table 6 Age structure of the UK population in 1999

Age group	Number of males (thousands)	Number of females (thousands)
0–4	1 845	1 752
5–9	1 970	1 897
10–14	1 980	1 872
15–19	1 900	1 797
20–24	1 804	1 716
25–29	2 118	2 045
30–34	2 374	2 318
35–39	2 350	2 292
40–44	2 030	2 006
45–49	1 883	1 884
50–54	1 993	2 005
55–59	1 552	1 589
60–64	1 389	1 452
65–69	1 218	1 349
70–74	1 042	1 275
75–79	823	1 195
80–84	420	761
85 and over	318	870

a Calculate the total size of the UK population in 1999.

b Draw a population pyramid using these data. You can construct it like a pair of histograms lying on their sides. You can either leave a space in the middle (as in Fig. 24 and Fig. 25) where you can label the axis, or you could use the left-hand edge as the axis and let the two halves of your pyramid join in the centre.

Fig. 24 Population pyramid for the UK in 2006

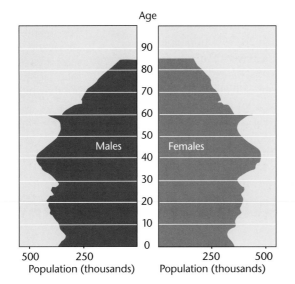

Fig. 25 Population pyramid for Mexico in 1980

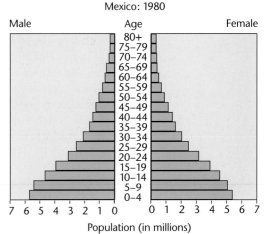

Source: US Census Bureau, International Data Base

Fig. 25 shows a population pyramid for Mexico in 1980. It is clearly a very different shape from the pyramid for the UK in 2006 (Fig. 24) or 1999 (which you may have drawn for Question 18). This pyramid is much wider at the base than higher up. This suggests a very high birth rate. There are a relatively small number of people of reproductive age in the population compared with the number of children.

As these young people grow up, their 'bulge' will steadily move up the pyramid. Assuming that these people reproduce at a similar rate to their parents, the pyramid will continue to have a very wide base. Indeed, you would expect the base to get even wider. This is a growing population. A bottom-heavy pyramid is a sure sign that a population is increasing in number.

Survival curves and life expectancy

Fig. 26 shows another way in which we can display information about population size and age structure. It is called a **survival curve**, and

Fig. 26 Survival curves for men and women in Great Britain in the early 21st century

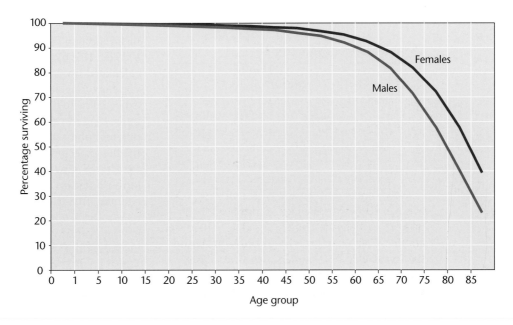

this one is for Great Britain. The graph shows two separate curves, one for males and one for females. The curve shows the percentage of people who are still surviving at each age. By definition, at age 0, the survival is 100%. You can see from the graph that, at age 20, 99% of people are still alive – in other words, fewer than 1% of us die in our childhood or teens. By age 60, 90% of men and 93% of women are still alive.

Survival curves can tell us a lot about what is happening to a population. Fig. 27 is a curve for Botswana, a stable and economically sound country in southern Africa that has been hit very badly by the HIV/AIDS epidemic.

19 This question is about the survival curves in Fig. 26 and Fig. 27.

a The steepest part of a survival curve shows the age group in which death rates are highest. When are death rates highest in Great Britain? When are they highest in Botswana?

b Suggest reasons for these differences.

Fig. 27   Survival curve for Botswana in the early 21st century

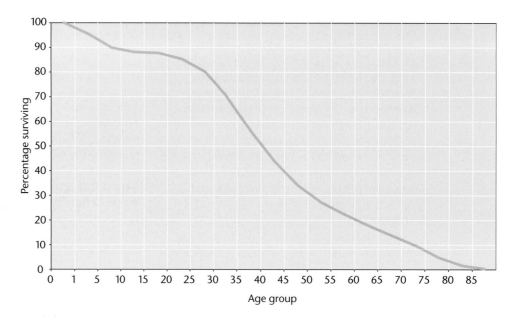

Fig. 28   Life expectancy in the UK from the age of 65 years

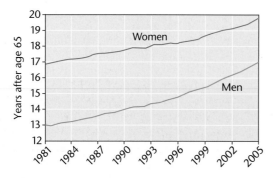

A simpler way of indicating how long a person might expect to live is to calculate the mean lifespan of everyone in a population. This is called **life expectancy**. It is usually calculated from birth – that is, the value tells you how long a newborn baby could be expected to live. Sometimes, though, it is interesting to calculate life expectancy from a different point in a person's life. Fig. 28 shows the life expectancy of a 65-year-old person in the United Kingdom. This tells us how much longer a person could expect to live, on average, assuming that they have reached the age of 65.

You can see from Fig. 28 that life expectancy for the over-65s in the UK has steadily increased. It is still increasing today. Things are not so good in many other countries, however. Table 7 shows the life expectancy at birth for

Table 7 Life expectancy at birth

Country	Life expectancy at birth		Infant mortality	Under-five mortality
	Males	**Females**		
Afghanistan	44	44	157	235
Austria	77	83	4	5
Botswana	50	51	46	68
Iceland	80	83	3	4
Russian Federation	59	73	17	21
Sierra Leone	41	44	160	278
UK	77	82	5	6
USA	76	81	6	8

eight countries, as well as infant mortality (the number of children of every 1 000 who die before their first birthday) and under-five mortality (the number who die before their fifth birthday). The very low life expectancies for sub-Saharan Africa are partly due to the HIV/AIDS epidemic, which tends to kill people in their thirties to forties.

The demographic transition

Earlier, we saw that population growth tends to be much greater in developing countries than in developed countries. We can get some idea of why this is so by looking at survival curves for developed countries from the past compared with today.

Fig. 29 shows survival curves for people living in England at various times since 1541. They are only approximations, because of course we do not have complete data for any of these years; they have been calculated by using many historical documents, and are thought to be reliable. You can see that there has been a steady change in the shape of the curve over the years.

The most obvious difference in the curves is the high childhood mortality in the past compared with today. In 1591, only six of every 10 lived to see their 10th birthday. The

Fig. 29 Survival curves for England

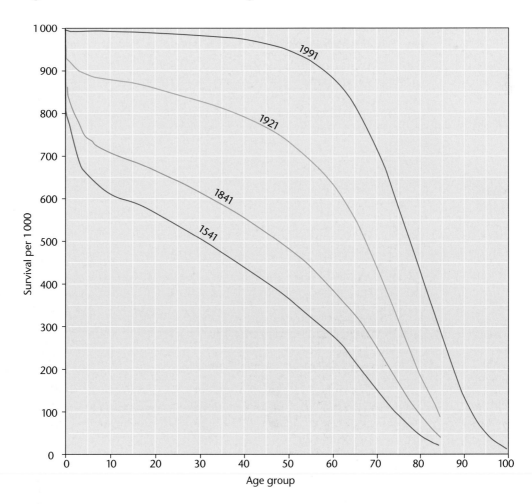

steepest part of the curve was over the 0–10 age range, whereas in 1991 it was in the 65-plus age range.

This change in shape of survival curves is typical of a country making the transition from being undeveloped to developed. The change is called the **demographic transition**. A country steadily shifts from high birth rates and high death rates to lower birth rates and lower death rates. In between there is often a stage where death rates have reduced but birth rates have not yet fallen significantly – this is the stage where population growth is greatest.

Fig. 30 illustrates the different stages of the demographic transition as it took place in Britain between 1700 and 2000.

In the 18th century, death rates were high. There were many reasons for this, including deaths from infectious diseases (which today we can treat with antibiotics), poor nutrition and poor sanitation. Birth rates were also high, but birth and death rates fluctuated. There were times when birth rate was higher than death rate (and population grew) and other times when the situation was reversed (and population fell). Overall, the population remained relatively constant during this period.

As we moved through the 18th and 19th centuries, industrialisation and advances in medicine brought improvements in standards of living, and death rates fell. You can see that, for some time, birth rates continued to be high. Only many years later did birth rates also begin to fall. This seems to be a reaction of a population to the fall in death rate. When death rates are high, people choose to have many children, expecting many of them to die. They want at least some of their children to live to see them into their old age. Only after a long period of relatively low death rates does there appear to be a response – whether conscious or not – of people having fewer children.

By the mid-20th century, both death rate and birth rate had fallen. Today, they are very close to each other and, as we have seen in Table 5 on page 29, we may soon be in a position where birth rate is lower than death rate.

Fig. 30   Birth rates and death rates in Britain between 1700 and 2000

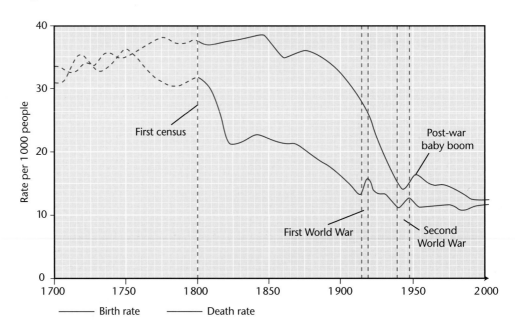

Fig. 31 The demographic transition

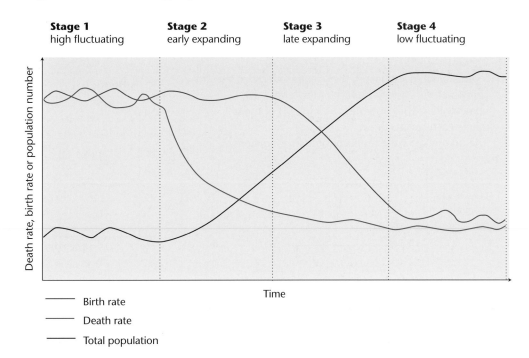

| **Stage 1** high fluctuating | **Stage 2** early expanding | **Stage 3** late expanding | **Stage 4** low fluctuating |

Death rate, birth rate or population number

Time

—— Birth rate

—— Death rate

—— Total population

It is intriguing to discover that this pattern seems to occur in *every* country that has been studied. If we take out the blips in the curves in Fig. 30, we arrive at something like Fig. 31. We can divide it into four stages:

- Stage 1, where birth and death rate are both high and population fluctuates but does not change overall
- Stage 2, where death rate falls but birth rate remains high, so population increases
- Stage 3, where death rate levels out and birth rate falls, so population continues to increase but at a decreasing rate
- Stage 4, where birth and death rate are both low and population remains approximately constant.

It is thought that there may also be a Stage 5, which we are entering now. Here, birth rate may fall *below* death rate, so population will actually decrease.

Looking around the world today, we can identify countries that are at different stages of the demographic transition. For example, the native people who live in the Amazonian rainforest are at Stage 1, as are some of the world's poorest countries, such as Bangladesh and Sierra Leone. Stage 2 can be seen in some countries in Latin America, including Peru; these are where the greatest current increases in population are taking place. Stage 3 is occurring in China and Australia. Most European countries are now in Stage 4, and some – including Austria – already seem to be moving into Stage 5.

key facts

- The human population is growing, and may eventually reach a point at which abiotic or biotic factors cause it to decrease. Growth rate is generally higher in developing countries than in developed countries.

- Population pyramids show the numbers (or percentages) of people in each age group at any one point in time. A pyramid with a wide base shows an increasing population.

- A survival curve shows the percentage of people who survive to each age within the total age range. A curve that falls near the beginning indicates high infant mortality. In a developed country, the curve is generally high and flat in the early years, with the sharpest downward gradient from age 60 or beyond.

- As a country develops, it gradually changes from high birth rates and death rates to low birth rates and death rates. This is called the demographic transition.

1 The pH of soil affects the growth of plants. Scientists tried to grow two species of plants on large areas of bare soil of different pH. They grew each species on its own.

In a different investigation, the scientists recorded the growth of the same two species in natural communities. They also measured the pH of the soil where each species grew.

The graphs show the results of both investigations.

Growth of common sorrel

3 4 5 6 7 8
pH of the soil

Growth of sheep's fescue grass

3 4 5 6 7 8
pH of the soil

Key
- - - Plant species growing on its own
—— Plant species growing in a community

a The results are very similar when common sorrel and sheep's fescue were growing on their own. Suggest **one** explanation. (2)

b The results obtained when these plants were growing in communities are different from when they were growing on their own. Suggest **one** explanation for the difference in the results. (3)

Total 5

AQA, June 2007, BYB678B, Question 3

2 Tree and canyon lizards are found in desert areas in Texas. Both species eat insects. In an investigation the population densities of both species of lizards were measured over a four-year period in:
- control areas, where both species lived
- experimental areas, from which one of the species had been removed.

The results for tree lizards are shown in **Figure 1** and the results for canyon island lizards are shown in **Figure 2**.

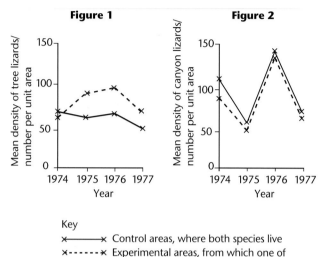

Figure 1   Figure 2

Key
×——————× Control areas, where both species live
×- - - - -× Experimental areas, from which one of the species had been removed

a Name the type of interaction being studied in the control areas. (1)

b Describe how the population density of each lizard species is affected by the presence of the other. Give evidence from **Figure 1** and **Figure 2** to support your answer.
  i Tree lizard (1)
  ii Canyon lizard (1)

c The investigators concluded that, during the four-year period, an abiotic factor had a greater effect on the canyon lizards than on the tree lizards. What evidence from **Figure 1** and **Figure 2** supports this conclusion? (1)

d Adult tree lizards were found to have shorter lives in experimental areas than in control areas. Using **Figure 1**, suggest an explanation for this. (1)

Total 5

AQA, January 2007, Unit 5, Question 8

3 The following table lists the numbers of adult butterflies in two areas of the same tropical forest. In the logged area some trees had been cut down for timber. In the virgin forest no trees had been cut down. The two areas were the same size.

Butterfly species	Logged forest		Virgin forest	
	Number	**$n(n-1)$**	**Number**	**$n(n-1)$**
Eurema tiluba	72	5112	19	342
Cirrochroa emalea	43	1806	132	17292
Partenos sylvia	58	3306	14	182
Neopithecops zalmora	6	30	79	6162
Jarmides para	37	1332	38	1406
Total	**216**	**11 586**	**282**	**25384**

a Describe a method for finding the number of one of the species of butterflies in the virgin forest. (2)

b What does the table show about the effects of logging on the butterfly populations? (2)

Total 4

AQA, June 2006, Unit 6, Question 3 (a & c)

4 The following table shows the population in different age groups in the UK in 1971 and in 2004.

Age group/years	Population/millions	
	1971	**2004**
0–15	14.0	11.4
16–64	34.7	38.9
Over 65	7.3	9.6
Total	**56.0**	**59.9**

a Calculate the percentage increase in the 16–64 age group from 1971 to 2004. Show your working. (2)

b For each age group, suggest **one** factor that could account for the change shown in the table. Give a different explanation in each case. (3)

Total 5

AQA, June 2007, BYB8A, Question 1

5

a The graph shows how the birth rate and death rate changed during a demographic transition in a human population.

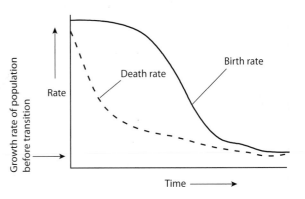

i Explain how **two** changes in social conditions could have reduced the death rate. (2)

ii Sketch a curve on the graph to show the growth rate of the population during the period shown. Start at the point indicated on the y-axis. (1)

b In some countries AIDS has resulted in a very large increase in the death rate among children and young adults. What effect will this have on the shape of the age pyramid of these countries? (1)

c There has been a rapid increase in the size of human populations during the past two centuries. Suggest **three** reasons why this has resulted in the reduction of populations of many other species. (3)

Total 7

AQA, June 2002, Unit 8, Question 7

hsw

how science works **assignment**

Changing population in New Zealand

New Zealand is famous for its wonderful landscapes, laid-back lifestyle and uncrowded places. The sheep population is 10 times greater than that of people.

Far from everywhere, New Zealand came relatively late to the population increases seen in virtually every part of the world. New Zealand today regularly tops the polls as a favourite holiday destination, but it was not inhabited by people until around 950 AD, when a few Polynesian people from the South Pacific islands arrived and settled. European explorers first arrived in New Zealand in 1642, but they did not stay. It was not until the late 1790s that European settlers set up their homes there. Although the population has greatly increased since then, it is still a relatively uncrowded country – with a land area of just under 270

000 km², it supports a population of around 42 million people. Compare this with Singapore, which has a land area of 683 km² and a population of 45.5 million people.

Age pyramids based on census data for 1901 and 2001, and a projection made in 2008 for 2101, tell a story of change in the age structure of the population of New Zealand, and give us information about how the size of the population is likely to change in the future. Today, the greatest contributor to the increase in population is immigration, especially of people from more crowded and less affluent countries in Southeast Asia.

Fig. 32   Age pyramids for New Zealand

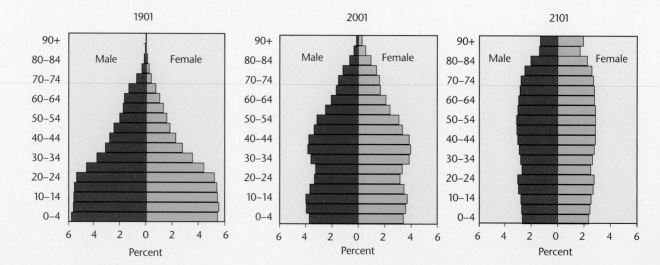

Fig. 33   Components of population change in New Zealand between 1982 and 2006

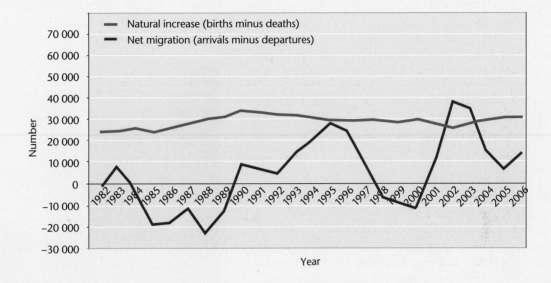

A1   Use the population pyramids in Fig. 32 to identify the stage of the demographic transition that New Zealand was at in:

a   1901

b   2001.

A2   From the pyramids, estimate what happened to the mean (average) age of the population between 1901 and 2001.

A3   How was the proportion of elderly people to younger ones predicted to change between 2001 and 2101?

A4   Fig. 33 shows how the population of New Zealand changed between 1982 and 2006.

a   What stage of the demographic transition was New Zealand at over this time period? Use the data in the graph and the age pyramids to provide evidence for your answer.

b   Which was greater during the period 1984 to 1989 – immigration or emigration?

c   Make a copy of the graph axes, and use the data on natural increase and net migration to sketch a line representing total change in population over this time period.

2 Photosynthesis

The atmosphere that surrounds the Earth is unique; no other planet that we know of has a blanket of gases that enables it to support life. For many living organisms the key component of this blanket is oxygen, a vital requirement for aerobic respiration. However, in the modern industrial world, it is carbon dioxide that is attracting our attention.

The oxygen in our atmosphere comes from photosynthesis – a metabolic reaction that takes place in plants and various microorganisms. Carbon dioxide, on the other hand, is used in photosynthesis and produced in respiration. There is currently great concern about the increasing concentration of carbon dioxide in our atmosphere. This increase appears to be at least partly due to the burning of fossil fuels. The carbon dioxide increase is causing global warming and climate change.

Rich rainforest regions are an important 'sink' for carbon dioxide, helping to absorb some of the excess that we produce. However, researchers have realised very recently that these huge forests don't always do what we think they should. We know that, under the right conditions, 10% of all the carbon dioxide in the Earth's ecosystems is locked up in the huge rainforests of the Amazon basin. In 1993, for example, the Amazon was storing a massive 0.7 trillion kilograms of carbon dioxide. However, this only happens when the climate in the Amazon is cool and wet. When it is warmer and drier, as has happened in the recent El Niño episodes, the Amazon basin actually produces more carbon dioxide than it stores. In 1992, for example, it added 0.2 trillion kilograms of carbon dioxide to the atmosphere.

Researchers are now trying to find out what makes the Amazon a sink or a source for carbon dioxide. This delicate balance is an important factor for countries to consider as they cooperate to try to manage carbon dioxide production.

Rainforests are a source of continuous research.

2.1   Energy and ATP

Think about the last time you were running about playing football or tennis, or the last time you were dancing at a party: you had plenty of energy, but where did it come from? Most people would answer a question like this with something like, 'From my muscles!' or 'From the pizza and chips I had at lunch time!' Both would be true to a certain extent but if you think about the question more deeply it starts to get really fascinating.

Ultimately, all our energy and all the energy used by all organisms on Earth comes from the Sun. But we can't gain enough energy to dance the night away by sunbathing: we need plants to do that for us. Plants have a wonderful collection of biochemical mechanisms and pathways that allow them to absorb the energy in sunlight and use it to convert water and carbon dioxide into sugars; they literally make food from thin air. These sugars contain chemical energy that was originally light energy. The plant can build up the basic sugars into more complex molecules such as starch, fats and proteins. When we eat plant foods, we digest the complex molecules back into simple ones. Later, our cells can strip out the energy that the plant put there using light. Respiration – the process that does this in the body – is essentially the reverse of photosynthesis.

In this chapter we look at photosynthesis; respiration is dealt with in Chapter 3.

Basic principles

Before going into the various stages of
photosynthesis and respiration, it is important
to cover some basic principles.

Fig. 1 gives a general overview of
photosynthesis and respiration. You should
remember the summary equations for
photosynthesis and respiration from your
GCSE studies.

Fig. 1 Photosynthesis and respiration

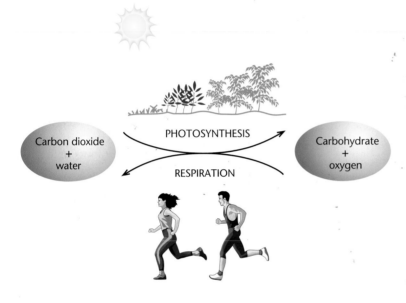

Photosynthesis:
$$6CO_2 + 6H_2O + [Energy] \rightarrow C_6H_{12}O_6 + 6O_2$$

Respiration:
$$C_6H_{12}O_6 + 6O_2 \rightarrow 6CO_2 + 6H_2O + [Energy]$$

As you can see, these two equations are
actually the reverse of each other.

Photosynthesis takes places inside
chloroplasts, which contain chlorophylls and
other pigments that absorb light energy. This
energy is used to power a series of **reduction
reactions** that reduce carbon dioxide and
water to glucose to provide the plant with an
energy store.

The reactions of respiration are **oxidation
reactions**. They oxidise glucose into carbon
dioxide and water, releasing the stored energy
and making it available for immediate use by
the organism.

Oxidation reactions take away electrons or
hydrogen atoms from molecules. Oxidising
agents take away electrons. Most oxidation
reactions are exothermic – energy is released.

Reduction reactions add electrons or
hydrogen atoms to molecules. Molecules that
supply electrons in a chemical reaction are
called reducing agents. Most reduction reactions
are endothermic – they require an input of
energy.

Whenever one chemical is oxidised, another
must be reduced, since electrons and hydrogen
atoms cannot just disappear or exist on their
own. This means that oxidation reactions and
reduction reactions always occur together; they
are often referred to as **redox reactions**.
Photosynthesis and respiration both consist of a
complex chain of reactions in which electrons
are passed from one molecule to another like
juggling balls.

Fig. 2 Redox reactions

Oxidation is:
- loss of electrons
- loss of hydrogen
- gain of oxygen.

Reduction is:
- gain of electrons
- gain of hydrogen
- loss of oxygen.

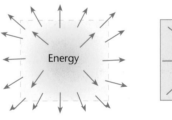

Energy is **released**   Energy is **trapped**

Cells carry out photosynthesis and respiration
to provide them with energy, so that they can
carry out all the metabolic reactions necessary
to sustain life. In photosynthesis, the overall
point of reduction is to capture light energy and
convert it to chemical energy in glucose and
other molecules. Oxidation in respiration allows
plants and other organisms to release this
chemical energy for use in the cell.

ATP

The energy in organic molecules could be
released in one go – rapid, uncontrolled

oxidation. This is what happens when substances burn. Burning is an oxidation reaction in which much of the energy in the burning molecules is released as heat.

This would be no use to the body: most of the energy would be lost as heat, and the high temperature would be lethal. Instead, the body uses enzymes to break down food fuels such as fats, proteins and carbohydrates step by step, releasing energy in small, usable amounts.

Although some of the energy is released as heat, most of it is not. It is used to make a substance called **adenosine triphosphate**, **ATP**. ATP stores energy within the cell, and releases it instantly for reactions that need it, such as during muscular contraction.

Each cell makes its own ATP, as and when it needs it. When energy is required, just enough ATP is broken down to release a suitable quantity of energy. An ATP molecule releases a useful-sized amount of energy to fuel a reaction in a cell. Breaking down a whole glucose molecule would be likely to release too much. ATP molecules are little 'packets' of energy. It is often known as the 'energy currency' of a cell.

ATP breakdown is usually a **coupled reaction**. This means that the ATP breakdown happens at the same time as another reaction that uses energy. The energy released from the ATP is used immediately to fuel the other reaction.

As its name suggests, adenosine triphosphate contains three phosphate groups. To make ATP, inorganic phosphate (Pi) is added to **adenosine diphosphate, ADP** (Fig. 3). This uses energy – the energy that will be stored in the ATP molecule.

$$ADP + Pi + [Energy] \rightarrow ATP$$

As you will see, ATP is formed during photosynthesis and during respiration.

About 33 kJ of energy is required to combine one mole of inorganic phosphate with one mole of ADP. This is just a little more energy than is needed to form bonds in most molecules. When ATP is broken down again into ADP and Pi, these 33 kJ of energy are released for use inside the cell. This means that ATP is a particularly good 'store' of energy, because the quantity of energy stored is neither too large nor too small to fuel a particular reaction in a cell.

So there are two main advantages of using ATP as an immediate energy source in a cell, rather than glucose.

- An ATP molecule contains a very useful amount of energy. A glucose molecule might contain too much to fuel a particular reaction, so a lot would be wasted.
- An ATP molecule can be broken down very quickly, in just one step, releasing its energy immediately. The breakdown of a glucose molecule (in respiration) requires many more steps and takes longer.

1   Energy is important in the human body in lots of ways. Give an example of:

a   a part of the body that generates a lot of heat

b   a part of the body where active transport is used to pump substances into or out of cells

c   a synthetic reaction that needs energy.

Fig. 3   The synthesis of ATP

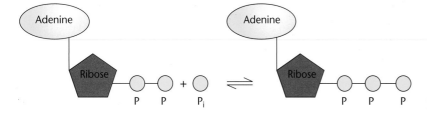

Energy in photosynthesis

In the rest of this chapter we will look in detail at exactly what happens inside a chloroplast during photosynthesis. The fundamental event is the transfer of light energy into chemical energy in glucose. This involves two main stages – the **light-dependent reaction** and the **light-independent reaction** – each of which is itself made up of many smaller steps. As you will see, ATP is involved in this process. ATP is made in the light-dependent reaction using energy from sunlight. It is then used in the light-independent reaction, releasing its energy to be stored in a glucose molecule.

Fig. 4 summarises what happens during these two stages. In the light-dependent reaction, energy from sunlight is used to split water into hydrogen and oxygen. The oxygen is given off as a waste product. The hydrogen splits into protons and electrons. These are picked up by a substance called NADP (nicotinamide adenine dinucleotide phosphate), thus reducing it. The energy in the electrons is used to make ATP.

The NADP and ATP are then used in the light-independent reaction. In this stage, carbon dioxide is converted to carbohydrates, using the reducing power and energy of the reduced NADP and ATP.

Both stages of photosynthesis take place inside **chloroplasts**. The light-dependent stage happens on the **thylakoids** – membranes in which chlorophyll and other pigments are embedded. The light-independent stage takes place in the **stroma**. Fig. 5 shows the structure of a chloroplast.

Both sets of reactions are kept separate from other reactions in the cell by the outer chloroplast membranes. There are two of these, forming an **envelope**. Carbon dioxide readily diffuses through the outer membranes of the chloroplasts into the stroma. Sugars that are made in photosynthesis can be transported out though these membranes. However, not all the sugar is exported from the chloroplast; some is converted into starch and stored in starch grains inside the stroma. During the night, carbohydrate may be released from the starch grains and translocated to other parts of the plant.

Fig. 4   The two stages of photosynthesis

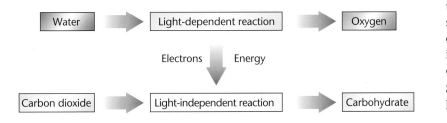

Fig. 5   Inside a chloroplast

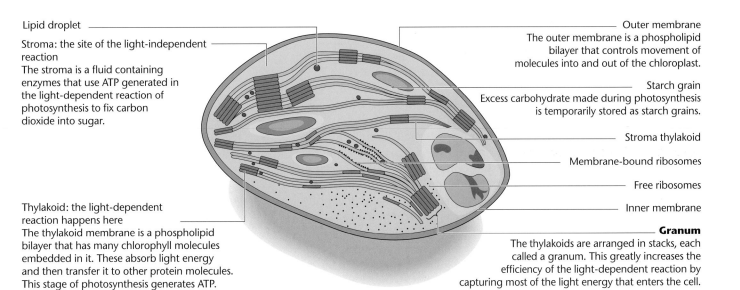

Lipid droplet

Stroma: the site of the light-independent reaction
The stroma is a fluid containing enzymes that use ATP generated in the light-dependent reaction of photosynthesis to fix carbon dioxide into sugar.

Thylakoid: the light-dependent reaction happens here
The thylakoid membrane is a phospholipid bilayer that has many chlorophyll molecules embedded in it. These absorb light energy and then transfer it to other protein molecules. This stage of photosynthesis generates ATP.

Outer membrane
The outer membrane is a phospholipid bilayer that controls movement of molecules into and out of the chloroplast.

Starch grain
Excess carbohydrate made during photosynthesis is temporarily stored as starch grains.

Stroma thylakoid

Membrane-bound ribosomes

Free ribosomes

Inner membrane

Granum
The thylakoids are arranged in stacks, each called a granum. This greatly increases the efficiency of the light-dependent reaction by capturing most of the light energy that enters the cell.

key facts

● ATP is the energy currency of the cell. It is made by combining ADP with inorganic phosphate, a reaction that uses energy. Its breakdown releases energy that the cell can use.

● All cells make ATP as a result of respiration, a series of reactions in which glucose or other energy-containing substances are oxidised.

● Plants photosynthesise, a series of reactions in which carbon dioxide is reduced to form carbohydrates. These carbohydrates contain energy that has been transferred from energy in sunlight.

● Photosynthesis takes place inside chloroplasts. It has two main stages: the light-dependent reaction, in which ATP, reduced NADP and oxygen are formed; and the light-independent reaction, in which carbon dioxide is converted to carbohydrate using some of the products of the light-dependent reaction.

2.2   The light-dependent reaction

In this stage of photosynthesis, energy captured from sunlight by pigment molecules is used to split water, to create ATP, and to produce reduced NADP.

The thylakoid membranes have a similar basic structure to other cell membranes. They are phospholipid bilayers with various proteins floating within them. However, thylakoid membranes also contain many **pigment** molecules. These are arranged in little groups called **photosystems**. Each photosystem contains numerous **chlorophyll** molecules, of two types – chlorophyll *a* and chlorophyll *b* – and other pigments known as **carotenoids**.

When light falls onto any of these pigments, the energy is passed through the photosystem and transferred to a chlorophyll *a* molecule. Chlorophyll *a* is one of a special group of compounds that can absorb light energy and use it to boost the energy level of electrons. Electrons can be compared to satellites orbiting Earth. To move a satellite into a higher orbit requires energy, for example, from firing a booster rocket. Similarly, energy is required to boost an electron into a higher orbit around the nucleus of an atom, from where it can be passed on to another atom more easily.

Chlorophyll *a* absorbs light energy and the excited electrons that it then contains are used to power three important redox reactions –

photolysis, photophosphorylation and the production of **reduced NADP**. Fig. 6 shows a summary of these processes.

Fig. 6   The light-dependent reaction

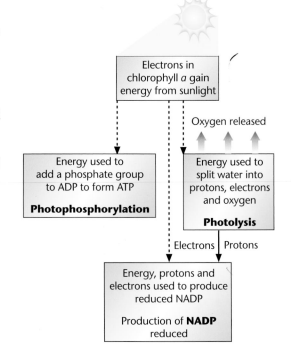

Colours and pigments

White light is a mixture of light of different colours; these are easily seen in a rainbow, or when light is passed through a prism. Each colour of light has a different wavelength. Blue light has a wavelength of approximately 450 nm, red light has a wavelength of approximately 650 nm, and the other visible colours have wavelengths in between.

Objects appear to be different colours because of the wavelengths of light they absorb and reflect. A white object reflects all wavelengths of light, while a black object absorbs all and reflects none. A yellow object reflects yellow light and absorbs other colours.

An **absorption spectrum** shows how much light specific objects or molecules absorb at each wavelength. Fig. 7a shows the absorption spectra of the different pigments in a chloroplast.

Fig. 7b shows a complete absorption spectrum for a leaf, involving all its different pigments. This graph also shows an **action spectrum** for photosynthesis – that is, the rate of photosynthesis at different wavelengths of light.

Fig. 7 Absorption spectra and action spectrum for photosynthesis

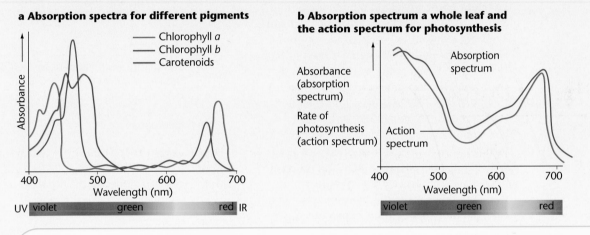

a Absorption spectra for different pigments

— Chlorophyll *a*
— Chlorophyll *b*
— Carotenoids

b Absorption spectrum a whole leaf and the action spectrum for photosynthesis

2

a What are the main colours of light absorbed by chlorophyll *b*?

b Explain why chlorophyll appears green.

c What colour would you expect carotenoids to appear?

3

a Copy the graph axes in Fig. 7b. On your copy, draw the action spectrum of a leaf if it contained only chlorophyll *a*.

b Suggest why it is an advantage to the plant to have several different pigments in its chloroplasts.

4

Light does not penetrate very far into water. Short wavelengths penetrate to greater depths than long wavelengths. Different types of seaweeds contain different photosynthetic pigments. Low down on the shore, where they are covered by deep water when the tide is in, seaweeds tend to be red or orange. Higher up the shore, where they are covered only by shallow water for short periods of the day, they are usually green. Suggest how these different colours may be useful adaptations to their environment.

Photolysis

Energy from excited electrons in chlorophyll *a* molecules is used to split water molecules into protons (H⁺), electrons (e⁻) and oxygen.

$$\underset{\text{energy from excited electrons}}{}$$

water $\rightarrow$ protons + electrons + oxygen
$2H_2O \rightarrow 4H^+ + 4e^- + O_2$

This reaction is known as **photolysis** (photo = using light, lysis = splitting).

The oxygen is released into the atmosphere. This reaction is the source of almost all the oxygen that makes up 20% of the atmosphere, which covers the Earth with a gas blanket 10 miles deep. That's a lot of oxygen. Millions of tonnes of oxygen are present in the atmosphere, only because of the work of photosynthetic organisms over millions of years. Without it there would be no animal life on Earth.

Photophosphorylation

Energy from excited electrons can also be used to produce ATP. Because the energy comes from light, this method of producing ATP is called **photophosphorylation** (photo = using light; phosphorylation = adding a phosphate group).

$$ADP + Pi + [\text{energy from excited electrons}] \rightarrow ATP$$

This process takes place as the energy-containing electrons are passed along a series of **electron carriers** in the thylakoid membranes. These carriers form an **electron transport chain**.

As the electrons are transferred from one carrier to the next, their energy is used to pump hydrogen ions across the membranes, so that the hydrogen ions accumulate inside the spaces in the thylakoids. The hydrogen ions then flow back through the membrane, down their concentration gradient. They have to pass through proteins in the membrane that act as ATPases, making ATP from ADP and inorganic phosphate. This process is very similar to the way that ATP is made during respiration; you can read more about it in Chapter 3.

Production of reduced NADP

The electrons released from water molecules by photolysis are eventually used to reduce carbon dioxide to form glucose. However, electrons cannot react directly with carbon dioxide. The excited electrons in chlorophyll a molecules are transferred to **NADP**, converting the NADP to **reduced NADP.**

With each electron that NADP takes up, it also accepts a proton, and so it does not become negatively charged. The transfer of electrons and protons to NADP requires energy, which is obtained from excited electrons in other groups of chlorophyll a molecules.

energy from excited
electrons

NADP + protons + electrons $\rightarrow$ reduced NADP

NADP + H$^+$ + e$^-$ $\rightarrow$ reduced NADP

---

5 Summarise what happens to each of these substances during the light-dependent reaction of photosynthesis:

a chlorophyll a

b water

c electron carriers in the thylakoid membranes

d ADP

e NADP.

---

key facts

- The light-dependent stage of photosynthesis takes place in the thylakoids, where photosynthetic pigments are embedded.

- Light energy excites electrons in chlorophyll.

- Some of this energy is used in the photolysis of water, producing protons, electrons and oxygen.

- The protons and electrons are picked up by NADP, producing reduced NADP.

- Energy from the excited electrons is used to generate ATP through photophosphorylation. This involves the transfer of electrons along a chain of electron carriers arranged in the thylakoid membranes.

2.3 The light-independent reaction

As we have seen, the light-dependent reaction of photosynthesis requires energy from sunlight. It can therefore occur during the day, but ceases at night. Although the light-independent reaction of photosynthesis carries on after dark, it does so for no more than a few seconds.

However, the next stage of photosynthesis does *not* need an input of light energy. It can therefore take place even in the dark. In practice, however, it cannot go on all night, because it needs the ATP and reduced NADP that have been made in the light-dependent reaction. Once these have been used, the light-independent reaction grinds to a halt.

In the light-independent stage of photosynthesis, the energy from ATP and the electrons from reduced NADP are used to reduce carbon dioxide from the atmosphere and build it into carbohydrate. This carbohydrate can provide a long-term store of energy. It can also be used as building blocks to produce all the other organic molecules – such as fats and proteins – that are needed for growth. Fig. 8 summarises the events that take place during the light-independent reaction. It is often known as the **Calvin cycle**, after the person who first worked out the series of reactions involved in the cycle.

Fig. 8   The light-independent reaction

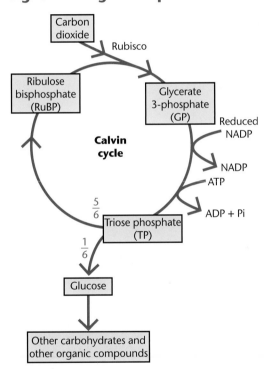

As this is a cycle, we could start anywhere, but let's begin at the top left, where carbon dioxide first comes into the picture. The carbon dioxide combines with a substance called **ribulose bisphosphate**, generally known as **RuBP** for short. The reaction is catalysed by an enzyme called RuBP carboxylase, or **rubisco**. Rubisco is the most abundant enzyme in the world.

An RuBP molecule contains five carbon atoms, and a carbon dioxide molecule contains one, so this reaction produces a six-carbon molecule. This instantly splits into two, forming two three-carbon molecules, called **glycerate 3-phosphate**, **GP**. The carbon dioxide is now

said to be **fixed**, meaning that it has become part of a compound within the plant.

GP is not actually a carbohydrate. It is converted to carbohydrate using energy from ATP and electrons from reduced NADP – both supplied from the light-dependent stage of photosynthesis. This is a reduction reaction. The GP is reduced to form **triose phosphate (TP)**. This *is* a carbohydrate – it is a three-carbon sugar.

All of these reactions take place in the stroma of the chloroplast.

Various things can now happen to the TP. Most of it – about five-sixths – is used to regenerate RuBP. If this did not happen, the chloroplast would quickly run out of RuBP and the reaction would stop. The rest is used to make other kinds of useful organic molecules. TP is a three-carbon compound, so two molecules of it can be combined to form the hexose six-carbon sugar, glucose. Other carbohydrates, such as sucrose, starch or cellulose, can be synthesised depending on the cell's needs. TP can also be converted to fats (lipids) and – with the addition of nitrate or ammonium ions – to amino acids and proteins.

6 Look at Fig. 8. Give an example of:

a an oxidation reaction

b a reduction reaction.

In each case, state which substance is being reduced and which is being oxidised.

7 Suggest what happens to the NADP, ADP and inorganic phosphate that are regenerated in the Calvin cycle. (Remember that this is all taking place inside a chloroplast.)

8 At the beginning of this chapter we saw that photosynthesis transfers energy from sunlight into energy in glucose. How does energy from sunlight enter the Calvin cycle and become incorporated into glucose molecules?

9 Using Fig. 5 and what you now know about the light-dependent and light-independent reactions of photosynthesis, discuss the ways in which the structure of a chloroplast is adapted for its functions.

hsw

how science works

Working out the Calvin cycle

In the 1950s, researchers led by Melvin Calvin were the first to find out what happened during the light-independent stage of photosynthesis. The researchers used apparatus similar to that shown in Fig. 9.

Fig. 9   Apparatus used to investigate Calvin cycle

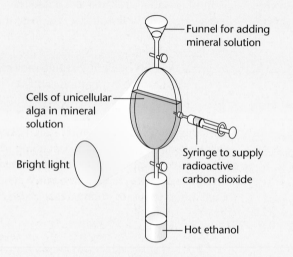

Funnel for adding mineral solution

Cells of unicellular alga in mineral solution

Bright light

Syringe to supply radioactive carbon dioxide

Hot ethanol

The glass-sided disc contained cells of a tiny one-celled photosynthetic organism called *Chlorella*. The apparatus was placed in a dark room and the *Chlorella* cells were supplied with carbon dioxide containing radioactive carbon.

The contents of the apparatus were mixed thoroughly and then a light was switched on. At five-second intervals Calvin extracted a few of the *Chlorella* cells from the apparatus and placed them into hot ethanol to kill them.

The cells were then homogenised (mashed up to form a liquid). The liquid was analysed to find out what it contained. This was done using two-way paper chromatography. This technique involves running the chromatogram with one solvent, then turning the paper through 90° and running it again with a different solvent. Although the substances being separated were not coloured, Calvin could find where they were on the paper because they contained radioactive carbon.

By using this technique, Calvin's team was able to investigate the sequence in which different organic compounds were produced. Fig. 10 shows two of the chromatograms he obtained. The dark spots contain radioactive carbon compounds. The researchers were able to collect samples of these compounds and find out exactly what they were.

In another series of experiments, Calvin's team used similar techniques to investigate the effect of light and dark periods on the compounds formed in the light-independent reaction. Fig. 11 shows the results of one of these investigations.

Fig. 10   Chromatograms

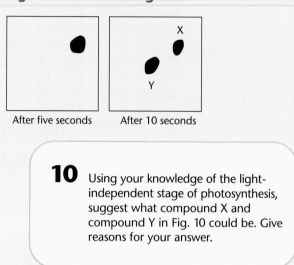

After five seconds   After 10 seconds

10 Using your knowledge of the light-independent stage of photosynthesis, suggest what compound X and compound Y in Fig. 10 could be. Give reasons for your answer.

Fig. 11   Effect of light and dark periods on the compounds formed in the light-independent reaction

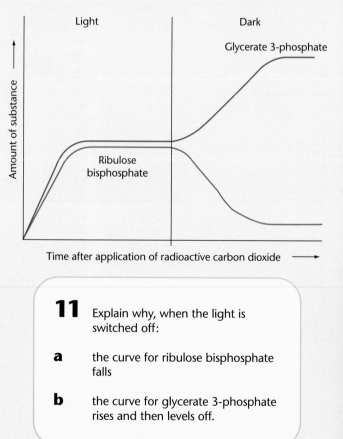

Light   Dark

Glycerate 3-phosphate

Amount of substance

Ribulose bisphosphate

Time after application of radioactive carbon dioxide

11 Explain why, when the light is switched off:

a the curve for ribulose bisphosphate falls

b the curve for glycerate 3-phosphate rises and then levels off.

- The light-independent stage of photosynthesis takes place in the stroma of the chloroplast.

- Carbon dioxide is accepted by the five-carbon compound RuBP to form two molecules of the three-carbon compound glycerate 3-phosphate (GP). This reaction is catalysed by rubisco.

- GP is reduced to triose phosphate (TP), with the help of reduced NADP and ATP generated in the light-dependent reaction.

- One-sixth of the triose phosphate is converted to glucose and other useful organic compounds. Five-sixths is used to regenerate RuBP in the Calvin cycle.

2.4 Limiting factors

What determines how fast a plant can photosynthesise? Obviously, light intensity will have an effect, as this is the energy source that drives the whole process. Carbon dioxide and water are the two raw materials, so these will be critical. Temperature is also likely to have an effect, because high temperatures speed up particle movement and therefore increase the frequency of collisions between reacting particles. On the other hand, very high temperatures denature proteins in living organisms and therefore stop enzyme-catalysed reactions from taking place at all.

The rate of photosynthesis is important in food production because it determines the crop yield. All the factors mentioned above can affect the rate of photosynthesis of a crop. Good crop management involves the manipulation of these factors so as to maximise photosynthesis and achieve good yields.

If one of these environmental factors falls below a certain level, it will start to limit the rate of photosynthesis. Although temperature, carbon dioxide and light may all affect photosynthesis, only the one that is in the shortest supply will limit the rate at any particular moment. This factor is called the **limiting factor**. The rate of photosynthesis can be increased by increasing that factor.

Light
The effect of light on photosynthesis depends on:

- light quality – that is, the wavelengths that it contains; plants can use only certain wavelengths for photosynthesis

- light duration – that is, the day length
- light intensity – that is, how strong the light is.

At low light intensities, an increase in the rate of photosynthesis is directly proportional to increasing light intensity (Fig. 12). But as light intensity increases, photosynthesis reaches a maximum rate and fails to increase further. This could be because:

- The photosynthetic reactions are proceeding as fast as is possible for the photosynthetic 'machinery' in that particular plant.
- Some other environmental factor is now limiting the rate, such as carbon dioxide concentration or temperature.

Fig. 12 Effect of light intensity on the rate of photosynthesis

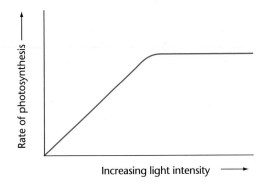

The maximum amount of light that can be used in photosynthetic reactions is estimated to be about 10 000 lux (the SI unit of illuminance). On a clear summer's day, solar illumination may reach 100 000 lux. At this time of year in a sunny field, therefore, light intensity is not the limiting factor. The plants have as much light as they can possibly use. However, at very high light intensities there may be damage to the chlorophyll molecules, resulting in a drop in the rate of photosynthesis.

Temperature

When light intensity is high, increasing temperature can have an effect on the rate of photosynthesis. Between the range 10 °C to about 35 °C, a 10 °C rise in the temperature will double the rate of photosynthesis (Fig. 13).

Fig. 13 Effect of temperature on the rate of photosynthesis

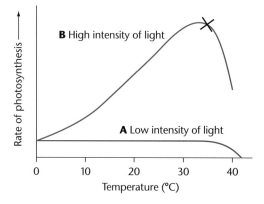

It is interesting to note that temperature does not have a direct effect on the light-dependent reactions of photosynthesis. This is because, unusually for a metabolic reaction, the energy that drives the reaction is light energy. The kinetic energy of the particles involved is almost irrelevant. However, temperature *does* affect the

rate of the light-independent stage, just as it does any other metabolic reaction.

Carbon dioxide

In tropical areas, temperature and light intensity are not usually the limiting factors of photosynthesis; carbon dioxide is. Carbon dioxide is the source of carbon atoms used to make all the organic products of photosynthesis. Carbon dioxide is needed in the light-independent reactions of photosynthesis, where it is reduced to carbohydrate and other organic compounds. Atmospheric carbon dioxide usually makes up only about 0.04% of the volume of the air. This is a tiny proportion. For most plants, this is lower than the optimum value for photosynthesis. So, under normal conditions for crops, carbon dioxide is often the limiting factor (Figs. 14 and 15).

Fig. 14 Effect of carbon dioxide concentration on rate of photosynthesis

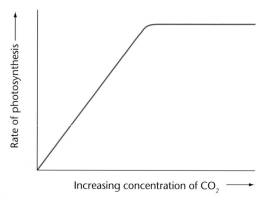

Fig. 15 Limiting factors and photosynthesis

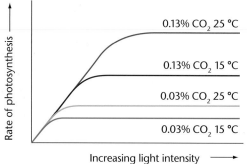

12

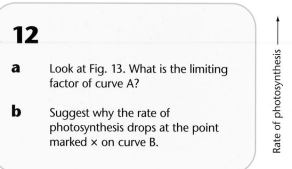

a Look at Fig. 13. What is the limiting factor of curve A?

b Suggest why the rate of photosynthesis drops at the point marked × on curve B.

13 Which curve in Fig. 15 shows carbon dioxide concentration to have the greatest effect as a limiting factor?

Glasshouse management

Imagine that a grower in the UK produces tomatoes to sell in bulk to supermarkets. In our climate, tomatoes can only be grown outdoors for a few months of the year, as they are killed by frost and the fruits grow poorly in low temperatures. So the grower produces the crop in glasshouses. Glasshouse cultivation allows:

- better yields to be achieved
- some crops to be grown out of season and so provide a better economic return
- some plants to be grown in regions where they would not normally grow.

In order to achieve maximum yields, possible limiting factors inside the glasshouse need to be controlled. The faster the plants can photosynthesise, the more carbohydrates they can make. This provides them with both the materials and the energy required to make new cells for growth, and for fruit formation. Rapid photosynthesis means rapid production of a large crop of tomatoes.

Over short periods, 0.5% carbon dioxide has been found to be the optimum concentration for photosynthesis. Over longer periods, however, this concentration may cause the stomata to close, resulting in a drop in photosynthesis. For glasshouse crops like tomatoes, 0.1% carbon dioxide is the optimum over long periods.

Tomato plants die when the temperature falls below about 2 °C. At the other extreme, temperatures above 30 °C can damage tomato plants. The grower aims to keep the temperature somewhere between 15 °C and 25 °C. This allows the plants to photosynthesise rapidly, without there being any danger of cells being damaged. It is just as important to stop temperatures from rising too high in the summer as it is to keep them from dropping too low in the cooler spring and autumn months.

At the ideal temperatures for tomato growth, water loss by transpiration is likely to be high, so the plants will need to be kept well watered.

Fig. 16 Glasshouse environment

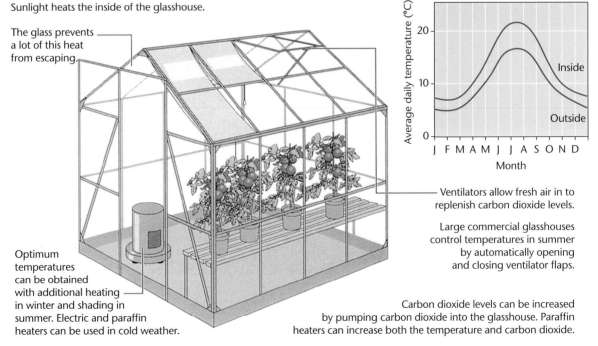

Sunlight heats the inside of the glasshouse.

The glass prevents a lot of this heat from escaping.

Optimum temperatures can be obtained with additional heating in winter and shading in summer. Electric and paraffin heaters can be used in cold weather.

Ventilators allow fresh air in to replenish carbon dioxide levels.

Large commercial glasshouses control temperatures in summer by automatically opening and closing ventilator flaps.

Carbon dioxide levels can be increased by pumping carbon dioxide into the glasshouse. Paraffin heaters can increase both the temperature and carbon dioxide.

Excessive water loss can lead to closure of the stoma, meaning that carbon dioxide will not be able to enter the leaves and photosynthesis will come to a halt. Many glasshouses have automatic watering systems with sprinklers and humidifiers. However, it is also important to regulate humidity in order to limit fungal diseases, which can increase when the humidity is too high.

Artificial lighting can be used in glasshouses when the natural light intensity falls too low. The light sources that are used provide a mix of the wavelengths that can be absorbed by the plant pigments – a spectrum similar to the absorption spectrum shown on page 45. This way, the grower is not spending money on electricity to produce light or wavelengths that will be wasted.

All of these factors can be controlled by computers. Sensors are used to monitor the level of each factor and the feedback is processed by the computer.

Keeping carbon dioxide concentration, temperature and light intensity at optimum levels is expensive. Carbon dioxide can be added from generators or from cylinders, which have to be bought in. Sometimes, it is generated by burning fuels such as paraffin, which has the added bonus that this also increases temperature. Lights run from electricity. The grower has to calculate the expense of controlling all of these factors, and balance it against the price he or she expects to get from the crop. The increase in yield has to be worth more than the costs of environmental control, if it is to be worth the investment.

key facts

● A factor that prevents photosynthesis taking place at a faster rate is known as a limiting factor. Light intensity, temperature and carbon dioxide concentration may all act as limiting factors.

● A knowledge of limiting factors can enable growers to increase the yield of crops grown in glasshouses. The costs of maintaining optimum temperature, light intensity and carbon dioxide concentration need to be outweighed by the greater income from the crop.

1 The diagram shows a cell organelle.

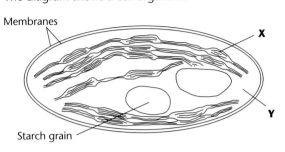

Membranes

X

Starch grain

Y

a Identify the parts labelled **X** and **Y**. (2)
  Light was shone on a suspension of these organelles while it was kept in an atmosphere of pure nitrogen. During this time the organelles made large amounts of two different substances, and gave off oxygen.

b Name the two substances that were made when the organelles were illuminated. (2)
  The membranes around the outside of the organelles were then broken, and the parts labelled **X** and **Y** were separated from each other. In the dark, part **Y** was supplied with a substance, which it converted into a carbohydrate.

c What is the name of the substance which was converted into a carbohydrate? (1)

d **i** Name the part of the organelle where the enzymes of the light-independent reaction are located. (1)
   ii Use the diagram to give the function of one other enzyme that is present in this part of the organelle. (1)

Total 7

AQA, June 2002, Unit 4, Question 1

3 The diagram shows a summary of the light-independent reaction of photosynthesis.

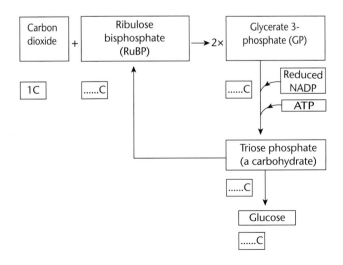

Carbon dioxide

+

Ribulose bisphosphate (RuBP)

→2×

Glycerate 3-phosphate (GP)

1C

......C

......C

Reduced NADP

ATP

Triose phosphate (a carbohydrate)

......C

Glucose

......C

a **i** Complete the boxes to show the number of carbon atoms in the molecules. (2)
   ii In which part of a chloroplast does the light-independent reaction occur? (1)
   iii Which process is the source of the ATP used in the conversion of glycerate 3-phosphate (GP) to triose phosphate? (1)
   iv What proportion of triose phosphate molecules is converted to ribulose bisphosphate (RuBP)? (1)

b Lowering the temperature has very little effect on the light-dependent reaction, but it slows down the light-independent reaction. Explain why the light-independent reaction slows down at low temperatures. (2)

Total 7

AQA, June 2004, Unit 4, Question 2

4 An investigation was carried out to find out the sequence of biochemical changes that occur during photosynthesis. Radioactive carbon dioxide was added to a suspension of algal cells, and they were allowed to photosynthesise. At intervals, samples of the suspension were removed into hot alcohol. These samples were analysed for different radioactively labelled compounds.

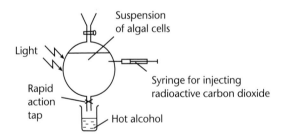

Suspension of algal cells

Light

Syringe for injecting radioactive carbon dioxide

Rapid action tap

Hot alcohol

a Explain how the use of radioactive carbon dioxide in this investigation allows the sequence of biochemical changes in photosynthesis to be followed. (2)

b Suggest a reason for the use in this investigation of
  i hot alcohol (1)
  ii a rapid action tap. (1)
  Samples were removed from the suspension at five different times, between 5 seconds and 600 seconds after the start of the experiment. In each sample, the radioactivity in four different organic compounds, **P**, **Q**, **R** and **S**, was measured. The table on the next page shows the results.

Organic compound	Amount of radioactivity present/arbitrary units				
	5s	**15s**	**60s**	**180s**	**600s**
P	0.01	0.02	0.08	0.17	0.67
Q	1.00	2.00	3.10	3.15	3.15
R	0.10	1.50	2.20	2.30	2.40
S	0.05	0.11	0.16	1.00	1.00

c Use this information to place the compounds **P**, **Q**, **R** and **S**, in the order in which they were formed in photosynthesis. (1)

d Using your knowledge of the light-independent reaction, explain why the level of radioactivity in compound **Q** remained steady after 180 seconds. (2)

Total 7

AQA, January 2002, Unit 4, Question 5

5

a Describe how NADP is reduced in the light-dependent reaction of photosynthesis. (2)

b In an investigation of the light-independent reaction, the amounts of glycerate 3-phosphate (GP) and ribulose bisphosphate (RuBP) in photosynthesising cells were measured under different environmental conditions. **Figure 1** shows the effect of reducing the carbon dioxide concentration on the amounts of glycerate 3-phosphate and ribulose bisphosphate in photosynthesising cells.

Figure 1

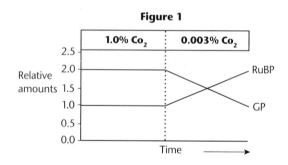

i Explain why there is twice the amount of glycerate 3-phosphate as ribulose bisphosphate when the carbon dioxide concentration is high. (1)

ii Explain the rise in the amount of ribulose bisphosphate after the carbon dioxide concentration is reduced. (1)

c **Figure 2** shows the results of an experiment in which photosynthesising cells were kept in the light and then in darkness.

Figure 2

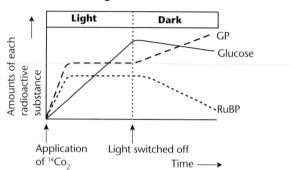

i In the experiment the cells were supplied with radioactively labelled $^{14}CO_2$. Explain why the carbon dioxide used was radioactively labelled. (1)

ii Explain how lack of light caused the amount of radioactively labelled glycerate 3-phosphate to rise. (2)

iii Explain what caused the amount of radioactively labelled glucose to decrease after the light was switched off. (1)

Total 8

AQA, January 2006, Unit 4, Question 3

6 Tomato growers have increased the yield of fruit from 100 to 400 tonnes per hectare by growing the tomato plants in automatically heated glasshouses and enhancing the carbon dioxide concentration. To control the nutrient supply to the roots, the plants are grown without soil in plastic troughs, as shown in the diagram.

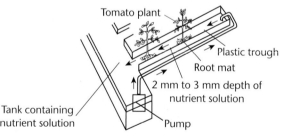

a Explain how enhancing the carbon dioxide concentration helps to increase the yield. (2)

b Maintaining a high temperature in a glasshouse in winter, when the light intensity is low, may reduce the yield. Explain how. (2)

c Tomato fruits have a high percentage of water. When making tomato ketchup, it is more economical to use fruits which have a low percentage of water. Growers can reduce the water content of the fruit by adding sodium chloride to the nutrient solution in the plastic trough. Explain how adding sodium chloride can reduce the water content of the fruit. (2)

Total 6

AQA, January 2004, Unit 6, Question 2

how science works **assignment**

Maximising the yield

Tomato plants.

Tomatoes are an important crop in many parts of the world. They may be used locally for food, but more often they are an important export crop, bringing in significant income for farmers. Growers have to decide whether it is best to use a low-tech approach, simply growing the tomatoes in open fields, or whether it may be worth making a substantial investment to build glasshouses where they can control the conditions in which the plants are growing.

An investigation was carried out into the effect of temperature on the growth of tomatoes. Tomato plants of the same variety were grown in an open field and also in three different glasshouses. The experiment was carried out in a subtropical country, where the outside temperature never dropped below 29.5 °C.

Even when not heated, the temperature inside a glasshouse is generally higher than outside. This is because of the greenhouse effect – short wavelength rays from the Sun pass through the glass into the glasshouse, and are reflected from the surfaces inside the glasshouse as longer wavelengths, which cannot escape. These longer wavelengths warm the air inside the glasshouse.

The glasshouses in the experiment had different coverings. Fig. 17 shows the variations in temperature in the four growing environments throughout the 10 months of the investigation.

A1

a Compare the temperature in growing environments T1 and T2.

b Explain the reasons for the differences you have described.

Equal numbers of one-month-old plants of the same tomato variety were planted into each of the growing environments in February. The table shows the yields of fruit from the tomatoes in each of the four growing environments.

Growing environment	Mean plant height (cm)	Mean fruit yield per plant (g)
T1	97.50	1 348.5
T2	105.83	2 145.2
T3	104.83	2 055.0
T4	75.40	981.0

A2 What appears to be the temperature range at which this variety of tomato plants produces the highest yield? Use the information in the graph and the table to support your answer.

A3 Suggest one factor, other than temperature, that may have been partly responsible for the differences in yield in T1 and T4. How significant might this factor have been? Do you think that it brings the reliability of the results into question?

A4 Discuss the extent to which the results of this investigation could be used by growers in a temperate country such as the UK to help them to determine the best environment in which to grow tomatoes.

Fig. 17   Variations in temperature

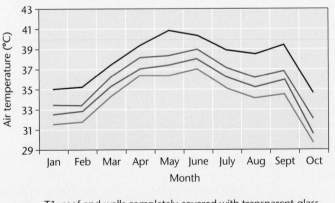

—— T1: roof and walls completely covered with transparent glass

—— T2: roof covered with transparent glass; all four walls covered with shading allowing in only 75% of the light

—— T3: roof and two walls covered with transparent glass; two walls covered with shading allowing in only 75% of the light

—— T4: open field

3 Respiration

Deep inside the underground passages leading from Peak Cavern in Derbyshire is a skeleton of a 20-year-old man, Neil Moss. He died there in March 1959, and his body was never removed.

Neil, a university student, was part of a group of experienced cavers intending to explore a recently discovered fissure deep in the cave complex. It was difficult enough getting to the fissure – the group had to squeeze through a wet and narrow tunnel called Mucky Ducks, followed by a long crawl through a passage where they often had to lie full length in order to wriggle through, ending with a very awkward bend. By the time they reached the unexplored opening, they were already a long way from the surface and in total darkness.

The group managed to manoeuvre a lightweight ladder into the fissure, and Neil climbed down it into the unknown space below. It was a very narrow cleft, part of it twisting like a corkscrew. Leaving the ladder, he worked his way down through the narrow, twisting crack. Eventually, he reached a point where he could go no further, and asked his friends to help pull him out.

They got a line down to him, but it broke when they tried to pull him up. They tried again – and again – but each time the line snapped. Neil was panicking now. He was somehow wedged into the cleft below the corkscrew so that he could not even bend his legs or elbows. He was totally stuck.

More help was needed. Experienced cave rescuers arrived, bringing a stronger rope with them. But by the time it arrived, Neil was unconscious. The rope dangled near his head, but he was totally unaware of it and could do nothing to help himself. The air around his face was now full of the carbon dioxide he had been breathing out, and the lack of oxygen caused his brain cells to run out of fuel and switch off. He never regained consciousness. It was decided that no-one else should risk their life in an effort to remove his body, and it is still there today.

Neil died because of a lack of oxygen in his blood. We can survive for a maximum of about six minutes without oxygen. Without water, we can live for about six days, and for about six weeks without food. Oxygen is easily the most urgent of our needs. Without it, respiration in our cells stops. The cells quickly run out of ATP, and can no longer carry out the energy-requiring metabolic reactions that maintain life. Brain cells are the most vulnerable to lack of oxygen because they can only make ATP through aerobic (oxygen-requiring) respiration.

3.1 The stages of respiration

In Chapter 2, we saw how photosynthesis transfers energy from sunlight into energy in glucose and other organic molecules. Respiration releases this energy again, using it to make ATP, which is the energy currency of all cells.

Respiration is a series of oxidation reactions. It is made up of many small steps, which we can group into four stages.

- **Glycolysis** – this word means 'sugar splitting'. In glycolysis, glucose molecules (each containing six carbon atoms) are split and converted into a 3-carbon compound called pyruvate. In the process, some of the

energy from the glucose is used to make a small number of ATP molecules. This stage happens in the cytoplasm of the cell.
- **The link reaction** – in this stage, the pyruvate made in glycolysis is transported into a mitochondrion. Here, the pyruvate is oxidised to produce acetyl coenzyme A.
- **The Krebs cycle** – still in the mitochondrion, the acetyl coenzyme A has electrons and carbon dioxide removed from it. This happens in the matrix of the mitochondrion.
- **The electron transport chain** – transport of the electrons down a series of electron

carriers transfers energy, which is used to produce ATP. This happens on the membranes of the cristae inside the mitochondrion, and is known as oxidative phosphorylation.

Fig. 1 shows the structure of a mitochondrion, and summarises the stages of respiration that take place in the different parts.

1 What similarities are there between the structure of a mitochondrion and a chloroplast (shown in Fig. 5 on page 43)?

Fig. 1   A mitochondrion

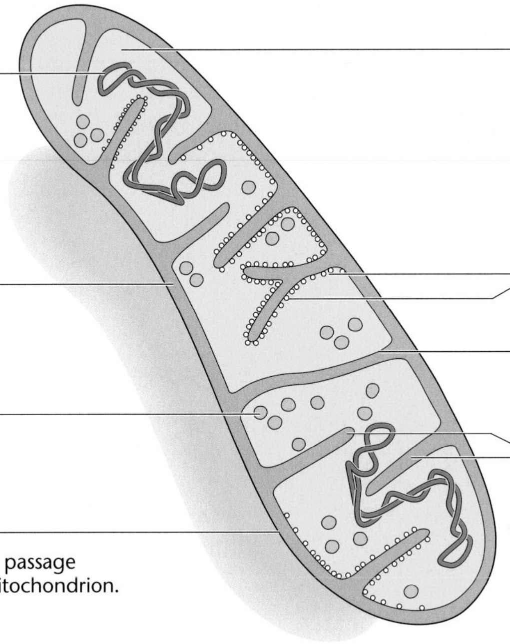

DNA
The DNA threads contain the information required for the mitochondria to replicate.

Matrix
The matrix contains the enzymes involved in the oxidative decarboxylation of pyruvate, and in the Krebs cycle

Intermembrane space

ATPase molecules that generate ATP

Inner membrane

Ribosome

Cristae
The inner phospholipid membrane is highly folded to form cristae. The membranes of the cristae contain the electron transport chain

Outer membrane
The outer membrane controls the passage of materials into and out of the mitochondrion.

key facts

● Respiration is the oxidation of glucose (or other substrates). During the reaction energy is released from glucose molecules. Much of this energy is used to synthesise ATP.

● Aerobic respiration involves four stages. Glycolysis takes place in the cytoplasm. The link reaction and the Krebs cycle take place in the matrix of a mitochondrion. Oxidative phosphorylation takes place in the inner membrane (on the cristae) of a mitochondrion.

3.2   Glycolysis

Glycolysis is the first stage of respiration. The overall process of glycolysis, which takes place in the cytoplasm, is illustrated in Fig. 2.

Sugars are not very reactive. Therefore, to start the process, two molecules of ATP are used to add two phosphate groups to the glucose molecule. This produces a phosphorylated 6-carbon sugar.

The phosphorylated sugar is then broken down into two molecules of the 3-carbon sugar **triose phosphate**. Each triose phosphate molecule is then converted to **pyruvate**. This is an oxidation reaction. Electrons are removed from the triose phosphate and transferred to a coenzyme called **NAD** (nicotinamide adenine dinucleotide). This produces **reduced NAD**.

Fig. 2   Glycolysis

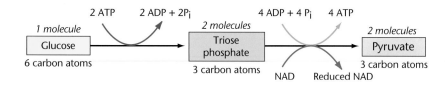

As we will see later, reduced NAD behaves in a similar way to reduced NADP (which you learned about in photosynthesis).

The conversion of two triose phosphates to pyruvate also releases enough energy to convert four molecules of ADP + Pi to four molecules of **ATP**.

So, two molecules of ATP were used at the start of glycolysis, and four have been made by the end. There is thus a net gain of two molecules of ATP from every molecule of 6-carbon sugar broken down during glycolysis.

The end products of glycolysis are pyruvate, reduced NAD and ATP. Reduced NAD is generated at several stages in respiration. In section 3.5 we will consider the outcome of all the reduced NAD produced.

> **2** Triose phosphate is involved in both photosynthesis and respiration. Compare the way in which triose phosphate is *produced* in photosynthesis with the way in which it is converted to pyruvate in glycolysis.

key facts

- In glycolysis, glucose is converted to pyruvate in a series of small steps.

- Each glucose molecule is phosphorylated, and then split to form two triose phosphate molecules. This uses two molecules of ATP for each molecule of glucose.

- NAD takes two electrons from each triose phosphate molecule, becoming reduced NAD.

- As the triose phosphate is converted to pyruvate, enough energy is released to make two molecules of ATP for every triose phosphate – that is, four molecules of ATP for every glucose molecule that we started with.

3.3   The link reaction

This stage of aerobic respiration links glycolysis, which occurs in the cytoplasm, with the Krebs cycle, the main ATP-generating phase that occurs in mitochondria. Fig. 3 gives an overview of the link reaction.

At the start of the link reaction, pyruvate produced by the process of glycolysis leaves the cytoplasm and enters the mitochondria. There,

NAD removes electrons from the pyruvate, oxidising it to a 2-carbon acetate molecule and carbon dioxide. The acetate is picked up by a carrier molecule – coenzyme A – and **acetylcoenzyme A** is formed. Because a carbon atom is removed from pyruvate in this process, and because the pyruvate is oxidised, this reaction is termed **oxidative decarboxylation**.

Fig. 3   The link reaction

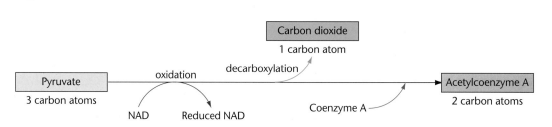

3.4 The Krebs cycle

Acetylcoenzyme A then enters a cyclic series of reactions called the Krebs cycle, named after Hans Krebs, who discovered the sequence. The reactions of the Krebs cycle, summarised in Fig. 4, occur in solution in the matrix of the mitochondria. The main goal of the Krebs cycle is to provide a continuous supply of electrons to fuel the next stage of respiration, the electron transport chain. The Krebs cycle

itself produces only a small amount of ATP directly.

At the start of the Krebs cycle, the 2-carbon molecule of acetylcoenzyme A combines with a 4-carbon organic acid to form a 6-carbon molecule. At two points a molecule of carbon dioxide is then removed; the resulting 4-carbon molecule is recycled, and it combines with acetylcoenzyme A so that the cycle keeps on turning.

During these reactions, several electrons are also removed from the reactants and transferred to NAD and FAD, forming reduced NAD and reduced FAD, respectively, and one molecule of ATP is formed. (FAD stands for flavine adenine dinucleotide; it is a similar substance to NAD and NADP and has a similar role.)

If you look at Figs. 3 and 4, you can see that three carbon dioxide molecules are eventually removed from the original pyruvate molecule. This accounts for all of the three carbon atoms in the pyruvate. This is where the carbon dioxide comes from, which is the waste product of respiration.

You may have noticed, however, that so far there has been no mention of oxygen. This is because there is no need for oxygen throughout the first three stages of respiration (glycolysis, the link reaction and the Krebs cycle). However, oxygen is vital for the next and final stage, which is the production of ATP in the electron transport chain.

Fig. 4   The Krebs cycle

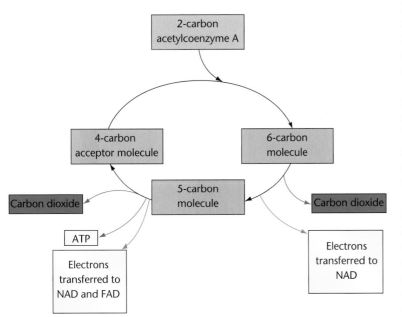

3.5 The electron transport chain

In either the link reaction or the Krebs cycle, all the electrons removed from the three carbon atoms in the pyruvate molecule are transferred to NAD or FAD to produce reduced NAD or reduced FAD. If you look back at Fig. 2, you will also see that some reduced NAD was produced in glycolysis.

This reduced NAD and FAD contains chemical energy that has been transferred from

the glucose molecule that started the whole process. Inside the mitochondrion, some of this energy is now used to make ATP.

A summary of the reactions that occur in the electron transport chain is shown in Fig. 5.

Reduced NAD and reduced FAD transfer their electrons to a chain of electron carrier molecules embedded in the phospholipid membranes inside the mitochondria. When electrons are transferred

Fig. 5   The electron transport chain

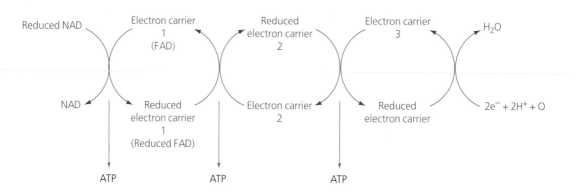

from reduced NAD or reduced FAD and pass along the chain of electron carriers, they provide the energy to power the active transport of hydrogen ions (protons) from the mitochondrial matrix, across the inner mitochondrial membrane and into the space between the two membranes.

The movement of hydrogen ions produces a high concentration of hydrogen ions between the two membranes, and a relatively low concentration in the matrix. There is therefore a steep concentration gradient, so the hydrogen ions have a tendency to diffuse back into the matrix. To do this, they have to get across the inner mitochondrial membrane, and this is not easy. There is usually only one route for them – through an enzyme that spans the membrane. As hydrogen ions pass through this enzyme, they provide enough energy for the enzyme to attach Pi groups to ADP molecules to form ATP. The enzyme is therefore called an **ATPase**.

The passage of one pair of electrons along the chain of carriers from reduced NAD provides enough energy to move sufficient hydrogen ions to produce three molecules of ATP. The first electron carrier is FAD; only two molecules of ATP are produced from reduced FAD.

So now we can see how respiration generates large amounts of ATP. But still no oxygen has been involved. It is not until the end of the process that its role becomes clear.

The final carrier in the chain transfers the electrons to oxygen atoms. Each oxygen atom picks up two protons and two electrons to produce a molecule of water, which is released back into the cell.

It may be surprising that this simple role is so vital; that, without oxygen, we cannot make enough ATP to stay alive. Without oxygen, there is nowhere for the electrons to go once they arrive at the end of the electron transport chain. So the electrons remain in the chain. This means that the reduced NAD and reduced FAD molecules cannot offload their electrons, so there is no unoccupied NAD or FAD to accept the electrons released by the Krebs cycle. Therefore, the Krebs cycle has to stop, and acetylcoenzyme A and pyruvate accumulate.

Cells do have a temporary way out of this fix. They are able to dispose of the accumulated pyruvate so that at least glycolysis can continue to run for a while, even when no oxygen is available. The way in which they do this is described in section 3.7 (Anaerobic respiration).

3.6 How much ATP is produced during aerobic respiration?

ATP is produced in two different ways during respiration – by **substrate-level phosphorylation** and by **oxidative phosphorylation**.

Substrate-level phosphorylation means that the energy released from a particular reaction is transferred directly to make ADP and Pi combine to form ATP. This chemical reaction happens:

- during glycolysis, when triose phosphate is converted to pyruvate, there is a net production of four molecules of ATP per molecule of glucose
- during the Krebs cycle, in the production of a 4-carbon molecule from a 5-carbon molecule, one molecule of ATP is produced per circuit.

Oxidative phosphorylation occurs when electrons from reduced NAD and reduced FAD flow along the electron transport chain.

- Three molecules of ATP are produced for every pair of electrons that are transferred from reduced NAD into the electron transport chain.
- Two molecules of ATP are produced from every pair of electrons that are transferred from reduced FAD into the electron transport chain.

Table 1 shows the number of molecules of reduced NAD and reduced FAD produced from one mole of glucose during the different stages in aerobic respiration.

3

a   Using Table 1, calculate the number of molecules of ATP that are theoretically produced during respiration. (Remember to count those that are produced by substrate-level phosphorylation as well.)

b   When measurements are made of the actual quantity of ATP made per glucose molecule in a cell, we find that it is less than this theoretical value. This is because some ATP has to be used to provide energy to move ADP into the mitochondrion, and also to move ATP out of it and into the cytoplasm. About 25% of the energy yield from oxidative phosphorylation is used for these purposes. Taking this into account, estimate the total actual yield of ATP from one molecule of glucose.

Table 1   Reduced NAD and reduced FAD produced during aerobic respiration

Stage	Number of molecules of reduced NAD formed per glucose molecule	Number of molecules of reduced FAD formed per glucose molecule
Glycolysis	2	
Link reaction	2	
Krebs cycle	6	2

key facts

- In the link reaction, pyruvate produced in glycolysis moves into a mitochondrion, where it is combined with coenzyme A and decarboxylated to form a 2-carbon compound, acetylcoenzyme A.

- In the matrix of the mitochondrion, acetylcoenzyme A is combined with a 4-carbon compound to make a 6-carbon compound. This is converted back to the 4-carbon compound in a series of enzyme-controlled steps. In several of these steps, carbon dioxide and/or electrons are removed. The electrons are picked up by NAD or FAD. In one step, ATP is synthesised from ADP and Pi. This cyclic series of reactions is called the Krebs cycle.

- The reduced NAD and reduced FAD pass their electrons to the first compound in a series of electron carriers situated in the inner membrane of the mitochondrion, called the electron transport chain. The electrons are passed along the chain, and are finally accepted by oxygen, forming water.

- The energy released from the electrons as they pass along the electron transport chain is used to pump hydrogen ions (protons) from the matrix, across the inner membrane and into the space between the two membranes. As the hydrogen ions move back down their concentration gradient, they pass through ATPases, and energy from them is used to synthesise ATP.

3.7 Anaerobic respiration

As we have seen, if oxygen is not available, the electron transport chain, the Krebs cycle and the link reaction grind to a halt. Pyruvate builds up in the cell. However, if the cell can get rid of this pyruvate, then it may at least be able to keep glycolysis going for a little while longer, even with no oxygen.

Human cells do this by converting pyruvate to **lactate**. This reaction uses reduced NAD, oxidising it to NAD once more (Fig. 6). This has the benefit that there is now NAD available to accept electrons from the glycolysis reactions. If this NAD were not regenerated, then even glycolysis would have to stop, because there would no unreduced NAD available to accept these electrons.

Fig. 6 The lactate pathway

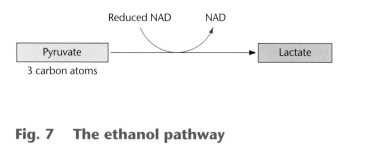

Fig. 7 The ethanol pathway

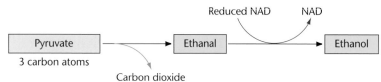

Plant cells and yeast have a different solution to the problem. They convert the pyruvate to **ethanol** and **carbon dioxide** (Fig. 7). The result, however, is similar. Not only is pyruvate removed, but NAD is regenerated.

These two sets of reactions – either glycolysis plus the lactate pathway, or glycolysis plus the ethanol pathway – make up **anaerobic respiration**.

Anaerobic respiration produces only a very small amount of ATP – just the molecules that are made during glycolysis. It cannot fuel our cells for very long, but it can help to tide them over for a short period when oxygen is not available, or to squeeze out an extra few ATP molecules when we are pushing our muscle cells to the limit by asking them to work particularly hard. However, the lactic acid that is produced is toxic and we cannot continue to produce it indefinitely. It has to be removed from the respiring cells and broken down.

The lactic acid produced in our anaerobically respiring muscle cells diffuses into the blood plasma, and is carried away in solution. As the lactic acid passes through the liver, the liver cells absorb it and metabolise it. To do this, the cells need oxygen. So carrying out anaerobic respiration did not mean we managed to do without oxygen – we just delayed the moment when it would be needed. Even after completing strenuous exercise, we continue to breathe heavily, taking in extra oxygen that the liver cells need to break down the lactic acid.

key facts

- Anaerobic respiration takes place when oxygen is in short supply, and the electron transport chain, the Krebs cycle and the link reaction cannot take place.

- Glycolysis continues as normal. To prevent the build-up of pyruvate, and to regenerate NAD, the pyruvate is converted either to lactate (in animals) or to ethanol and carbon dioxide (in plants and yeast organisms).

- Glycolysis plus either the lactate pathway or the ethanol pathway make up anaerobic respiration.

- Anaerobic respiration generates a very small amount of ATP compared with aerobic respiration.

3.8 Energy transfer in photosynthesis and respiration

A comparison of some of the fundamental features of photosynthesis and respiration is shown in Fig. 8.

In photosynthesis, reduced NADP acts as an electron donor to carbon (in carbon dioxide). In respiration, a similar compound, NAD, accepts high-energy electrons from carbon (in glucose) and becomes reduced NAD in the process. An easy way to remember which of these is involved in photosynthesis and which in respiration, is if you remember that the one with the P (NADP), is used in photosynthesis and the other (NAD) is used in respiration.

During respiration, carbon is oxidised. Four electrons are removed from each atom of carbon, which is exactly the opposite of photosynthesis. Whereas energy is required to add electrons to organic molecules in photosynthesis, removing electrons in respiration releases energy.

Fig. 8   Comparison of photosynthesis and respiration

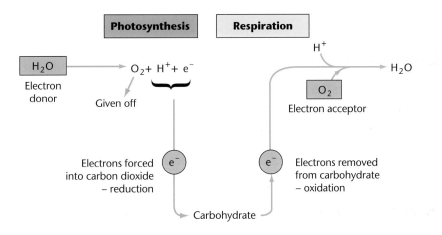

key facts

● In photosynthesis, energy from sunlight is used to generate reduced NADP. This donates electrons to carbon dioxide, producing carbohydrates.

● In respiration, carbohydrates are broken down to release electrons. These are picked up by NAD, which donates the electrons to the electron transport chain. The energy from the electrons is used to produce ATP, the energy currency of every cell.

1 The boxes in the diagram represent substances in glycolysis, the link reaction and the Krebs cycle.

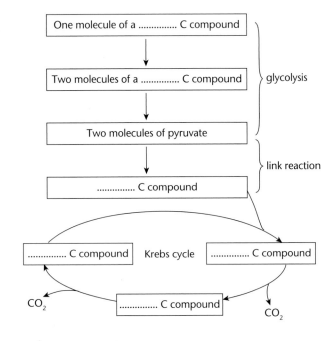

a Complete the diagram to show the number of carbon atoms present in **one** molecule of each compound. (2)

b Other substances are produced in the Krebs cycle in addition to the carbon compounds shown in the diagram. Name **three** of these other products. (3)

Total 5

AQA, January 2005, Unit 4, Question 2

2

a Mitochondria in muscle cells have more cristae than mitochondria in skin cells. Explain the advantage of mitochondria in muscle cells having more cristae. (2)

b Substance **X** enters the mitochondrion from the cytoplasm. Each molecule of substance **X** has three carbon atoms.

i Name substance **X**. (1)

ii In the link reaction substance **X** is converted to a substance with molecules effectively containing only two carbon atoms. Describe what happens in this process. (2)

c The Krebs cycle, which takes place in the matrix, releases hydrogen ions. These hydrogen ions provide a source of energy for the synthesis of ATP, using coenzymes and carrier proteins in the inner membrane of the mitochondrion.

Describe the roles of the coenzymes and carrier proteins in the synthesis of ATP. (3)

Total 8

AQA, June 2004, Unit 4, Question 7

3

a The table contains some statements relating to biochemical processes in a plant cell. Complete the table with a tick if the statement is true or a cross if it is not true for each biochemical process.

Statement	Glycolysis	Krebs cycle	Light-dependent reaction of photosynthesis
NAD is reduced			
NADP is reduced			
ATP is produced			
ATP is required			

(4)

b An investigation was carried out into the production of ATP by mitochondria.

ADP, phosphate, excess substrate and oxygen were added to a suspension of isolated mitochondria.

i Suggest the substrate used for this investigation. (1)

ii Explain why the concentration of oxygen and amount of ADP fell during the investigation. (2)

iii A further investigation was carried out into the effect of three inhibitors, **A**, **B** and **C**, on the electron transport chain in these mitochondria. In each of three experiments, a different inhibitor was added. The table shows the state of the electron carriers, **W–Z**, after the addition of inhibitor.

Inhibitor added	Electron carrier			
	W	**X**	**Y**	**Z**
A	oxidised	reduced	reduced	oxidised
B	oxidised	oxidised	reduced	oxidised
C	reduced	reduced	reduced	oxidised

Give the order of the electron carriers in this electron transport chain. Explain your answer. (2)

Total 9

AQA, June 2006, Unit 4, Question 7

4 The diagram shows glycolysis and the Krebs cycle.

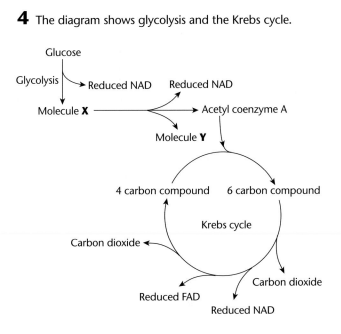

a Name
  i molecule **X** (1)
  ii molecule **Y**. (1)
b Where, in a cell, does glycolysis occur? (1)
c High concentrations of ATP inhibit an enzyme involved in glycolysis.
  i Describe how inhibition of glycolysis will affect the production of ATP by the electron transfer chain. (1)
  ii Explain this effect. (3)
  Total 7

AQA, June 2007, Unit 4, Question 1

5 In an investigation of aerobic respiration, isolated mitochondria were added to a prepared medium containing succinate and inorganic phosphate. Succinate is a 4 carbon compound, which occurs in the Krebs cycle, and can be used as a respiratory substrate. The medium was saturated with oxygen. Equal amounts of ADP were added at one-minute intervals, and measurements were taken of the oxygen concentration in the medium. The graph shows the results.

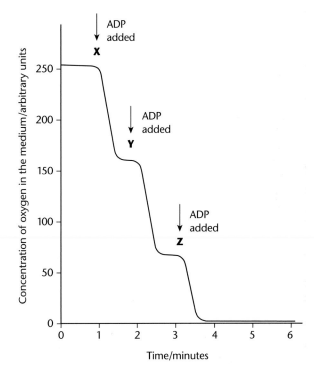

a Why was inorganic phosphate added to the medium? (1)
b Explain why the oxygen concentration in the medium decreased after adding ADP at **X**. (3)
c Explain why the fall in oxygen concentration was the same following the addition of ADP at **X** and at **Y**. (1)
d Explain why the fall in oxygen concentration, following the addition of ADP, was less at **Z** than at **Y**. (1)
e Fresh mitochondria were isolated from cells and a similar experiment was carried out. This time the medium contained glucose instead of succinate. Again, the medium was saturated with oxygen, and excess ADP was added. However, there was almost no fall in oxygen concentration, even after 10 minutes.
  i Suggest and explain a reason for this observation. (2)
  ii Explain, in outline only, how you could test your suggestion. (1)
  Total 9

AQA, January 2002, Unit 4, Question 10

Aerobic training

Many sports events require muscles to use large quantities of energy very quickly. The muscles therefore break down large amounts of ATP, which has to be replaced via respiration. The faster an athlete's muscles can generate ATP, the more power he or she can generate.

Paula Radcliffe, British world record-holder for long distance running; from the start of each race, she sets a punishing pace.

Aerobic training – that is, training that requires your muscles to respire rapidly, so that you have to breathe faster and deeper – produces changes in the structure and physiology of the fibres that make up our muscles. These fibres contain a dark-red pigment called myoglobin. This is similar in structure to one of the four polypeptides that make up haemoglobin; both of these pigments are able to combine with oxygen and then release it again. Myoglobin, however, only releases oxygen when the oxygen concentration has become very, very low. It therefore acts as an oxygen store in muscle tissues, providing the last few molecules of oxygen when the rate of supply by the blood is not enough. Aerobic training increases the quantity of myoglobin stored in muscles by up to 75%.

Another change brought about by aerobic training is an increase in the number and size of mitochondria in the muscles, and also in their ability to use oxygen rapidly and make large quantities of ATP at great speed.

This latter improvement is due to increased production of the enzymes involved in the various stages of respiration. Fig. 9 shows how the activity of one of the enzymes of the Krebs cycle, called succinate dehydrogenase, changed in an athlete undergoing aerobic training for a period of seven months.

Fig. 9   Changes in the activity of succinate dehydrogenase

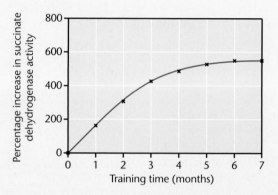

It can also be useful to measure the rate at which a sample of muscle from an athlete is able to use oxygen. In a laboratory, a small specimen of muscle is obtained by a needle biopsy. The sample is ground into a solution. This releases the mitochondria, which start to use oxygen and generate ATP. Fig. 10 shows the maximum rate at which muscle extracts taken from three people were able to use oxygen.

Fig. 10   Maximum rates at which muscles use oxygen

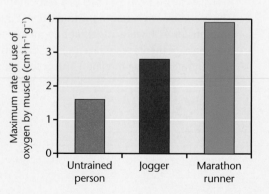

A1 Suggest which of these activities could be part of an aerobic training regime: swimming, weight lifting, running. Explain your answer.

A2 Think back to your AS work (or look back at section 12.5 in *AS Biology*). Sketch a dissociation curve for haemoglobin. On the same axes, sketch a curve for myoglobin.

A3 Explain how an increased quantity of myoglobin in a muscle, such as is produced by aerobic training, could give a 1500 m runner an edge over a rival with less myoglobin.

A4 Explain how the changes in succinate dehydrogenase activity shown in Fig. 9 could help to improve an athlete's performance.

A5 Discuss the changes in muscle structure and physiology that could account for the improved rate of oxygen use by a muscle sample taken from a marathon runner compared with an untrained person, shown in Fig. 10. (Some are mentioned on this page, but you may be able to think of others as well. Remember, though, that we are looking at the performance of a *sample* of muscle in a test tube, not in the person's body.)

4 Energy in ecosystems

Early humans were hunter-gatherers. We know something about their lifestyle because there are still some relatively small groups of people who live in a similar way today, for example, the Hadza tribe in Tanzania. In many ways it is a laid-back lifestyle; generally, the men hunt to kill animals for meat and skins, while the women collect plant foods. The human population density remains low, because a lot of land is needed to supply food to support them.

In Britain, the major sources of food for hunter-gatherers were fruits, nuts and seeds and also a variety of wild animals. Mammals such as deer and wild boar were hunted and killed. Animal foods also came from the sea, and fish and shellfish formed an important part of the diet of these early people. We still behave as large-scale hunter-gatherers today, for example, when we catch fish from the sea. And many of us enjoy picking blackberries in late summer.

About 10 000 years ago, people in Britain began to take control over the sources of their food. They became agriculturalists. They had already domesticated dogs, which must have helped with hunting. Now they also began to domesticate grazing animals, including cattle, sheep, goats and pigs. Crops were planted, tended and harvested, to provide food and other resources such as clothing and building materials.

Agriculture has allowed our population to grow much larger than hunting and gathering could ever do. We are able to use a greater proportion of the Sun's energy to produce large quantities of food in a relatively small area. Modern technology has allowed us to do this using relatively few people-hours, so that most of us now have nothing to do with food production and can spend our time doing other things. In the distant past, before farming began, finding food would have been the main activity of almost everyone in society. Without agriculture, human culture could never have developed to produce the complex society we see today.

A tribe member of the Hadza people (one of the last hunter-gatherer societies that exists) sometimes hunts with the help of the tribe's domesticated dogs.

Grains and sheep are two forms of agriculture that have allowed our population to grow.

4.1 Energy transfer

In Chapters 2 and 3 we saw how plants transfer energy from sunlight into organic molecules such as glucose, and how all organisms use respiration to transfer energy from glucose into ATP. The ATP is then used as the energy currency of the cell – the immediate source of energy for the myriad of energy-consuming metabolic reactions that maintain life.

Within any ecosystem, the community contains organisms that produce energy-containing organic molecules, using an external energy source to make inorganic molecules react together. These organisms are called **producers**. Usually, this is done by photosynthesis. However, there are a few ecosystems where other sources of energy are used, for example, in the communities found around deep-sea volcanic vents, where bacteria use energy from minerals to synthesise organic molecules.

The rest of the community relies on the producers. All the other organisms are **consumers**, so-called because their energy source comes from the organic nutrients they consume – which originally came from the producers.

Knowledge of how energy is passed from one organism to another is of great value in helping us to understand the interrelationships between the different species in a community. It is also vital to us in producing our own food. How can we harness more of the Sun's energy to make more food to feed everyone in the world? And can we do this without destroying the habitats and livelihoods of other species? As the human population continues to grow, these are questions that we urgently need to address.

Food chains

Organisms that feed directly on producers are called **primary consumers**, and they are usually **herbivores** (because producers are usually plants). Animals that eat primary consumers are **secondary consumers**. Above them are the **tertiary consumers** and so on. Secondary and tertiary consumers are **carnivores**. This sequence forms a **food chain**. A food chain is the sequence in which energy is passed from one organism to another,

in the form of chemical energy in organic nutrients. Fig. 1 shows two food chains; one based on land (terrestrial) and one in water (aquatic).

Each step in a food chain is a **trophic level**. 'Trophic' means 'feeding'. In both food chains in Fig. 1, there are five trophic levels.

One group of organisms that is often overlooked when constructing food chains is the **decomposers**. These are organisms that feed on the dead bodies and waste materials of organisms from all the trophic levels. Organisms that feed on corpses and excreta don't have a very good public image, but they play a vital role in recycling the materials that they were made from, as we shall see in Chapter 5.

1 If decomposers were added to the food chains in Fig. 1, at what point, or points, would they appear?

Fig. 1   Two examples of food chains

The arrows show the direction of energy flow from one trophic level to the next.

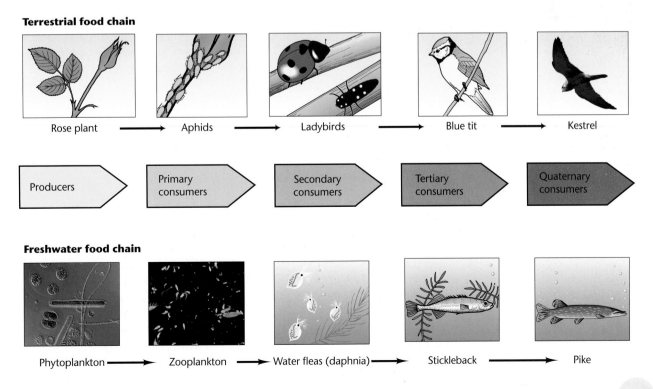

Terrestrial food chain

Rose plant → Aphids → Ladybirds → Blue tit → Kestrel

Producers → Primary consumers → Secondary consumers → Tertiary consumers → Quaternary consumers

Freshwater food chain

Phytoplankton → Zooplankton → Water fleas (daphnia) → Stickleback → Pike

Food webs

Food chains are a simplification of the actual feeding relationships in an ecosystem. Many animals feed on both plants and other animals, so they may feed at two or more trophic levels. In most communities it would be almost impossible to work out all the feeding relationships that exist.

Food web diagrams show how different food chains within an ecosystem interact with one another. Fig. 2 shows *some* of the possible interrelationships in a British wood. In reality, there are thousands of species in this ecosystem (including hundreds of different kinds of insects, and thousands of species of bacteria) so even this diagram is far too simple an example of the true situation. Notice, too, that the diagram should really have arrows going from every single organism to the decomposers – and that there are hundreds of different species of decomposers, each of which really deserves its own individual listing in the food web.

Pyramids of numbers

We have seen that the arrows in a food chain or food web represent the direction of the flow of energy. All the consumers depend on the amount of energy that is incorporated by the producers.

However, not all the energy available at a particular trophic level is transferred to the next, because some is lost to the environment at each level. This is always the case when energy is transferred from one form, or one place, to another. Some of it is always wasted during the transfer process.

Therefore, as a general rule, there is more energy available to primary producers than to primary consumers, and more energy available to primary consumers than to secondary consumers. All the way along the chain, the total amount of energy available to each trophic level decreases. As a result, the number of organisms at each trophic level also decreases. There are more producers than primary consumers, more primary consumers than secondary consumers, and so on. This can be shown in diagrammatic form as a **pyramid of numbers** (Fig. 3).

Fig. 2   Interrelationships in a British wood

British woodland ecosystem

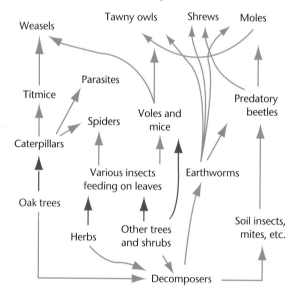

> **2** Use the information in Fig. 2 to draw complete food chains containing:
>
> **a** four organisms
>
> **b** five organisms
>
> **c** six organisms.
>
> **3** List all the secondary consumers in the British woodland food web.

Fig. 3   A pyramid of numbers

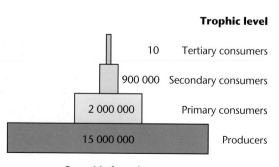

Pyramid of numbers
for 1 hectare of grassland

However, pyramids of numbers are not always this shape. It depends on the relative sizes of the different organisms in the food chain. Think, for example, of an oak tree on which caterpillars are feeding. Now the number of organisms at the first trophic level is one, but the number of organisms at the next trophic level (the caterpillars) is hundreds or thousands. A pyramid of numbers for one hectare of oak woodland looks more like a Christmas tree than a pyramid (Fig. 4.).

Fig. 4 A woodland pyramid of numbers

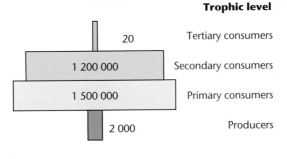

Trophic level

20	Tertiary consumers
1 200 000	Secondary consumers
1 500 000	Primary consumers
2 000	Producers

Pyramid of numbers
for 1 hectare of woodland

Other exceptions to these patterns in trophic levels may occur when parasites are part of the food chain. For example, a few *Buddleia* plants may support a large number of caterpillars. Tiny ichneumon wasps lay several eggs in each caterpillar and these eggs hatch into larvae that feed as parasites inside the living caterpillars. Each ichneumon larva may be infested with large numbers of still smaller larvae of another species of tiny wasps.

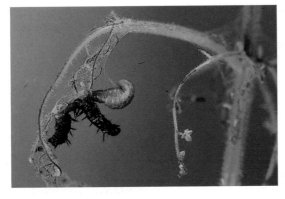

Larva of an ichneumon wasp emerging from the caterpillar of a peacock butterfly.

4 Draw a pyramid of numbers for the food chain:

Buddleia → caterpillars → ichneumon larvae → wasp larvae

5 Draw a pyramid of numbers from the following information. A copse of sycamore trees is infested with aphids feeding on sap from the leaves. Blue tits are small birds that feed on the aphids. The blue tits have mites living among their feathers and feeding on dead skin.

Pyramids of biomass

Pyramids of numbers have their limitations when used to compare different ecosystems. It is not very useful to equate the number of grass plants in a meadow with the number of oak trees in a wood, or the number of sticklebacks in a pond with the number of whales in the Atlantic. One way of overcoming this is to measure the **biomass** of all the living organisms in each trophic level (Fig. 5). The biomass of the grass plants would be the total mass of the grass plants in a given area.

The pyramid of biomass in Fig. 5 is based on the same woodland ecosystem that was used as a basis for the pyramid of numbers in Fig. 4.

Fig. 5 A pyramid of biomass

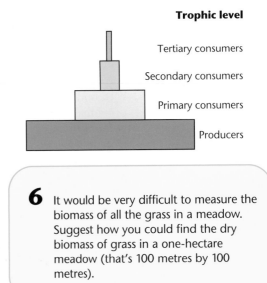

Trophic level

Tertiary consumers

Secondary consumers

Primary consumers

Producers

6 It would be very difficult to measure the biomass of all the grass in a meadow. Suggest how you could find the dry biomass of grass in a one-hectare meadow (that's 100 metres by 100 metres).

The mass of water in living organisms is highly variable. Therefore ecologists usually want to measure the **dry biomass**, which is the mass after all the water has been removed by gentle heating. However, this would kill the organism, so it is not something one would want to do to any animal.

In practice it is not easy to measure the biomass in a habitat. It is often not practical or desirable to find the dry biomass of the consumers. Just how to measure the dry mass of whales in the Atlantic is one of the enduring mysteries of ecology! Nevertheless, estimates of biomass can give a much more realistic comparison of the relative sizes of the trophic levels in a food chain.

You might expect that all pyramids of biomass would have a clear pyramid shape, since the producer biomass must be greater than that of primary consumers in order that they have enough to feed on. Although this is true in most cases, there are some ecosystems that give misleading results. For example, in the open sea, most of the producers are unicellular protoctists, the phytoplankton. Samples often show a lower biomass of photosynthesising organisms than of consumers. This is because the phytoplankton are eaten almost as quickly as they are produced and have a very short life, often less than 24 hours, whereas the animals that feed on them survive for much longer. If you were to measure the total biomass of phytoplankton produced in, say, a month and then compared this with the increase in mass of consumers, the mass of producers would be found to be much greater than the mass of primary consumers.

Pyramids of energy

You can run into strange-shaped pyramids with both pyramids of numbers and pyramids of biomass, but if you calculate the total amount of energy that flows through each trophic level, you always get an upright pyramid, even in communities with phytoplankton. The largest box is always at the base, as shown in Fig. 6.

When drawn to scale, pyramids of energy enable us to compare the amount of energy that enters the producers in an ecosystem, as well as the amount that flows through to each trophic level.

Fig. 6   A pyramid of energy

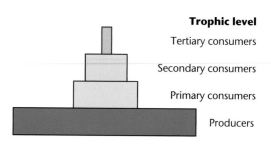

Energy losses between trophic levels

Only a very small proportion of the solar energy that reaches the Earth's surface is actually stored in producers and is therefore available to other trophic levels in food materials. Much of it is not even used in photosynthesis because only a small percentage of the Earth's surface is covered with producers, so a lot of the energy is absorbed by rock or water. Of the solar energy that does fall on plant leaves:

* Some solar energy is reflected.
* Some passes straight through the leaves.
* Some is the wrong wavelength; only visible light in the wavelength range of 380 nm to 720 nm excites electrons in the chlorophyll and starts the light-dependent reaction.
* Some is converted to heat during the reactions of photosynthesis.

At best, less than 10% of the solar energy is incorporated into carbon compounds by photosynthesis.

Moreover, the producers do not use all of these carbon compounds to make new materials for growth. They use some of the carbon compounds to provide energy for other processes. This energy is released from the carbon compounds in respiration. It is rare for more than 3% or 4% of the solar energy that falls on a plant to be incorporated into its growth. It is often much less. Only the small proportion of solar energy that is incorporated into the structure of the producers can be passed on to primary consumers.

Even then, most of the energy that is theoretically available to the primary consumers

Fig. 7   Heat losses from each trophic level

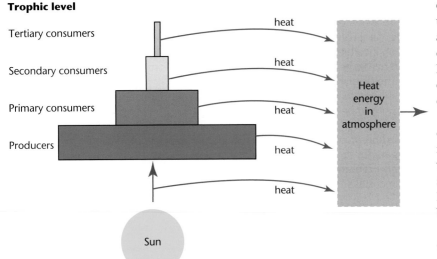

Trophic level

Tertiary consumers

Secondary consumers

Primary consumers

Producers

heat

heat

heat

heat

heat

Heat energy in atmosphere

Sun

will never get into their cells. Imagine, for example, a herd of cows grazing in a field. The grass in the field contains energy in its cells, which the cattle eat to obtain their energy. But the cattle will not eat all the grass, for example, the roots. They will not be able to digest all of the grass that they do eat; some of it will pass through their digestive systems and out of the body as faeces, never having been part of the cow's body. (This energy becomes available to the decomposers.)

Now think about the next organisms in the food chain – us. Imagine that the cows are killed to produce beef. The cows will not have incorporated all the energy they obtained from the grass into their cells – a lot of it will have

been lost as heat energy as the cows' cells respired. So only a proportion of the total energy they took in is theoretically available to us. And, once again, we only obtain a small amount of this energy. For example, humans do not eat every part of a cow, such as the hooves, teeth or intestines; we are unable to digest every molecule of the beef that we consume.

At each step, only about 10% of energy in one trophic level is passed on to the next. This is known as the **efficiency of energy transfer**. As similar losses happen at each successive trophic level, there can be only a limited number of trophic levels in a food chain before the energy 'runs out' (Fig. 7).

Most communities are so complex that it is difficult to obtain accurate data about the efficiency of energy transfer. Fig. 8 shows a simplified pyramid of energy for an aquatic ecosystem in Florida. Fig. 9 shows the energy data that the researchers found.

Fig. 8   Pyramid of energy for an aquatic ecosystem in Florida

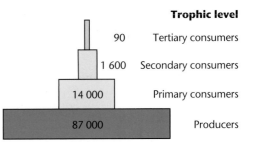

Trophic level

90   Tertiary consumers

1 600   Secondary consumers

14 000 Primary consumers

87 000 Producers

Fig. 9   Energy flow through an ecosystem in Florida

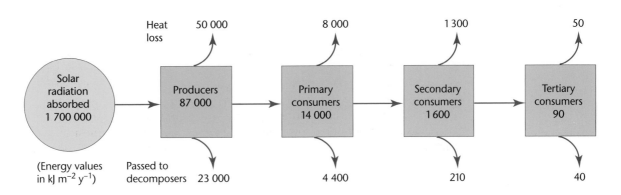

Heat loss   50 000   8 000   1 300   50

Solar radiation absorbed 1 700 000

Producers 87 000

Primary consumers 14 000

Secondary consumers 1 600

Tertiary consumers 90

(Energy values in kJ m^{-2} y^{-1})

Passed to decomposers 23 000

4 400   210   40

7 Flow diagrams can also be constructed for individual animals. Fig. 10 shows energy flow through four types of animals: an invertebrate herbivore, an invertebrate carnivore, a mammalian herbivore and a mammalian carnivore. The diagrams are drawn to scale, so that the widths of the arrows enables comparisons to be made.

a Suggest why the proportion of energy lost in respiration is greater in the mammals than in the invertebrates. (Clue – think about homeostasis.)

b Suggest explanations for the differences in the proportions of energy lost in the faeces of herbivores and carnivores. (Clue – think about digesting plant cells compared with animal cells.)

c Using all the information in the diagrams, describe and explain the differences between these four organisms in the efficiency of energy transfer to the next trophic level.

Fig. 10   Energy flow diagrams

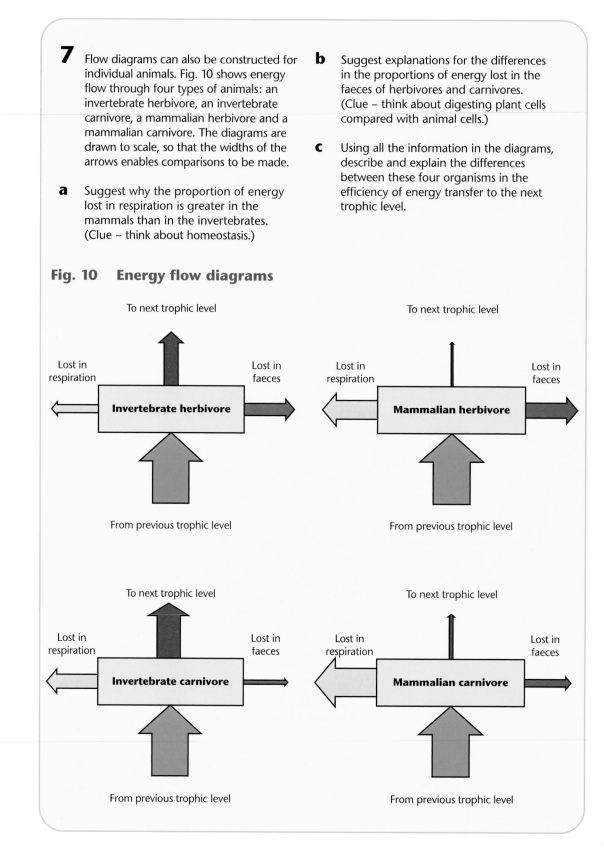

- Energy enters most ecosystems as sunlight, which is transferred by photosynthesis into chemical energy in organic compounds.

- Food chains and food webs show the direction of energy flow from producers and through each level of consumers.

- Pyramids of numbers, biomass and energy represent the quantity of these parameters at each trophic level.

- Energy is lost as heat at and between each trophic level. The proportion of the energy that is passed on from one trophic level to the next is known as the efficiency of energy transfer. It is generally about 10%.

4.2   Productivity

When crop plants are grown for food, we are essentially using the plants to convert energy in sunlight into chemical energy in carbohydrates and other nutrients. The rate at which plants do this is known as **primary productivity**.

Primary productivity is often measured in terms of the amount of energy that is converted from light energy into chemical energy, over an area of one square metre, during one year. Its units are therefore kJ m^{-2} year^{-1}. It is important to remember that productivity is a *rate*, and therefore its units must include time.

Another way of measuring primary productivity is the mass of new plant matter that is produced over an area of one square metre per day. In this case, its units are g m^{-2} day^{-1}. Table 1 shows productivity per day in g m^{-2} for six important food crops. (Only five of these are grown in the UK – rice cannot grow in our temperate climate.)

You can see that the average productivity for each crop is much lower than the highest value that can be obtained. If we could somehow move the world average closer towards the maximum, we could produce much, much more food each year – which perhaps could be one way of helping to feed the millions of people who do not get enough to eat. Later in this chapter we will look at some of the ways in which good farming practices can help to improve productivity.

Table 1   Crop productivity

Crop	Productivity per day of growing season (g m^{-2})	
	World average	**Highest productivity in ideal conditions**
Wheat	2.3	8.3
Oats	2.4	6.2
Maize	2.3	4.4
Rice	2.7	4.4
Potatoes	2.6	5.6
Sugar beet	4.3	8.2

Gross and net productivity

The rate at which a plant produces organic materials through photosynthesis is its **gross primary productivity** (**GPP**). However, as we have seen, during respiration plants use up some of these materials and release energy from them. The net gain of dry mass stored in the plant after respiration is known as the **net primary productivity** (**NPP**). It is the NPP that represents the potential food available to primary consumers.

Net productivity	=	Gross productivity	−	Respiratory loss

NPP = GPP − respiration

Table 2 gives some values for NPP in a range of different ecosystems. You can see that intensive agriculture (where the land is farmed to get as high a yield as possible, by using inputs such as pesticides and inorganic fertilisers) comes very high in the list, second only to tropical rainforests.

Table 2 Net primary productivity (NPP) in different ecosystems

Ecosystem	NPP (kJ m^{-2} year^{-1})
Extreme desert	260
Desert scrub	2 600
Subsistence agriculture	3 000
Open ocean	4 700
Shallow seas over continental shelf	13 500
Temperate grasslands	15 000
Temperate deciduous forest	26 000
Intensive agriculture	30 000
Tropical rainforest	40 000

Factors affecting primary productivity

Primary productivity is affected by a wide range of abiotic and biotic factors. Productivity is related directly to the rate of photosynthesis.

The rate at which plants can photosynthesise is determined by:

- the intrinsic capabilities of the particular species and variety of plants
- the intensity of sunlight that falls onto them
- the duration of light each day
- the amount of water that is available
- the temperature
- the concentration of carbon dioxide in the atmosphere
- the availability of inorganic ions such as nitrate in the soil
- competition for light and other resources
- damage to the plants by fungi, insects and other pests.

8

a Which of the factors listed above are abiotic factors, and which are biotic factors?

b Suggest which factors are likely to contribute to the high productivity in the two most productive ecosystems as listed in Table 2.

c Suggest which factors are likely to be reducing productivity in the two least productive ecosystems in Table 2.

hsw

how science works

The mean energy value of solar radiation during daytime is approximately 1 kJ m^{-2} s^{-1}. Fig. 11 shows what happens to each 1 000 units of solar radiation that falls on the leaves of actively growing plants.

9 Calculate the overall percentage efficiency of the use of the solar radiation by the plant to produce materials for growth.

10 What percentage of the energy absorbed by chloroplasts is actually used to synthesise carbohydrate?

11 What percentage of the energy in the carbohydrate produced by photosynthesis is incorporated into growth compounds?

Fig. 11 Photosynthetic efficiency

The amount of solar energy intercepted by green plants depends a great deal on geographical location. In Britain, this is estimated as approximately 1×10^6 kJ m^{-2} year^{-1}, but at least 95% of this is unavailable to plants for photosynthesis.

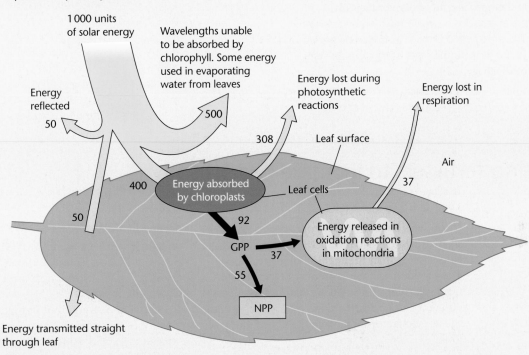

Source: adapted from The Open University Science Foundation course and data taken from ABAL, Cambridge University Press

12 Use your knowledge of leaves and chloroplasts to explain the features that:

a keep the amount of light transmitted to a minimum

b enable the chlorophyll to make use of the green wavelengths of light

c maximise the efficiency with which chlorophyll molecules absorb light.

13 The efficiency of net primary productivity (NPP) given in the diagram is close to the maximum ever found in nature and much higher than the average in most ecosystems. The mean percentage efficiency is rarely more than about one-fifth of this value. Suggest reasons for the efficiency being much lower than the possible maximum.

14 We make better use of the solar energy falling on an area of land if people feed on crops rather than on animals such as cattle.

a Use your understanding of efficiency and energy flow to explain why.

b Suggest why it may sometimes be an economical use of land and energy to use animals as a source of food.

key facts

- Primary productivity is the amount of energy per unit area per unit time that is transferred from sunlight to chemical energy in plants. This is net primary productivity, NPP.

- All organisms use some of the chemical energy in their organic molecules for their own purposes. This energy is released by respiration. The energy that is still available to be passed on to the next trophic level is the gross primary productivity, GPP.

  GPP = NPP – respiration

- Anything that limits the rate of photosynthesis, such as low carbon dioxide concentration, low temperature or low light levels, also limits primary productivity.

4.3   Using fertilisers to increase productivity

If you refer back to Table 1 on page 75, you will see that the average productivity for each crop is much less than the highest yields recorded. The productivity of a crop may be prevented from reaching its potential maximum by many different factors, both abiotic and biotic. Among the abiotic factors is the availability of inorganic ions.

Inorganic ions from fertilisers

Plants require many different inorganic ions (sometimes called minerals), which they obtain from the soil, taking them into their root hairs by diffusion or active transport. Of these ions, those that are most likely to be in short supply are **nitrate, phosphate** and **potassium**.

Nitrogen-containing ions are needed for the manufacture of amino acids, and therefore proteins, in the plant. Although air contains almost 80% nitrogen, this gas (N_2) is far too unreactive to be useful to a plant. Plants can only use nitrogen from a more reactive compound, such as nitrate (NO_3^-) or ammonium (NH_4^+) ions. Supplying plants with nitrogen means supplying nitrates or ammonium salts.

Phosphorus is needed for making nucleic acids – DNA and RNA – and also ATP. It is important in many enzyme-catalysed reactions in plants. It is also vital for cell division and is needed in areas of rapid early growth. If phosphorus is lacking, root growth is stunted. Nitrogen and phosphorus interact to affect crop growth. Plants absorb phosphorus in the form of phosphate ions (PO_4^{2-}).

Potassium is important in maintaining the balance of negative and positive ions inside and outside cells, and is involved in protein metabolism. Efficient photosynthesis and active transport rely on an adequate supply of potassium. If it is not available, leaves turn yellow and cereal crops produce less grain.

To achieve high productivity and therefore high yields, farmers apply fertilisers containing these ions to the soil. They may use an **artificial fertiliser** or a **natural fertiliser**. Inorganic or artificial fertilisers consist of inorganic compounds like ammonium nitrate. Organic or natural fertilisers consist of organic materials such as animal manures, composts and sewage sludge.

Good farming practice involves measuring the concentration of inorganic ions in the soil and calculating the amount that will be required by whatever crop is to be grown. Different crops require a different balance of minerals. This is sometimes expressed as the NPK value for the crop – the ratio of nitrogen: phosphorus: potassium that it needs. Spring barley (a cereal crop sown in spring) usually has an NPK value of 2 : 1 : 1. Beans, however, which are legumes, have a value of 0 : 1 : 1. This is because legumes have nitrogen-fixing bacteria in their roots, which make ammonium ions using nitrogen gas from the air spaces in the soil (see page 104).

When an ion such as nitrate is in short supply, adding some of it to the soil will increase productivity and growth. Up to a point, the more you add the better the growth. However, beyond a certain level, there is no more improvement. Indeed, if you go on adding more and more nitrate, you may actually *reduce* the productivity of the crop (Fig. 12). This can happen for a variety of reasons, one of which is that a very high concentration of inorganic ions in the soil produces such a low water potential that water is drawn out of the plant roots by osmosis.

Fig. 12   Effect of nitrogen-containing fertiliser on grain yield

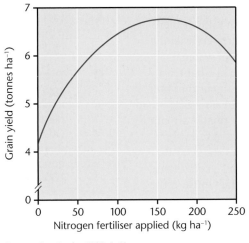

Source: after Cooke, 1980, in Harper, *Principles of Arable Crop Production*, Granada

15 With reference to Fig. 12, recommend the most appropriate quantity of nitrogen-containing fertiliser to provide for this crop. Explain your decision.

16 Suggest the limiting factor on yield at each of these points on the curves in Fig. 13.

a Nitrogen fertiliser application of 100 kg ha⁻¹, low phosphorus.

b Nitrogen fertiliser application of 21 kg ha⁻¹, high phosphorus. (You will have to make an educated guess about what might happen to the 'high phosphorus' curve in that region of the graph.)

Fig. 13   Effect of nitrogen and phosphate on yield of sugar beet

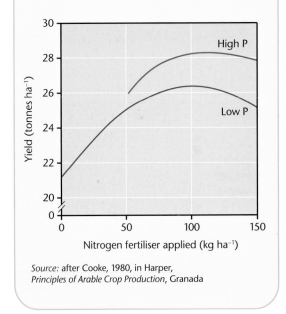

Source: after Cooke, 1980, in Harper, *Principles of Arable Crop Production*, Granada

The availability of different inorganic ions often interacts in terms of their effects on plant growth. For example, if a sugar beet plant is short of phosphorus, it may not be able to make maximum use of the nitrate ions that are available to it (Fig. 13). Giving it more nitrate ions will not increase its productivity if the limiting factor is actually phosphate ions.

Choosing and applying fertilisers

In an artificial fertiliser made up of a mixture of inorganic ions, the precise amounts and ratios of the different inorganic ions are known, and so they can be exactly matched to the needs of the crop. In contrast, the nutrient content of a natural fertiliser such as animal manure varies depending on the animal species and its diet, and it can be very difficult to match the nutrient needs of the crop. You do not know exactly what you are adding.

If soil is very low in nitrogen and a nitrogen-demanding crop is to be grown, then the speed of release of ions from the fertiliser used may influence a farmer's choice as much as the actual quantity of ions that the fertiliser contains. Ammonium nitrate, often the major component in artificial fertilisers, is soluble and releases nitrates to the soil easily. Urea, found in farmyard manure and other natural fertilisers, gives a much slower rate of release. In most natural fertilisers, there are some ions that are in a soluble, readily available form, and others that are released over a longer time by the decomposition of organic matter by microorganisms. So one application of natural fertiliser can have a much longer-lasting effect than artificial fertiliser.

Fertilisers need to be added to the soil, ready for when the crop's demand for them is at its greatest. Nitrate ions and ammonium ions are highly soluble; during periods of rainfall there is a risk that they will be lost through leaching (drainage of nutrients dissolved in water through the soil). As we will see in Chapter 5, this not only wastes money for the farmer but can also seriously damage nearby aquatic ecosystems. Potassium and phosphates are not so soluble, so problems with losses from the soil by leaching are not as great.

Artificial fertilisers are expensive, so farmers do not want to use more than they must. If the farmer is to make a profit, then the economic value of the increased yield of the crop must be greater than the cost of buying and applying the fertiliser.

Fertiliser can be applied in controlled amounts.

17 Table 3 shows the increase in yield when applying different quantities of nitrogen-containing fertiliser to a cereal crop.

Table 3

Quantity of fertiliser applied (kg ha^{-1})	Yield of grain (tonnes ha^{-1})
0	2.9
50	3.4
100	4.3
150	4.7
200	4.8

a Draw a graph of the data in Table 3.

b Explain why your curve does not pass through the origin.

c Explain why the curve does not show a proportional relationship between fertiliser application and grain yield.

d The farmer can sell the grain for approximately £200 per tonne. The fertiliser costs approximately £400 per tonne. What quantity of fertiliser would you recommend the farmer should use, in order to obtain a good profit from his crop? (There are 1000 kg in 1 tonne.)

Farmers also have to consider the costs of machinery and labour. Many artificial fertilisers are available in granules or pellets, and specialised machinery is needed for spreading them over the land. However, the machinery is light, and the fertiliser is easy to store and clean to handle. If kept in moisture-proof conditions, the fertiliser can be stored for long periods of time.

Natural fertilisers such as farmyard manure are bulky and difficult to store. There may be insufficient organic material available on the farm, so it may have to be transported from livestock areas into arable areas used for crop production. Heavy machinery is needed to

Applying farmyard manure can be a 'hit or miss' task, as far as knowing which nutrients are being added to the soil.

handle the manure, and it can be difficult to apply evenly over a field. Weed seeds and fungal spores that cause plant diseases may be present in animal manures. Sewage sludge may contain heavy metals such as lead, zinc and nickel, which can be toxic to plants. On the other hand, organic matter binds the soil particles together. This improves the overall soil structure by improving aeration and drainage in clay soils, and water retention in light, sandy soils. Organic material also acts as a food resource for soil organisms, and the activity of animals such as earthworms also improves soil aeration and drainage. Organic matter releases nutrients over a longer period of time as a result of the action of microorganisms.

Recycling organic waste makes good environmental sense. The nutrients in the organic material are added to soil where they will be used up by crops. If organic material is just left, for example, in landfill sites, then there can be problems of uncontrolled leaching.

18 Make lists of:

a the advantages

b the disadvantages

of using natural fertilisers (such as farmyard manure) compared with using artificial fertilisers.

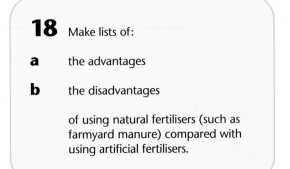

key facts

● Farmers use fertilisers to ensure that productivity of crop plants is not limited by lack of inorganic ions such as nitrate, potassium or phosphate. The quantity of fertiliser to be applied must be judged carefully, as over a particular level the extra yield obtained may not outweigh the costs of the fertiliser.

● Natural fertilisers, such as farmyard manure, have some advantages and some disadvantages compared with inorganic fertilisers. They may improve soil structure, and are a good way of dealing with otherwise unused waste materials. However, their ion content is not usually known and cannot be controlled, so it is not possible to calculate exactly how much should be applied for a particular crop growing in a particular soil type.

4.4 Pest control

Besides the abiotic factors affecting productivity and therefore crop yields, there are many biotic factors to take into consideration. These include:

- intraspecific competition; that is, competition between individual crop plants growing together for light, water and inorganic ions

- interspecific competition; that is, competition between the crop plants and plants of other species growing in the field
- grazing by herbivores such as rabbits or insects
- parasitism and disease, for example, fungi growing on the crop plants, or infection with disease-causing bacteria or viruses.

Plants of other species growing among crop plants are known as **weeds**. They can be controlled using chemicals called **herbicides**, or by non-chemical methods such as hoeing, or by using biological control (see page 84).

Insects or fungi that feed on crop plants are **pests**; the chemicals that are used to control them are **pesticides**. These include **insecticides** and **fungicides**. Non-chemical methods of control can be also used, including biological control.

Many people dislike the idea of eating food from crops that have been treated with pesticides. But the quantity of potential food lost to pests each year is enormous. Worldwide crop losses caused by insects, weeds and diseases are shown in Fig. 14. If we did not use pesticides, far less food would be produced than at present, and even more people would go hungry. However, as we will see, we can often greatly reduce pesticide use by combining it with other methods of pest control.

Fig. 14 Crop losses due to insect pests, weeds and plant diseases

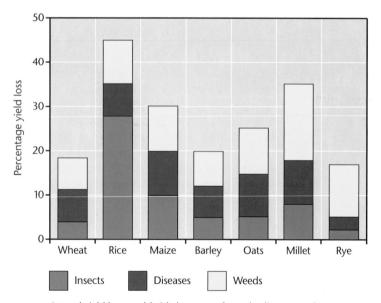

Annual yield loss worldwide because of weeds, diseases and insects as estimated by FAO, based on data pre-1990.

Source: Freeland, *Food for Life*, British Agrochemical Association

Chemical control of weeds

Weeds are plants that grow in the wrong place, at the wrong time. They compete with crop plants for light, water and inorganic ions, thus reducing productivity and crop yield. Weeds

also reduce yields because they act as hosts for pests and disease-causing organisms. For example, fungi can survive in many grass species and then infect cereal crops. Seeds from weeds can get mixed up in harvested seed, and be sown along with the crop seed.

A wide range of herbicides is available for farmers to use in order to reduce competition from weeds in fields of crops. Some herbicides are non-selective, or **broad-spectrum**. This means they kill any plant, including crop plants. They can therefore only be used before a crop has been sown or before the seeds have germinated. They can be used to clear areas before cultivation. Other herbicides are **selective**. They have been developed to affect only certain types of plant. For example, they may kill broad-leaved plants – which includes most weeds – but do not kill plants with narrow leaves such as wheat and other cereals. They can be applied after a crop has germinated.

An example of a selective herbicide is **2,4-D**. This is a synthetic **auxin**. Auxins are plant hormones that affect the growth and development of plants; you will find out more about them in Chapter 12. The auxin 2,4-D has a similar molecular structure to the most common naturally produced auxin, known as IAA (indole acetic acid). When 2,4-D is sprayed onto a field of wheat, it makes the broad-leaved weeds grow very quickly for a short while, and then die. The wheat is not affected.

Chemical control of insect pests

The use of pesticides (chemicals to protect crops from pests such as insects and fungi) is not new. Three thousand years ago, sulfur was used by the Greeks to kill pests, and the Chinese used arsenic in AD 900. More recently, naturally occurring chemicals such as nicotine from ground tobacco leaves, and pyrethrum from certain Kenyan daisies have been used. Since 1950, more than 500 chemical substances have been registered for use as pesticides in the UK. Some of these do not occur naturally but are made in laboratories. These are known as synthetic pesticides.

Like herbicides, insecticides can be broad-spectrum or specific. A broad-spectrum insecticide kills all insects with which it comes into contact, including beneficial species such as bees (which help to pollinate flowers) and ladybirds (which are hungry predators of insect pests such as aphids). So very careful timing is needed when applying a broad-spectrum insecticide, to try to limit its effects on beneficial insects. For example, you would not spray a broad-spectrum insecticide on an apple tree when it was in flower; if you did, you would kill the bees visiting the flowers, and so would get little or no crop of apples. Unfortunately, however, there aren't many insecticides that target only a particular species, and they are considerably more expensive than broad-spectrum insecticides.

Insecticides can be classified as **contact** or **systemic**. Contact insecticides kill insects with which they make direct contact. Systemic insecticides are absorbed by the plant, and transported to every part of it in the phloem. These are much more effective against sap-sucking insect pests such as aphids (greenfly) because every aphid that is feeding on the sap will be killed. A contact insecticide, however carefully sprayed on to the plant, would be likely to miss many individual aphids.

Insecticides differ in their **persistence**. This is a measure of how quickly they break down. Many natural insecticides are quickly decomposed in the environment. The early types of synthetic insecticides, such as the infamous DDT (Dichloro-Diphenyl-Trichloroethane), may not be decomposed as quickly. However, food regulations now stipulate that food crops must have no more than a very low quantity of pesticides on them when offered for sale. As a result, most pesticides are formulated to ensure that they break down quickly enough to meet the regulations.

Table 4 shows the **half-lives** of several pesticides; that is, the length of time it takes for half of the original quantity of the pesticide to break down. In practice, half-lives vary enormously according to environmental conditions. The half-life for DDT, for example, can be anything between two and 15 years. Pesticides break down more quickly when it is warm and wet than when it is cold and dry. This is because the microorganisms that act on them are most active in these conditions.

Another problem with using chemicals to kill insects is that populations that are resistant to the insecticide may evolve. All members of a pest species are not equally likely to be affected

Table 4   Half-lives of pesticides

Pesticide	Use	Legal in the UK?	Half-life
DDT	killing insects	no	minimum 5 years
paraquat	killing weeds	no; banned in 2008	3 years
benomyl	killing fungal pests on crops	yes, although there are questions about its safety	240 days
pyrethum	killing insects	yes	12 days
rotanone	killing insects	yes	2 days
Bt toxin	killing insects	yes	2 days
malathion	killing insects	no; banned in 2006	1 day

Synthetic pesticides are shown in blue; natural pesticides are shown in green.

19 Suggest one advantage and one disadvantage of an insecticide that takes a long time to be broken down by decomposers.

20 Choose two of the pesticides listed in Table 4 and find out:

- the particular uses that they have
- any evidence that they may be a threat to human health.

by a particular insecticide. Some individuals are genetically less susceptible and may survive to pass on the alleles for resistance to the next generation. This could lead to the development of significant pest resistance. Most insects have rapid reproductive rates, so the evolution of resistance can be quite quick (Fig. 15). Female aphids (greenfly) are capable of breeding two weeks after birth, so large numbers of resistant individuals could appear in the population in a single growing season. The more frequently an insecticide is used, the more likely resistance in the pest insect species is to evolve.

Fig. 15   Resistance of budworms to insecticide

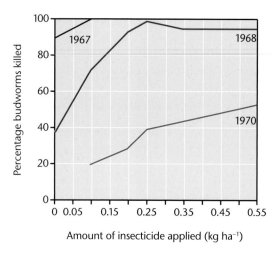

Source: ABAL, Ecology, Cambridge University Press, 1985

Fig. 16   The ideal pesticide

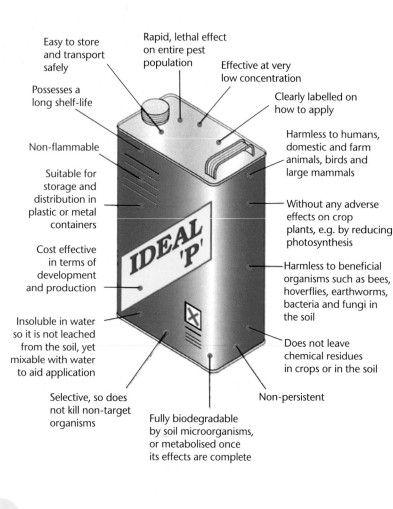

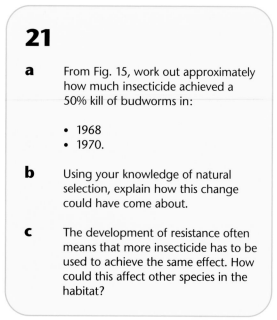

21

a   From Fig. 15, work out approximately how much insecticide achieved a 50% kill of budworms in:

- 1968
- 1970.

b   Using your knowledge of natural selection, explain how this change could have come about.

c   The development of resistance often means that more insecticide has to be used to achieve the same effect. How could this affect other species in the habitat?

The number of pesticide-resistant species increases every year. Many species have developed resistance to more than one type of pesticide. Using different insecticides for the same insect pest can delay the development of resistance.

The development, manufacture and application of pesticides are now strictly regulated, and certain ideal characteristics are aimed for (Fig. 16). However, the environmental and long-lasting impact of pesticides must not be overlooked. They can cause harm to humans, and to beneficial animals and plants, thus upsetting the ecosystem.

Biological control

Biological control involves the introduction of a natural predator or parasite, which will feed on the pest species and reduce its numbers. Biological control is not a new idea. In the 13th century, Chinese farmers put ants on citrus trees to protect the trees from pests such as aphids.

Introducing a predator to the pest reduces the level of the pest population by increasing the death rate. It does not get rid of the pest completely. This helps to prolong the effect of the control agent, because if no pests survived the predator population would also die out. Fig. 17 shows typical changes in population sizes of the pest and its control agent, before and after the control agent is introduced.

Fig. 17   Biological control

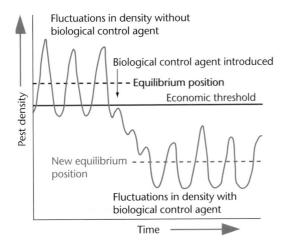

Source: Biological Sciences Review, Vol. 7, 1995

22 Look at Fig. 17. Using your knowledge of factors that affect population size, suggest why the numbers of pests fluctuate both before and after the introduction of control measures.

A biological control programme consists of the following stages:

- finding where the pest species comes from originally
- finding predators of the pest, to become the control species
- testing the control species to make sure no unwanted diseases are introduced and that only the target species is attacked
- finding out whether the control is likely to work on a large scale
- breeding or mass culture of the control species
- releasing the control species
- monitoring and evaluating the success of the programme.

If properly chosen, biological control agents are very specific and attack only one pest, with few if any adverse effects on other organisms. Biological control can be cheaper than chemical insecticides if the control species can reproduce in the habitat where it is needed. Insecticide would need to be applied regularly but it is possible that just one introduction of the biological control agent would be sufficient to keep the pest numbers down to an acceptable level. A successful biological control method would replace the use of insecticides. The levels of insecticides in food chains and ecosystems would then fall.

However, some ecological disasters have resulted from the ill-considered introduction of biological control agents. One of the worst of these occurred in Australia, where cane toads were introduced to kill beetles that were feeding on crops of sugar cane. The cane toads now have a huge and growing population, and have spread over much of Australia. They kill and eat small (and often rare) marsupials such as the rare Mulgara. Cane toads produce poisonous secretions on their skin, and predators often die after killing them. The toads are now more of a problem than the original sugar cane pests.

A cane toad, eating its recently caught prey.

Integrated pest management

Biological control will not achieve the same level of pest control as will the use of chemical pesticides. Although many biological control products are being developed, they account for less than 1% of the total crop protection market. Biological control is unlikely to be the only control measure used. It often takes some time for the population of the control species to increase to such a level that it has an impact on the pest population level. In the meantime, a lot of crop damage may have occurred. Moreover, if pest numbers have built up to very high levels

Fig. 18   Integrated pest management

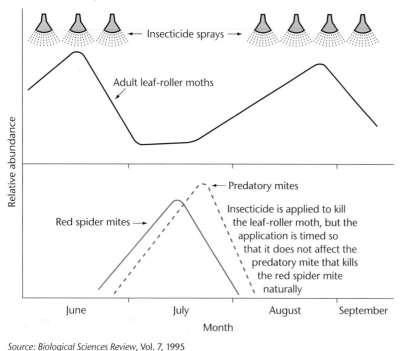

Leaf-roller caterpillars feed on apples and red spider mites blemish the apples. Both pests need to be controlled

Insecticide sprays

Adult leaf-roller moths

Relative abundance

Predatory mites

Red spider mites →

Insecticide is applied to kill the leaf-roller moth, but the application is timed so that it does not affect the predatory mite that kills the red spider mite naturally

June   July   August   September

Month

Source: Biological Sciences Review, Vol. 7, 1995

by the time the biological control agent is introduced, then it may prove impossible for the control species to have any significant effect on the pest population. There is also an appropriate reluctance to introduce any new species into a country, as it is impossible to predict whether or not it may become a pest itself.

In many instances, a system of control measures has been worked out that uses both pesticides and biological control, as well as other methods such as the development of strains of crop plants that are naturally resistant to a particular pest. Such a system is called **integrated pest management (IPM)**.

IPM can greatly reduce the quantity of pesticides that need to be used, so reducing costs for the farmer and possible harmful effects on plants or animals other than the pest species.

One example of the use of IPM is shown in Fig. 18. Leaf-roller caterpillars and red spider mites are significant pests of apples. In this IPM programme, a combination of spraying with insecticide to kill the caterpillars, and using predatory mites to control the red spider mites, can be very successful.

key facts

● Insect and other pests cause immense amounts of damage to crops each year, greatly reducing yields. This can be a particular problem in developing countries, where it contributes to food shortages.

● Pesticides are chemicals that kill pests. Insecticides and fungicides are widely used. They may be specific (killing only the pest species) or broad-spectrum (killing many different species). There are few specific pesticides, and they tend to be very expensive.

● In developed countries, there are strict regulations about the amounts of pesticide traces that may be present on food offered for sale. Many pesticides used in the past have now been banned, as evidence has emerged of their possible harmful effects on human health.

● In the past, widespread use of persistent insecticides such as DDT and dieldrin led to their bioaccumulation, and subsequent damage to organisms at the top end of food chains. These are now banned in many countries, but are still used in some developing countries as the only practical and affordable method of controlling malaria and other insect-borne diseases.

● Natural selection may produce strains of the pest that are resistant to a pesticide.

● Biological control uses natural predators or parasites of the pest to keep its numbers at a low level. This avoids the use of potentially harmful pesticides, but care has to be taken that the control agent does not itself become a pest.

● Integrated pest management uses a carefully worked-out combination of biological control and pesticides to control the pest.

4.5 Intensive rearing of domestic livestock

Farming practices differ enormously in different countries, and even in different regions of one country. In some countries, and in some areas of Britain, land is farmed using very few or no inputs of fertilisers or other chemicals. This is called **extensive farming**. Farmers spend relatively little money on the land. Sheep grazing extensively farms upland areas such as the Pennines in northern England and southern Scotland. No fertiliser is used on the land, so productivity of the grass is quite low and not many sheep can be supported on one hectare.

In contrast, much of Britain is farmed intensively. **Intensive farming** is geared to getting the maximum crop, or the maximum amount of milk or meat, from an area of land (Table 5). In order to achieve high outputs, high inputs of fertilisers and pesticides are needed. There is also high energy input in terms of the fossil fuels used to drive farm machinery and in the manufacture of fertilisers and pesticides. Farmers spend much more money than they would if they farmed extensively, but this is offset by the higher income received from the larger crop.

Table 5 Extensive and intensive sheep farming

Extensive farming	Upland grazing on infertile land	No inputs maximum 0.5 sheep per hectare
	Upland grazing on improved land	High inputs (fertiliser for the vegetation and winter feed for sheep) maximum 10 sheep per hectare
	Lowland grazing	High inputs (fertiliser for the vegetation and winter feed for sheep) maximum 14 sheep per hectare
Intensive farming	Keeping sheep in buildings	Very high inputs (processed feed for the sheep) and 1 sheep per 2 m² (equivalent to 5 000 sheep per hectare)

Sheep farming

The range of methods by which sheep are farmed in the UK is a good example of the differences between extensive and intensive farming. Many hill farmers simply let their sheep range over large areas of land, where the sheep find what they can to eat from the poor-quality vegetation. The sheep are left to their own devices for much of the warmer months. The farmer uses little or no fertilisers on the land. His input costs are therefore low. The poor-quality land can support only a very low density of sheep.

Extensive sheep farming takes place on large areas of land.

If the sheep are stocked too densely, then they may **overgraze** the land. This means that the vegetation is eaten more rapidly than it can regenerate. Some species disappear completely. Patches of bare ground appear, which are prone to erosion when it rains. Overgrazing does great damage to biodiversity. Soil that is lost through erosion may take hundreds of years to reform.

If the farmer wants to stock more sheep on the land, then the inputs must be increased. Fertiliser can be applied to increase the rate of growth of the vegetation, allowing more sheep to be kept in the same area without signficant problems of overgrazing. The farmer can give the sheep extra food in the form of hay or commercial pelleted food. The extra cost of the inputs may be outweighed by the extra returns the farmer gets when selling the meat or wool.

Lowland soils tend to be more fertile than in the Britain's uplands, so here it is possible to graze more sheep per hectare. Generally, though, lowland farmers tend to farm fairly intensively, using fertilisers to increase the productivity of the grass. The higher densities of sheep can lead to health problems, as parasites spread more easily, so there will be more inputs in the form of medicines to treat the animals.

The most intensive type of sheep farming involves keeping the animals indoors. They do not graze at all, but are fed on concentrated commercial foods. The sheds may be heated, so that the sheep use more of the energy in the food they eat to put on body mass, rather than respiring it to generate heat to keep warm. They do not move around as much as they would if outdoors, so more of the energy in their food goes into making biomass, rather than being lost as heat to the environment. All of this means that the farmer gets more product (meat) for a given area of land and a given input of energy – but the inputs are expensive, and it is difficult to make a profit with this system in the UK, where prices for lamb and mutton scarcely meet the costs of production.

Intensive sheep farming is geared to producing the maximum yield.

Egg production

In the UK, there are three methods by which hens are kept for egg production.

Battery farming involves keeping a large number of hens in small cages in a relatively small area, inside a heated building. Automated systems supply the hens with water and carefully formulated food and collect the eggs that they lay. There are quite strict regulations about how the hens are kept – for example, the minimum size of the cages. This is the most intensive method of egg production.

These chickens are used for battery egg production.

Barn eggs come from hens that are also kept indoors, in a heated building, but with more space to roam around. The hens are given perches, which is where they often prefer to rest. The system is much less automated than battery farming, and the hens lead more natural lives.

Free-range eggs come from hens that are free to walk out of the building in which they are housed. They have access to an outdoor area – although generally this will be fenced off to stop them wandering too widely, and also to keep out predators such as foxes. This is the least intensive method of egg production.

These chickens produce free-range eggs.

Free-range eggs are usually more expensive than battery-produced eggs. This is because the farmer's costs are high for free-range egg production, for example, the system cannot be automated, so more people-hours are put in, and more land is required.

However, many people are concerned about the conditions in which battery hens are kept. They believe that welfare considerations should outweigh the need to produce cheap food. Hens kept in battery cages have very little space to move around. They can do nothing except sit down, stand up, eat, drink and lay eggs. These issues are being recognised by an increasing number of people, and battery egg production is likely to be banned in the UK by 2012.

For many people, however, the cost of the food they buy is more important than ethical, animal-welfare issues associated with its production. Until 2008, more battery-produced eggs were bought than free-range or barn eggs. March 2008 saw the first occasion on which sales of free-range eggs overtook those of battery-farmed eggs. Several supermarket chains no longer sell battery-produced eggs.

key facts

- Intensive farming involves the use of high levels of inputs – such as fertiliser, processed feeds, medicines – in order to achieve high productivity and therefore high outputs. Extensive farming uses few inputs, and achieves low productivity and low outputs.

- Intensive farming may result in cheaper food, but this is often at a cost of damage to the environment, or poor living conditions for animals.

1 Starfish feed on a variety of invertebrate animals that are attached to rocks on the seashore.
The diagram shows part of a food web involving a species of starfish.

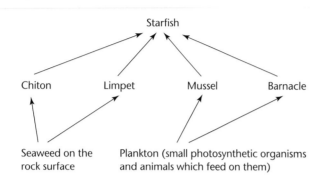

a Explain why a starfish can be described as both a secondary and a tertiary consumer. (1)
b When starfish feed on mussels they leave behind the empty shell. Explain how quadrats could be used to determine the percentage of mussels that had been eaten by starfish on a rocky shore. (3)
c The table shows the composition of the diet of starfish.

	Prey species			
	Chitons	Limpets	Mussels	Barnacles
Percentage of total number of animals eaten	3	5	27	65
Energy provided by each species as a percentage of total energy intake	42	5	38	15

i The percentage of barnacles in the diet is much higher than the percentage of energy they provide. Suggest **one** explanation for this difference. (1)
ii The table shows that the amount of energy provided by chitons is greater than the amount of energy provided by limpets. Calculate the number of limpets a starfish would need to eat in order to obtain the same amount of energy as it would obtain from one chiton. (1)

Total 6

AQA, June 2006, Unit 5, Question 4

2 The diagram shows the flow of energy through trees in a woodland ecosystem. The numbers represent mean inputs and outputs of energy in kJ m^{-2} day^{-1}.
a Use information in the diagram to
i give the amount of energy incorporated into tree biomass; (1)
ii calculate the percentage of solar energy that is fixed by photosynthesis. (2)

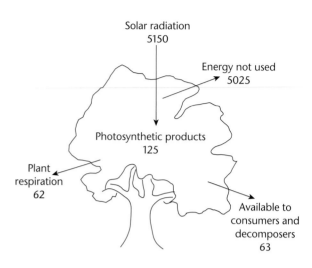

b Not all the solar radiation reaching the leaves of the tree is used in photosynthesis. Give **two** explanations for this. (2)
c The graph shows the rate of photosynthesis and the rate of respiration in tree leaves at different temperatures.

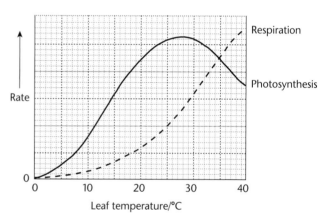

Give the range of temperatures over which the leaves will show the greatest increase in biomass. Explain your answer. (2)

Total 7

AQA, June 2003, Unit 5, Question 1

3 The diagram shows a pyramid of energy for an ecosystem.

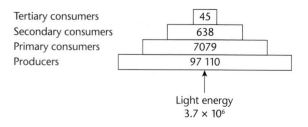

a Suggest suitable units for the measurement of energy transfer in this pyramid of energy. (1)

b **i** Calculate the percentage of energy transferred from primary consumers to tertiary consumers. (1)

   ii Give **one** reason why the percentage of energy transferred between consumers is generally low. (1)

c **i** Give **one** reason why all the light energy reaching the producers cannot be used in photosynthesis. (1)

   ii Explain how light energy is used to generate ATP in plants. (3)

Total 7

AQA, January 2007, Unit 5, Question 2

4 The diagram shows the flow of energy through a marine ecosystem.

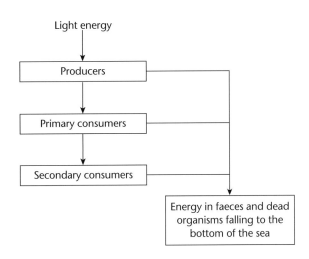

a Give **one** reason why not all the light energy falling on the producers is used in photosynthesis. (1)

b Describe what happens to the energy in faeces and dead organisms which fall to the bottom of the sea. (3)

c The producers in this ecosystem are seaweeds, which have a large surface area to volume ratio. Give **two** advantages to seaweeds of having a large surface area to volume ratio. (2)

d Some species of seaweed are submerged in water for most of the time. Explain how being under water might affect the rate of photosynthesis. (3)

Total 9

AQA, June 2005, Unit 5, Question 2

5 A popular lake in the United States was affected by large swarms of midges (small biting insects) in the summer. The lake was sprayed with insecticide to kill the midges. The effect of spraying on the following food chain in the lake was investigated.

Plant plankton → sunfish → western grebe

a Shortly after spraying the concentration of insecticide in the water of the lake was 2×10^{-5}g dm^{-3}. After four weeks the concentration in the plant plankton was equivalent to 5×10^{-3}g dm^{-3}.
By how many times was the insecticide concentrated? (1)

b After a few months the concentration of insecticide in the grebes was more than six times the concentration in the sunfish. Explain why. (2)

c Another lake, which contained dead trout, was found to be contaminated by waste containing mercury ions.

   i In an investigation, the lethal concentration of mercury ions for trout was 42 parts per billion over four days. Explain what is meant by *lethal concentration*. (1)

   ii Explain how heavy metal ions are toxic to animals. (3)

Total 7

AQA, January 2004, Unit 6, Question 4

6 Organic and inorganic fertilisers increase crop yield.

a Other than cost, give **one** advantage and **one** disadvantage of using an organic fertiliser. (2)

b Scientists investigated the effect on weed growth of sowing oats at two different densities, **J** and **K**. The graph shows their results.

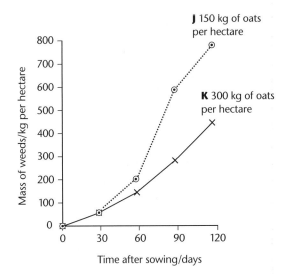

   i Describe the results of the investigation. (1)

   ii Use your knowledge of photosynthesis to explain the results. (4)

   iii Over the 120 days, the leaf area index increased in both the oat crop and the weeds. What is meant by *leaf area index*? (1)

Total 8

AQA, June 2007, Unit 6, Question 6

hsw

how science works **assignment**

Balancing the benefits

When insecticides are not broken down, they may persist in food chains. As the chemicals pass from one trophic level to another, they become concentrated, particularly if they are fat-soluble, in which case they build up in fat deposits of top carnivores such as birds of prey. This is called **bioaccumulation**. The effect can be dramatic. For example, the broad-spectrum, persistent, fat-soluble insecticide DDT is now found in virtually all animal tissue, in every food chain, and even in Antarctic snow. At the kind of concentration in which it is applied to insects, DDT is not at all toxic to other animals. However, by the time it reaches the top carnivores in a food chain, the concentrations are high enough to weaken and sometimes kill these animals.

The first major use of DDT was in 1943, when it was used to kill body lice and halt the spread of a very unpleasant disease that they transmitted, called typhus. Since then, it has been widely used in many parts of the world to kill the mosquitoes that spread malaria. The World Health Organization (WHO) estimates that the use of DDT for these purposes has probably saved at least 30 million human lives. DDT is also used as an insecticide on crops such as cotton, to kill pests such as the cotton boll weevil. It was seen as the perfect insecticide – cheap to make, very toxic to insects but not to mammals, and very stable – so you only needed to spray a small amount in your house and there would be no mosquitoes for a very long time.

The 1950s decline in sparrowhawk and peregrine populations in the UK has been linked to the use of persistent insecticides such as DDT and dieldrin (which was developed as a possible replacement for DDT (Fig. 19). These insecticides were used to treat cereal seeds before they were sown. They caused the death of many wildlife species, including long-tailed field mice, which are prey for sparrowhawk and peregrines. Table 6 shows the results obtained by ecologists who trapped field mice and determined the concentrations of dieldrin in body residues. Two traps were used for four days before seeds treated with the chemicals were sown, and four traps for six days after the seeds had been sown.

Table 6   Results of long-tailed field mice study

Period	Trap	Number of mice caught	Number of mice analysed	Mean concentration of dieldrin/ ppm
Before seeds were treated	1	11	9	0.15
	2	15	4	0.23
After seeds were treated	1	18	2	6.49
	2	18	7	10.96
	3	9	5	8.70
	4	12	5	12.06

Fig. 19   Peregrine populations in the UK

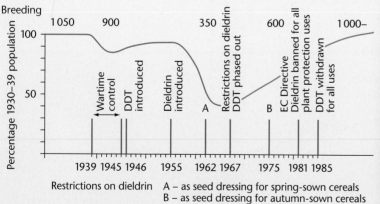

Restrictions on dieldrin   A – as seed dressing for spring-sown cereals
                           B – as seed dressing for autumn-sown cereals

Source: after Ratcliffe 1980, in Cadbury *et al.*, *Birds of Prey Conservation in the UK*, RSPB Conservation Review, Vol. 2, pp. 9–16, 1988

A peregrine with a chick.

There is evidence that the decline in peregrine populations resulted from a thinning of the shells of their eggs, making them less likely to develop and hatch. If the eggs fall below a certain mass, they break during laying. In general, the egg shells were thinner in the intensively farmed south east compared with the north and west where DDT was less widely used. Experimental evidence on captive birds of prey confirmed that DDT causes shell thinning in various species.

The use of DDT and dieldrin has been banned in Europe, including Britain, and in the USA since the early 1970s. The Worldwide Fund for Nature (WWF) would like to see a global ban. However, DDT is still a major weapon in the fight against insect-borne diseases such as malaria, and there is as yet no replacement for DDT that is as effective and cheap.

A1   Discuss the evidence in Fig. 19 that dieldrin and/or DDT were responsible for the fall in the UK peregrine population in the 1950s to 1960s. What other evidence would you like to see to convince you that there is a causal link between them?

A2   Summarise the conclusions that can be drawn from the results in Table 6.

A3   As far as you are able, without more detail of exactly how the investigation was carried out, evaluate the reliability of the results shown in Table 6. Suggest how the design of the investigation could have been improved.

A4   Use the Internet to research the current situation with regard to the use of DDT to fight against insect-borne diseases in the developing world. What are the positions of the Worldwide Fund for Nature and the World Health Organization on this issue? Put forward you own opinions on whether or not DDT should be banned globally, backing up your suggestions with evidence wherever possible.

5 Nutrient cycles

An area of York city, York (above) and of Thatcham, Berkshire (below) during the 2007 floods.

In the summer of 2007 it rained and rained, resulting in the wettest May to July since 1766.

Worst affected were the areas around the rivers Don, Severn and Thames. So much rain fell, and so quickly, that river channels overflowed. Flood defences could not cope. Around 49 000 houses were seriously affected by the flooding and many people lost irreplaceable possessions. Power supplies and clean water supplies were cut off.

Although this was by far the worst flood event in memory, it was by no means the only flood event. There have been many floods in recent years, far more than in the past. Insurance premiums are increasing, and some householders have difficulty getting house insurance at all, because their houses have such a high risk of flooding.

Are things changing? Are the floods a result of climate change? It is difficult to be sure, but it is a possibility. It is important to be aware of the difference between weather – what happens day to day – and climate, which is the long-term pattern of weather that we experience on a regular basis. Most researchers believe that they have seen enough evidence to suggest that climate change really is happening, and that we should expect to get more extreme weather events such as the heavy rain of the 2007 so-called summer.

So, if climate change really is happening, is it our fault? Again, it is difficult to be sure, but most researchers believe that the evidence linking an increasing carbon dioxide concentration in our atmosphere with rising global temperatures – and therefore climate change – is strong. Whether or not it is our activities that are responsible is still up for discussion, but few scientists doubt that we are at least contributing to the problem, even if our activities are not the only cause.

5.1 The carbon cycle

Producers make carbohydrates and other organic compounds from inorganic substances. When plants photosynthesise, they use carbon dioxide from the air (or dissolved in the water in which they live) and water to build simple sugars. These sugars may then be used to make structural carbohydrates such as cellulose, or combined with other elements to produce carbon-containing compounds such as proteins, phospholipids and nucleic acids.

Most of the carbon is eventually returned to the atmosphere in the form of carbon dioxide, by respiration. This is respiration by:

- producers
- primary consumers, which have eaten producers
- secondary and tertiary consumers, which have eaten carbon compounds passed along food chains
- decomposers, which have absorbed carbon compounds from dead organisms or from excretory products.

These processes can be summarised as the **carbon cycle**, shown in Fig. 1 opposite.

Fig. 1 The carbon cycle

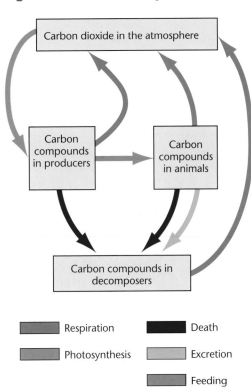

Respiration

Photosynthesis

Feeding

Death

Excretion

Decomposers and detritivores

Decomposers play an important part in the carbon cycle, and in recycling other elements. When an organism dies, or when a consumer excretes organic compounds, other organisms soon get to work on breaking down the organic constituents. A range of **scavengers** such as foxes, crows and beetles are often the first to feed on the bodies of dead animals, thus preventing the countryside and roads from being littered with more dead animals than there are.

Invertebrate animals that feed on partly broken down dead matter (detritus) are called **detritivores**, and these play an important role in disposing of dead plants. Earthworms and woodlice are detritivores. They eat the dead plant material and take it into their guts, where digestive enzymes break down the organic compounds. However, plant tissue contains a large percentage of cellulose, and few organisms can make cellulase, the enzyme needed to digest it. Most detritivores simply eliminate the cellulose in their faeces. Only a few, such as termites, can make use of the cellulose. They do this by acting as hosts to microorganisms in their gut that do produce cellulase and so can break down cellulose into sugar monomers.

Most of the cellulose has to await the attention of fungi and bacteria that live in the soil. These are the true **decomposers** that complete the process of breakdown. Decomposers differ from detritivores in that they secrete enzymes outside their bodies. They then absorb the products of digestion. This is called **saprobiotic nutrition**. The detritivores help by breaking dead matter into much smaller pieces. This makes decomposition more efficient because there is a much larger surface area for enzymes to work on. Soil that contains many earthworms is very fertile because decomposition is much more rapid.

An example of a polypore, a mushroom without a stalk, growing on a dead log.

The decomposers use the substances they have absorbed, breaking them down by respiration to form carbon dioxide and water. The carbon dioxide is released into the atmosphere and is then available again for photosynthesis. This is essential to keep the carbon cycle going. If the carbon dioxide were not returned to the air, then plants would be unable to photosynthesise and the entire cycle would stop.

1 Which monomer is the product of cellulose digestion?

2 Decomposition is much more rapid in tropical rainforests than in British woods. Explain why.

Carbon sources and carbon sinks

The carbon cycle should, in theory, keep the carbon dioxide concentration in the atmosphere constant, since all living organisms eventually die and are decomposed. However, some dead organisms do not fully decompose, for example, because they fall into deep water or marshes where there is very little oxygen for the decomposers. Over many millions of years their remains have accumulated and produced oil, coal and natural gas. Carbon was locked away in these deposits. Such deposits are sometimes called **carbon sinks**.

Humans are now plundering these fossil fuels as sources of energy, with the result that an extra six billion tonnes of carbon dioxide are being released into the atmosphere each year from their combustion. Rapid deforestation may also be adding another two billion tonnes, especially where land is being cleared by burning. Fig. 2 shows the global carbon dioxide economy. The figures are very approximate. Moreover, they do not remain steady – for example, the quantity of carbon dioxide added to the air from the combustion of fossil fuels is increasing year on year.

The greenhouse effect

As you will be aware, carbon dioxide is one of the gases that contribute to the **greenhouse effect**. This is an entirely natural effect that affects global temperatures. Fig. 3 illustrates how this effect works.

Much of the radiation arriving at the Earth from the Sun is short-wavelength radiation. This passes freely through the atmosphere to the ground. Some of it is emitted as radiation with a longer wavelength, towards the infrared range of the spectrum. This long-wavelength radiation has a heating effect on the atmosphere.

If we had no greenhouse gases, then the long-wavelength radiation would simply travel out through the atmosphere into space. However, the atmosphere contains several gases that absorb a lot of this radiation. Instead of escaping, the radiation is trapped.

Fig. 2   Global carbon dioxide economy

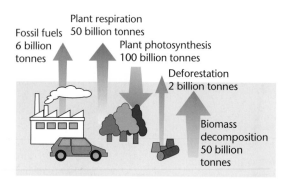

Plant respiration
50 billion tonnes

Fossil fuels
6 billion
tonnes

Plant photosynthesis
100 billion tonnes

Deforestation
2 billion tonnes

Biomass
decomposition
50 billion
tonnes

Human activities may be affecting the global distribution of carbon dioxide.

Fig. 3   The greenhouse effect

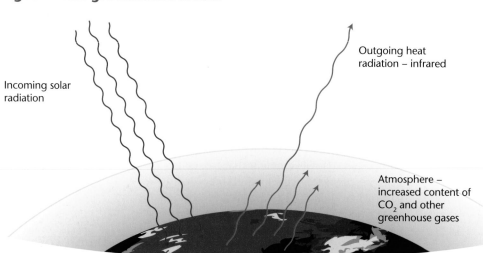

Outgoing heat
radiation – infrared

Incoming solar
radiation

Atmosphere –
increased content of
CO_2 and other
greenhouse gases

Increasing levels of carbon dioxide and other gases reduce the loss of heat to space, so increasing the temperature of the atmosphere

This warms the Earth. Without this effect, the Earth would be 30 °C colder than it is. It would be so cold that there would be no liquid water, and probably no life at all, so we need the greenhouse effect.

Several gases contribute to the greenhouse effect. Carbon dioxide is the main naturally occurring gas. At present, about 0.04% of the atmosphere is made up of this gas. Methane is another greenhouse gas. Although there is a lot less of it than carbon dioxide, it is more efficient at trapping the long-wavelength radiation, so its effect is proportionately greater than carbon dioxide.

The enhanced greenhouse effect

Fig. 4 shows what has happened to mean global carbon dioxide concentrations since 1750. Some of these measurements have been made from ice cores. Hugely thick layers of ice have built up in Greenland and other parts of the world as snow has been compacted. By drilling into these and taking out a core, we can gain access to tiny bubbles of air trapped long ago. By analysing the gases in these bubbles, we can find out how much carbon dioxide there was in the atmosphere at different times in the past.

Fig. 4 Carbon dioxide concentrations since 1750

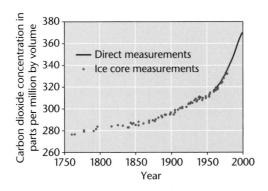

Fig. 5 shows a 'closer view' of how carbon dioxide concentration has changed since 1960. You can see how the concentration rises and falls each year. This happens at different times in the northern and southern hemispheres, as each is affected by the rate of photosynthesis.

Fig. 5 Carbon dioxide concentrations since 1960

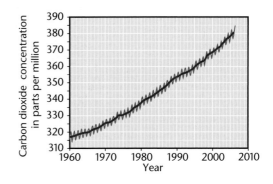

3 In the northern hemisphere, during which season would you expect atmospheric carbon dioxide concentrations to be lowest? Explain your answer.

Fig. 6 gives us an even closer look at what happened to carbon dioxide concentrations between 2004 and 2008. The rise in 2007 was worrying, because it was the third highest since measurements began in 1958. And this is at a time when the world is trying to reduce carbon dioxide emissions. (To find out what has happened to carbon dioxide concentrations since 2008, visit the NOAA website: www.noaa.gov.)

Fig. 6 Carbon dioxide variations between 2004 and 2008

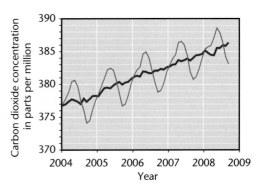

Fig. 7 Atmospheric methane concentrations since 1984

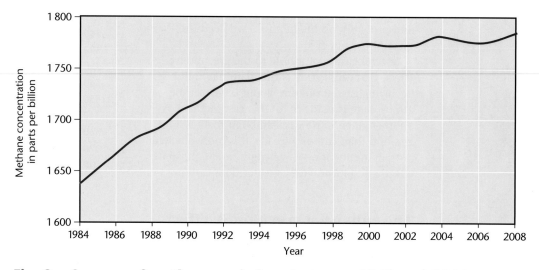

Fig. 8 Sources of methane emissions between 1860 and 2000

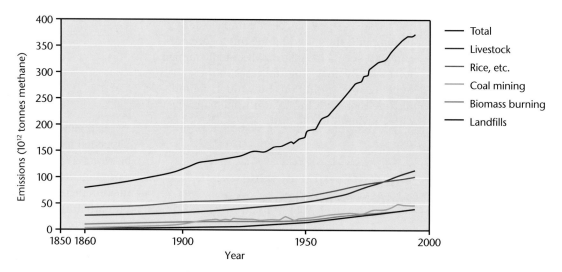

Carbon dioxide is not the only greenhouse gas whose atmospheric concentration has been increasing. Fig. 7 shows changes in the concentration of methane since 1984. Fig. 8 shows the main sources of these methane emissions. Methane is emitted by the anaerobic breakdown of organic waste material by microorganisms. Landfill sites and paddy fields (where the ground is flooded with water, creating anaerobic conditions in the mud) are major sources of methane. Coal mining also liberates methane that has been trapped deep in the Earth, generated from the same biomass that eventually turned into coal.

Fig. 9 Global warming

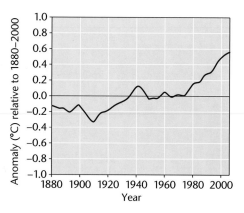

This graph shows the difference between the mean global temperature each year (the line from '0' on the y-axis)and the mean global temperature calculated for the years between 1880 and 2000.

Global warming

The rise in greenhouse gases is causing the mean global temperature to increase. This is known as global warming. While there is no doubt that this is happening (Fig. 9), it is debatable just how much of it is due to human activities – there may well be other contributory factors. Regardless of whether we are the only cause, we must still reduce our contribution if we are to slow down the rate of global warming.

Global warming matters because it will have – and is probably already having – major effects on climate. It is impossible to predict what these will be, and they will no doubt be different in other parts of the world. However, all researchers agree that higher temperatures in the atmosphere mean that there is more energy that can contribute to the intensity of weather events such as hurricanes and typhoons. These are likely to become more frequent and violent. In general, weather could become more extreme in many places – more droughts, floods, high winds and storms.

We will look at three aspects of these effects – the yield of crop plants, the life cycles and numbers of insect pests, and the distribution and numbers of wild animals and plants.

Crop plant yields

We have seen that carbon dioxide concentration is often a limiting factor for photosynthesis (Chapter 2). Increased carbon dioxide concentration should therefore increase crop yields. Numerous investigations have shown that this is likely to be true for most crop plants, including all of the world's main cereal crops – wheat, rice, maize, sorghum and millet. Table 1 shows the effect of increased carbon dioxide concentration on the yield from several varieties of soya beans.

4

a Calculate the percentage increase in seed yield at higher carbon dioxide concentration for each of the varieties of soya beans in Table 1.

b Explain why an increased carbon dioxide concentration results in increased seed yield per plant.

c Suggest why the percentage increases in yield of different varieties of soya bean are not the same.

In a glasshouse, a grower can increase the carbon dioxide concentration but still maintain the temperature at an ideal level for plant growth. Crops growing outside, however, will be subject to whatever climate change takes place as a result of raised carbon dioxide levels. In many places this will mean that the temperatures are a little higher, and in some parts of the world – such as northern Europe – this could also increase crop productivity. In others, however, such as many parts of Africa, the higher temperatures may reduce productivity, as they may go above the optimum level for photosynthesis in the plants.

The increased temperatures may make it possible to grow crops in Britain that we have not been able to grow before, or that there will be a longer growing season for those that need warm temperatures. Many crop plants, such as melons, tomatoes and potatoes, are killed by frost. At the moment, they can only be planted in late spring when danger of frost is over, or in climate-controlled glasshouses. Climate change may mean that our winters become milder and spring arrives earlier, enabling growers to plant crops earlier and perhaps also get higher yields over a longer growing season.

Climate change and insect pests

Although most insects are harmless, and many are beneficial, there are a few species that we would prefer to be without. Some are significant pests of crop plants. These include various species of aphids, which feed by sucking sap from phloem tubes. Aphids often transmit viral diseases from one plant to another. Other insects may transmit human diseases – for

Table 1 Effect of increased carbon dioxide concentration on soya bean yield

Variety	Seed yield (g per plant)	
	Normal carbon dioxide concentration	**Higher carbon dioxide concentration**
Arksoy	30.8	42.2
Harrow	32.7	45.1
Mandarin	31.3	58.4
Mukden	41.4	56.5
Williams	46.0	64.8

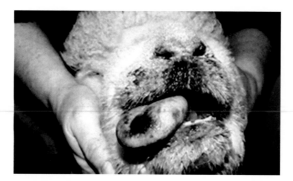

Bluetongue disease in a sheep.

example, malaria and dengue fever, which are both transmitted by mosquitoes. Insects that carry disease-causing organisms from one host to another are known as **vectors**.

If our climate warms, then it is probable that some insect pests will become more of a problem in the UK. Some species that were formerly not able to live here may move northwards and take up residence. For example, we may find that we are sharing our country with insect vectors that previously were unable to live here. In 2007, a disease of cattle and

hsw

how science works

Invasion of the green spruce aphids

The green spruce aphid, *Elatobium abietinum*, is a serious pest of spruce trees. Like most aphids, this species is parthenogenic, meaning that females produce young (often in large quantities) without needing to mate with males. The aphids feed on several different species of spruce trees, *Picea*. Spruce is not native to the British Isles but it is an important tree in many forestry plantations, where it is grown for timber. Spruce trees infested with the green spruce aphid can lose more than one-third of their leaves and grow by only three-quarters as much as uninfested trees. The aphids also disrupt the formation of

buds, so that the next year's growth is also greatly reduced.

Green spruce aphids migrate in spring each year. Winged adults fly during a two-month period. The earlier this period is, then the longer the aphids have to feed and breed on the spruce trees during the late spring and throughout the summer.

A long-term study into the timing and size of the spring migration of the green spruce aphid, and also into the climate in Britain, took place between 1966 and 2006. The measurements relating to climate used an indicator called the North Atlantic Oscillation, or NAO. The more positive the NAO index, the warmer the weather in the UK. The researchers used the winter NAO index, which reflects the temperatures in winter and early spring.

Aphids were caught using a trap placed at a height of 12 m, so they were only caught during their spring migration period.

Fig. 10 shows the measurements of the population of the green spruce aphid and the NAO index between 1966 and 2006. The y-axis scale for the aphid population is a log scale, not a linear one. A value of 0.0 means that there were no aphids at all. Each unit represents a 10-fold increase in population, so a value of 2.0 means that there were 10 times as many aphids in the population as a value of 1.0. Statistical calculations were used to draw the lines, rather than estimating where a best-fit line should go.

Fig. 11 shows the day of the year on which the first flying green spruce aphids were captured, and the day on which the last one was captured – in other words, the timing and the duration of their spring migration. Day 90 is the 90th day of the year – that is, April 1st.

Green spruce aphids, *Elatobium abietinum* (circled in the photograph above).

sheep called bluetongue appeared for the first time in Britain (see photograph opposite). This serious disease is caused by a virus, and its vector is a midge belonging to the genus *Culicoides*. The midge is usually killed in a hard winter, but it seems that with the mild winters in the mid 2000s more of these midges than usual survived, and they began to breed earlier, producing larger populations.

Conservationists are also concerned that insect species that are not native to Britain will find our warmer climate to their liking. If new species arrive, they are likely to compete with our native species for food or breeding places. It is probable that the populations of some of our native species will fall, and possible that some will become extinct as a result.

Fig. 10

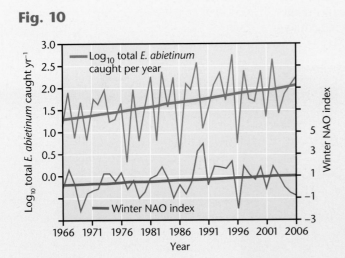

Fig. 12

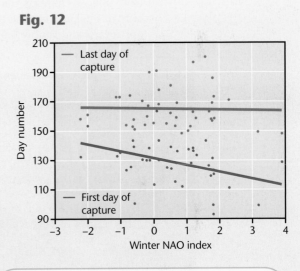

Fig. 11

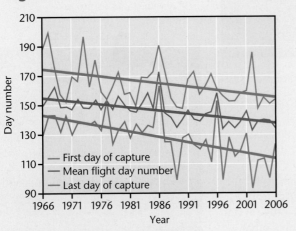

Fig. 12 shows the 41 values for the first day of the spring migration flight and the last day of the spring migration flight plotted against NAO index. Statistical calculations were used to draw the lines through the points.

5 What evidence is there in Fig. 10 that climate change is occurring in Britain?

6 Using the data in Fig. 10, discuss the relationship between the NAO and the population of the green spruce aphids. Suggest reasons for this relationship.

7 Using the data in Fig. 11 and Fig. 12, discuss the relationship between the NAO and the dates of the spring migration of the aphids.

8 It is expected that similar patterns will be found in the population size and behaviour of other species of aphids, including those that infest major agricultural crops such as wheat. Suggest how the results of this study could be used to help farmers make the most efficient use of pesticides.

Distribution of wild animals and plants

As climate change happens, wild animals and plants may find their usual habitats changing. Sometimes, these changes may be obvious. Melting of Arctic ice, for example, means that polar bears may struggle to find enough places to live and enough food to eat. Some people think that these bears may become extinct if global warming continues, as many researchers believe that it will.

It is likely that many species will be able to move from where they currently live to other places where the climate is better suited to them. As we have seen, we expect some insect species to move northwards into cooler places, as their current habitats get warmer. This is already happening in Britain. For example, the Holly Blue, a butterfly that was formerly found only in England, has increased its range northward and is now found in Scotland.

The Scottish primrose, *Primula scotica*.

Fig. 13 Distibution map for *Primula scotica*

The Holly Blue, *Celastrina argiolus*, is a butterfly that is extending its range northwards into Scotland, probably as a result of global warming.

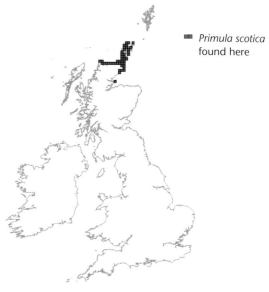

■ *Primula scotica* found here

However, for some species rising temperatures may bring disaster. Like polar bears, some species will have nowhere else to go. For example, several British plant species grow in the far north of Scotland or only on our highest mountain tops. The beautiful Scottish primrose, a wild flower, is already confined to the most northern parts of our country (Fig. 13), and we may lose it altogether if temperatures continue to rise.

The tufted saxifrage, *Saxifraga cespitosa*, grows near the tops of high mountains in Scotland. If global warming continues, it may not be able to find any suitable habitats and may become extinct.

Even if native species are able to find new places in which to live, they may face a threat from invaders. As our climate warms and new species are able to live here, they will compete with the ones already here, if their ecological niches are similar. Interspecific competition could result in the extinction of native species. Another problem is that the invaders may bring diseases with them, which could spread to the native species.

key facts

- Carbon is removed from the atmosphere by photosynthesis and returned by respiration.

- Incomplete decomposition can result in the formation of fossil fuels, which have locked up large amounts of carbon for millions of years. The combustion of these fuels returns carbon dioxide to the air.

- Carbon dioxide is a greenhouse gas, helping to prevent long-wave radiation from leaving the Earth. This is essential, as without it the Earth would be too cold to sustain life.

- The concentration of carbon dioxide, and also of other greenhouse gases such as methane, has been increasing. This has led to an enhanced greenhouse effect, which is causing climate change, including global warming. Human activities such as the burning of fossil fuels and deforestation are contributing to these changes.

- Increased carbon dioxide concentrations may result in increased rates of photosynthesis, and therefore increased productivity and yield of crop plants. However, in some parts of the world, increased temperatures may reduce yields, or even mean that some types of crop plants can no longer be grown there.

- Warmer winters and longer summers are likely to affect the life cycles of insects and increase their populations. This could mean that insect pests may become a greater threat to the health of humans and other animals, and may also reduce crop yields.

- Animals and plants adapted to living in our current climate may not be able to survive in the same areas in the future. For some, this may mean extinction if they cannot move to cooler habitats.

5.2 The nitrogen cycle

Decomposition is not only important for maintaining the supply of carbon dioxide in the atmosphere. All the other components of living organisms have to be recycled. Mineral ions are released from cells as tissues break down. Vital mineral nutrients are also released by the digestion of certain substances, such as magnesium ions from chlorophyll and phosphate ions from phospholipids and nucleic acids. Particularly important is the recycling of nitrogen compounds, since nitrogen is an essential element in all proteins and nucleic acids. Nitrates are often in short supply in soil and water, so it is vital to keep the cycle going.

The cycling of nitrogen is rather more complex than that of most ions because digestion of proteins does not release the nitrate ions that producers require to synthesise their own proteins. There are three main stages in the production of nitrate ions from proteins. These are:

- digestion of proteins to amino acids by decomposers – **saprobiotic** digestion
- formation of ammonium compounds from amino acids – **ammonification**
- conversion of ammonium compounds to nitrates – **nitrification**.

Saprobiotic digestion

We have already seen how decomposers break down dead animals and plants and their waste products. They secrete enzymes onto these organic materials, which break down large molecules to smaller ones. The smaller molecules are then absorbed into the bodies of the decomposers. Fungi and bacteria use this method of nutrition to break down proteins to amino acids, some of which are absorbed into their bodies. Some, however, remain outside them.

Ammonification

Some ammonifying bacteria can continue the process of decomposition by removing the nitrogen-containing amino groups ($-NH_2$) from amino acids, converting them into ammonia, NH_3, or ammonium ions, NH_4^+. This is very similar to the process of deamination that takes place in our livers, and which produces urea. In fact, urea excreted by animals is also converted to ammonia by some bacteria. If you have ever been in a horse's stable, you may have smelt the pungent ammonia resulting from this reaction.

Nitrification

The ammonia and ammonium ions are oxidised by another group of bacteria, called **nitrifying bacteria**. This takes places in two stages, and different species of bacteria are involved at each stage as follows:

$$NH_4^+ \longrightarrow NO_2^- \longrightarrow NO_3^-$$
ammonium   nitrite   nitrate
ions   ions   ions

These reactions are oxidations because hydrogen is removed from the nitrogen and oxygen is added. As in the oxidations that occur during combustion and respiration, energy is released in the reactions. The bacteria use this energy to carry out chemosynthesis, in the same way as green plants use light energy in photosynthesis.

An overview of the nitrogen cycle

The core of the nitrogen cycle is summarised in Fig. 14.

In theory, this cycle could maintain the supplies of nitrates for plants indefinitely. However, there is a complication in the real world. Where soils are waterlogged and there is little or no oxygen available for respiration, **denitrifying bacteria** use nitrate ions as electron acceptors for respiration instead of oxygen. This reaction produces nitrogen gas, which is released into the atmosphere, thus reducing the amount of nitrates in the soil.

Nitrogen fixation

The air contains enormous quantities of nitrogen – almost 80% of the air is nitrogen gas, N_2. However, this is so unreactive that it is useless to plants. Nitrogen gas diffuses in and out of plant leaves all the time, and in and out of their cells, but absolutely nothing happens to the nitrogen gas. As far as the plant is concerned, it might as well not be there.

Plants can only use nitrogen when it is combined with another element to form a compound. In particular, they can use nitrate ions and ammonium ions.

Some species of bacteria can use nitrogen gas to synthesise nitrogen compounds. These bacteria can reduce nitrogen (N_2) to ammonia (NH_3) in a process called **nitrogen fixation**. In this process, nitrogen gas is combined with hydrogen ions obtained from water. Some of these nitrogen-fixing bacteria live freely in the soil or water, but others can fix nitrogen only when living inside green plants.

For example, *Rhizobium* bacteria live inside small lumps (nodules) on the roots of leguminous plants such as clover, peas and beans. The bacteria have nitrogenase enzymes that fix nitrogen and produce ammonia. This is converted to amino groups that are used to synthesise amino acids inside the cells of the nodules.

Fig. 14   The circulation of nitrogen

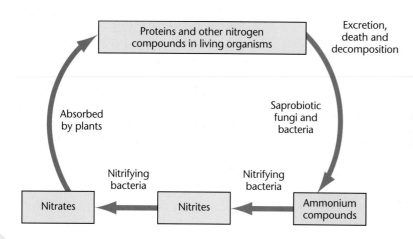

The root nodules of clover are an excellent example of nitrogen fixation.

Fig. 15   The complete nitrogen cycle

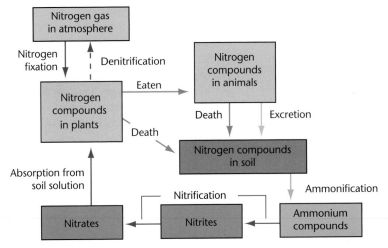

Nitrogen fixation is an example of mutualism, because both the bacteria and the plants benefit from the relationship. The leguminous plants obtain amino acids without having to absorb nitrates, which are often in short supply in soil. The bacteria get sugars from the plants, which they use as a source of energy to synthesise the ammonia.

The full nitrogen cycle

Once these additional processes are added to the diagram, the nitrogen cycle becomes more complex (Fig. 15). You can now see how vital microorganisms are and why it is important to maintain healthy populations of microorganisms in the soil.

9 Use your knowledge of the biochemistry of aerobic respiration to suggest how denitrifying bacteria might use nitrate and nitrite ions as electron acceptors.

10 With reference to the nitrogen cycle, explain how good soil drainage may help to maintain a high concentration of nitrate ions in the soil.

11 Farmers and vegetable growers may use crop rotations, in which they do not grow the same kind of crop in the same ground each year. Clover, peas or beans are often grown for one year in the rotation. Suggest how this may help to maintain the fertility of the soil.

Pollution from fertilisers

In Chapter 4 we saw how fertilisers are used to increase the productivity of crop plants. These fertilisers often contain nitrogen-containing ions, such as ammonium ions or nitrate ions.

Nitrogen-containing fertilisers may be added to grassland. Here, they promote the growth of the grass plants to such an extent that other plants may be out-competed for light. This is one of the main reasons why meadows today tend to contain far fewer species than in the past. The use of fertilisers reduces biodiversity.

A meadow like this can only continue to support such a wide range of species if the use of fertiliser is avoided.

Ammonium nitrate is highly soluble, and will readily wash out of the soil when it rains. This is called **leaching**. If the ammonium and nitrate ions get into water, they can largely affect the communities that live in the water.

The population sizes and rates of growth of aquatic algae and plants are often limited by a shortage of useable nitrogen. When nitrate or ammonium ions are added to the water, these organisms undergo rapid population growth. Often, it is algae that respond most rapidly, turning the water into something looking rather like pea soup. Their huge population prevents light from reaching plants that live deeper in the water, so these plants may die.

The dead plants – and eventually also the algae – provide a greatly increased supply of food for the decomposers in the water. Many of these are aerobically-respiring bacteria. As they have so much more food, their populations grow rapidly.

Fig. 16   How eutrophication occurs

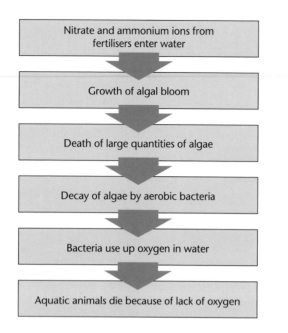

These aerobic bacteria use up most of the dissolved oxygen in the water. The lack of oxygen produces anaerobic conditions, in which many invertebrates and almost all vertebrates are unable to live. They move away, or die.

This sequence of events is called **eutrophication**. It is not only caused by pollution from fertilisers, but can also occur if untreated sewage or factory waste containing organic substances gets into the water. The process is summarised in Fig. 16.

12   The concentration of nitrogen compounds in aquatic ecosystems is not influenced only by pollution. There is natural variation in concentration at different times of year. Fig. 17 shows the nitrate and ammonium ion concentrations in a small lake over a 12-month period.

a   Suggest explanations for each of the following:

i   the decline in nitrate concentration during the spring and summer

ii   the rise in concentration during autumn and winter

iii   the steep increase in ammonium concentration during early autumn

iv   the slow decline in ammonium concentration through the late autumn and winter.

b   From the evidence in the graph, suggest whether this lake is seriously polluted by nitrogen-containing fertilisers from surrounding farms.

Fig. 17   Variations in nitrate and ammonium concentrations

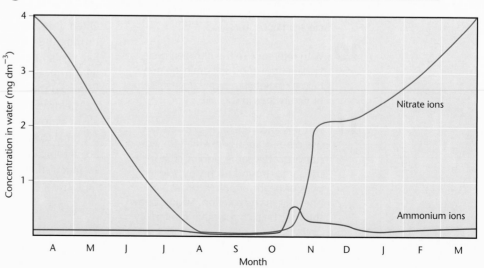

- Nitrogen gas is too unreactive to be used by plants, and must be converted to a compound such as nitrate or ammonia in order to become available to them. This is called nitrogen fixation. Some bacteria are important nitrogen fixers.

- Animals obtain their nitrogen by feeding on plants.

- Remains of dead plants and animals, and their waste products, contain proteins and other nitrogenous substances. These are broken down by saprobiotic decomposers to form ammonia.

- Nitrifying bacteria oxidise ammonia to nitrite and nitrate ions.

- Denitrifying bacteria reduce nitrate ions to nitrogen.

- Pollution of waterways by nitrogen-containing fertilisers leads to eutrophication, in which excessive growth of the populations of oxygen-requiring bacteria depletes the water of dissolved oxygen.

1

a Name the type of bacteria which convert
  i nitrogen in the air into ammonium compounds;
  ii nitrites into nitrates. (2)

b i Other than spreading fertilisers, describe and explain how **one** farming practice results in addition of nitrogen-containing compounds to a field. (2)

  ii Describe and explain how **one** farming practice results in the removal of nitrogen-containing compounds from a field. (2)

Total 6

AQA, June 2006, Unit 5, Question 1

2

a Nodules on the roots of some plants enable them to survive in soils with a low concentration of nitrate ions. Explain how. (2)

b Waterlogged soil may have a low concentration of nitrate ions. Explain why. (2)

Total 4

AQA, June 2007, Unit 6, Question 1

3 Arctic tundra is an ecosystem found in very cold climates. The diagram shows some parts of the carbon and nitrogen cycles in arctic tundra.

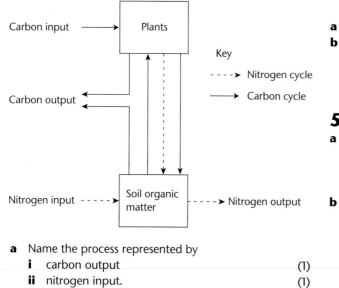

a Name the process represented by
  i carbon output (1)
  ii nitrogen input. (1)

b An increase in temperature causes an increase in carbon input. Explain why. (2)

c Fungi obtain their nutrients from the organic matter in soil. Explain how. (3)

Total 7

AQA, June 2007, Unit 5, Question 2

4 The diagram shows part of a river into which raw sewage is discharged. The table shows the results of a chemical analysis of water samples taken at sites **X**, **Y** and **Z** along the river.

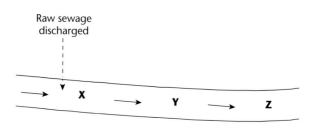

	Site X	Site Y	Site Z
BOD (biological oxygen demand) (mg dm^{-3})	38.0	17.0	3.0
Ammonium ions (mg dm^{-3})	0.3	0.7	0.1
Nitrate ions (mg dm^{-3})	4.0	60.0	7.0

a Explain the decrease in BOD from site **X** to site **Z**. (3)

b Explain the increase in the concentration of ammonium ions and nitrate ions from site **X** to site **Y**. (4)

Total 7

AQA, June 2007, Unit 5, Question 8

5

a The availability of nitrogen-containing compounds in the soil is often a limiting factor for plant growth. Explain **two** ways in which a shortage of nitrogen-containing compounds could limit plant growth. (4)

b Farmers apply nitrate fertilisers to improve crop growth.
  i Explain why plants may fail to grow if high concentrations of nitrate are applied to the soil. (2)
  ii Streams and rivers running through farmland can also be adversely affected by application of high concentrations of nitrate fertiliser. Fish cannot

survive when the oxygen levels of water are reduced. Explain how high concentrations of nitrate applied to farmland may result in the reduction of the numbers of fish present in aquatic ecosystems. (5)

c Malonate is a substance that inhibits the enzymes of the Krebs cycle. In an investigation, plant roots were placed in a solution containing nitrate and malonate. The solution had air bubbled through it. Explain why these plant roots took up nitrate more slowly than those in a control solution which contained nitrate but no malonate. (3)

Total 14

AQA, June 2003, Unit 5, Question 8

6 Pea plants are leguminous and have nodules on their roots which contain bacteria that are able to fix nitrogen. The diagram shows some of the processes involved in nitrogen fixation by these bacteria.

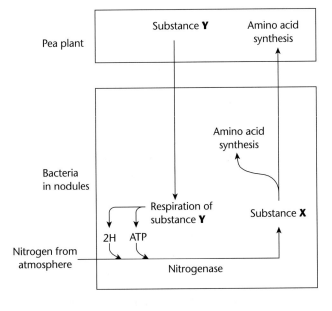

a Name
   i substance **X**; (1)
   ii substance **Y**. (1)
b Pea plants respire aerobically, producing ATP which can be used for amino acid synthesis. Describe the role of oxygen in aerobic respiration. (2)

c The bacteria respire anaerobically. This produces hydrogen and ATP used in nitrogen fixation. The hydrogen comes from reduced NAD. Explain how the regeneration of NAD in this way allows ATP production to continue. (2)

d The enzyme nitrogenase is specific to the reaction shown. Explain how **one** feature of the enzyme would contribute to this specificity. (2)

e Sodium ions act as a non-competitive inhibitor of the enzyme nitrogenase. Explain how the presence of a non-competitive inhibitor can alter the rate of the reaction catalysed by nitrogenase. (3)

Total 11

AQA, June 2005, Unit 5, Question 3

7 The diagram shows a river system in an area of farmland. The numbers show the nitrate concentration in parts per million (ppm) in water samples taken at various locations along the river. Concentrations above 250 ppm encourage eutrophication in the river.

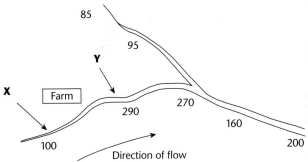

a **i** Explain how farming practices might be responsible for the change in nitrate concentration in the water between point **X** and point **Y**. (2)
   ii Describe the effect the nitrate concentration may have in the river at point **Y**. (5)
b Single-celled organisms were cultured from samples of river water. Give **three** characteristics of the cells that would enable you to distinguish prokaryotes from eukaryotes. (3)

Total 10

AQA, June 2005, Unit 5, Question 7

BOD and indicator species

In the UK, the Environment Agency has the responsibility of monitoring water quality in our rivers. Each year, it takes water samples from each river, and it also responds to reports of pollution incidents. The source of the pollution is tracked down, and the people responsible can be fined. Water pollution incidents decreased significantly between 2000 and 2006. If you would like to find out whether this decrease has continued, visit the Environment Agency's website at: www.environment-agency.gov.uk.

One of the tests that Environment Agency scientists do is to measure the biochemical oxygen demand, **BOD**, of the water. The greater the pollution with organic material or fertilisers, the more aerobic bacteria there will be in the water. And the more aerobic bacteria, the faster the oxygen is used up. BOD is a measure of the rate of use of oxygen.

Fig. 18 Water pollution incidents in the UK between 2000 and 2006

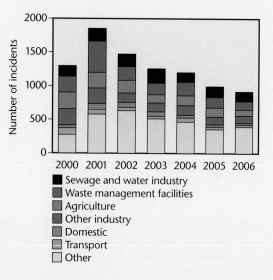

To measure BOD:

- Two samples of water are collected from the same site.
- The oxygen content of one sample is measured using an oxygen meter.
- The second sample is sealed in an air-free container and incubated in darkness at 20 °C for five days.
- At the end of this time, the oxygen content of the second sample is measured.
- The difference between the oxygen concentration on the first and last days is used to calculate the BOD.

Water pollution can also be detected using an even simpler method, which does not involve using test tubes or

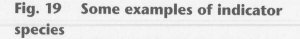

Fig. 19 Some examples of indicator species

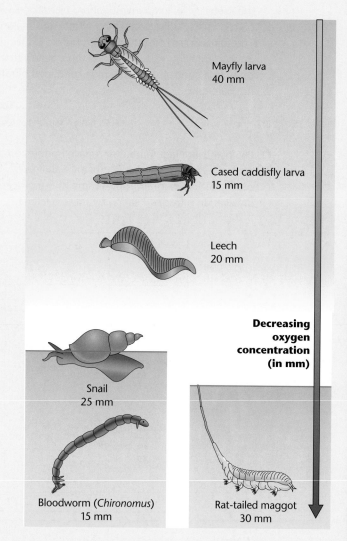

water samples at all. The organisms that live in a river or stream are greatly affected by the abiotic factors in the water, including the oxygen concentration. Some species, such as rat-tailed maggots and sludgeworms, have adaptations that allow them to survive in very anaerobic conditions. Others, such as mayfly larvae, can only live in water containing a high concentration of dissolved oxygen. These are **indicator species** – their presence gives an indication of the oxygen concentration in the water, and therefore the degree of pollution.

Each indicator species is allocated a 'biotic index'. The higher the biotic index, the more dissolved oxygen it requires. Researchers can get a good idea of the degree of pollution in a river by collecting samples in a standard way, and counting the number of each organism that they find. Table 2 shows the species present in samples taken from two points on a river, one upstream of a suspected pollution incident and one just downstream from it.

Table 2   Data from two sites on a river

Species	Biotic index	Number of individuals at site 1	Number of individuals at site 2
Stonefly larvae			
Species 1	10	4	
Species 2	10	7	
Species 3	10	7	
Mayfly larvae			
Species 1	4	2	8
Species 2	10	7	
Caddisfly larvae			
Case made of stones	10	2	
Case made of leaves	7	1	
Freshwater shrimp	6	35	
Leeches			
Species 1	3	11	1
Species 2	3	3	2
Snails			
Species 1	3	45	30
Fly larvae			
Species 1	2		30
Species 2	2		65
Worms	1		120

A1 Use the data from the Environment Agency (Fig. 18) to determine the relative importance of water pollution from agriculture compared with other sources. Did this change between 2000 and 2006?

A2 Explain how measuring BOD can give a good indication of whether pollution has occurred.

A3 Explain the importance of each of the following measures that must be taken when measuring BOD:

a ensuring there are no air bubbles or air spaces in the water samples in their containers

b keeping the second sample in the dark, not in the light, for five days.

A4 Suggest how BOD measurements could be used to help to track down the source of pollution.

A5 Using the data in Table 2, determine which of the two sites was above the source of the pollution. Explain how you used the data in making your decision.

A6 Discuss the advantages and disadvantages of using indicator species to detect water pollution.

A7 Bloodworms and rat-tailed maggots are perfectly able to survive in a relatively high oxygen concentration, but they are almost never found in well-oxygenated water. Suggest which types of environmental factors might cause their absence.

6 Succession and conservation

Left to its own devices, this house and the land around it would become woodland after another 25 years or so.

Oak woodland is the natural vegetation in wetter parts of Britain. This one is in Derbyshire.

What would happen if your school playing fields, your garden or the local park were suddenly abandoned? If no one mowed the grass any more? If no one ever killed the weeds? If no one ever walked there?

It would not take long for nature to take over. First, the grass and other plants would grow taller and taller. Seedlings of shrubs would begin to appear. In a few years time, it would probably be covered with bushes and young trees. It would not be long before there was no trace of any human activities. Eventually, the land would probably develop into a wood.

The natural vegetation in most parts of Britain is woodland. Given a chance, this is what would grow on the land. Human activities prevent this from happening. We cover things with concrete, dig up allotments, pull up weeds, mow grass, cut hedges, plough fields, and allow sheep and cattle to graze on hillsides.

Our activities, over thousands of years, have greatly changed the appearance of our landscape. Without this, there would be less variety of habitats in Britain. If we want to keep all of these varied habitats – chalk downlands, flower-rich meadows, freshwater ponds – we have to take steps to stop them all gradually turning into woodland. Conservation of habitats has to be active. If we just let things be, we would live in a land covered by oak, ash and beech trees.

6.1 Succession

The change from a grassy park to an oak wood is an example of **succession** (Fig. 1). Succession is a gradual change in the communities in a particular place. At the end of the process, the final community is called **the climax community**. The climax community tends to be fairly stable. In Britain, most places – other than those that are especially wet, have very poor soil or are exposed to very high winds – develop woodland as their climax community. The woodland will not change very much over time. Individual trees will die, and new ones will grow, but the overall composition of the community will remain pretty much the same over long periods of time.

Fig. 1 Succession: a grassy park gradually turning into woodland

As grazing is reduced ⟶

Grassland with small flowering plants, like daises.

Taller herbaceous plants, like willowherb and foxgloves, grow and cut off light to small plants, allowing tree seedlings to take root.

Bushes and shrubs, such as hawthorn and bramble, grow. Most herbaceous plants die out.

Fast-growing trees, such as birch, grow up, forming dense, low forest.

Larger, slower growing, but stronger oak trees grow above the birch and establish the climax community.

Why succession occurs

Succession occurs because conditions in the habitat change. Abiotic factors and biotic factors are altered, so that different species are better adapted than the current species to live there.

As each species takes possession of the habitat, it changes many different factors – for example, the amount of light reaching the ground, the quantity of nutrients in the soil, the food sources for animals. New species are able to move in once the habitat becomes suitable for them. They in turn change the habitat, making it suitable for yet more species. We generally find that, during succession, the number of different species, known as **species richness**, steadily increases.

Sometimes, we are able to study succession occurring on a completely blank canvas – a piece of ground where there are absolutely no living things to start with. This can happen on a new island that has appeared as a result of volcanic eruption beneath the sea, as happened when Surtsey suddenly appeared off the coast of Iceland in November 1963. Retreating glaciers left behind areas of ground completely devoid of life. In 1980, a massive volcanic eruption of Mount St Helens took place, destroying all life over huge areas – the succession that subsequently took place there is described on page 117.

The first organisms to live on empty land are called **pioneer species**. They are often lichens – strange mutualistic combinations of an alga and a fungus that are able to grow in the harshest of conditions such as on bare rocks with no soil. They arrive as spores, carried by the wind or perhaps on the feet of a briefly visiting bird or insect. Their presence begins to provide small amounts of humus in which wind-dispersed plant seeds can be trapped, and which hold water and minerals. After some years, plants may be able to grow. These pioneer

1

a In the example of a grassy park gradually turning into woodland in Fig. 1, what is the initial change in environmental factors that kick-starts the process of succession?

b At the start of this process, the ground is covered with meadow grass and small plants such as daisies and plantains, *Plantago*, a small herb. Explain why these are not present in the woodland community that has developed by the end of the succession.

plants generally have adaptations that allow them to survive in difficult conditions.

At the early stage of succession, abiotic factors tend to be the most important factors in determining which species can live there.

Lichens are often the first species to grow on empty ground. They are a combination of a photosynthetic alga and a fungus which is able to extract inorganic ions from even the most inhospitable substrates.

These yellow verbascums and pink willowherbs have managed to colonise the abandoned railway track. They are pioneer plants. Their seeds blew in on the wind and they are able to cope with the lack of water and minerals in the almost non-existent soil between the rails.

There is little or no soil, so plants have access to very few minerals, and have difficulty in anchoring themselves to the ground. Water runs off the bare ground, because there is little or no humus to be able to hold water for a time after rainfall. There is no shelter from wind or the hot sun.

However, as soil gradually builds up over the years, conditions become suitable for other species to survive. Conditions are much less hostile. After a while, many different species will live there, all competing with each other for food, water, light and living space. This is when biotic factors such as competition, predation and grazing become very important in determining which species live there, and how big their populations will be.

Succession on sand dunes

We don't need to stand in one place for a few hundred years to study succession. If we can find a place where succession has been happening for a while, we may be able to see different stages of succession at one glance.

Sand dunes give us a good opportunity to see the stages. Next to the sea, the sand is an inhospitable place for plants to live. It is sharp-draining, so it does not hold water for any length of time. It is constantly shifting, giving no opportunity for plants to establish their roots. It holds very few nutrients, so plants cannot get the inorganic ions – nitrates, for example – that are essential for their growth.

Not much further away from the sea, however, we can often see sand dunes that have been there for a short while. And further behind these, we can find older dunes. If we set out a transect from the sea to behind the inland dunes, we can study what species live in this environment at different stages of succession – the earliest stages nearest the sea, and the latest stages furthest away from the sea. Fig. 2 illustrates the process of succession on sand dunes, as described in the following text.

The pioneer plant on the developing sand dunes is often marram grass, which is superbly adapted for life on shifting sands. Marram grass has a dense network of branching underground stems, which hang on tightly to the sand and help to stabilise it. It has tough leaves, able to withstand the high winds to which it is often exposed. The leaves can roll up, sheltering the stomata and reducing water loss by transpiration. Fig. 3 shows the many adaptations of marram grass for reducing water loss by transpiration.

Once the marram grass has taken root, wind-blown sand tends to collect around it, which is how dunes form. Within 10 years, the dunes may be up to three metres high, and well covered by marram grass.

Fig. 2 Succession on sand dunes

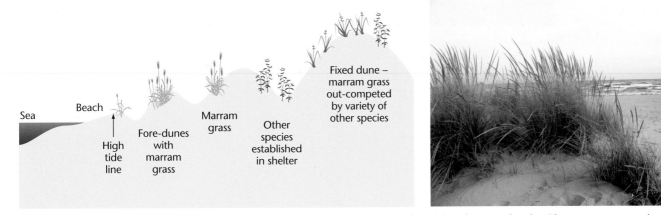

A fore-dune on a beach, with marram grass growing on it.

Fig. 3 Transverse section of a rolled-up leaf of marram grass, *Ammophila arenaria*

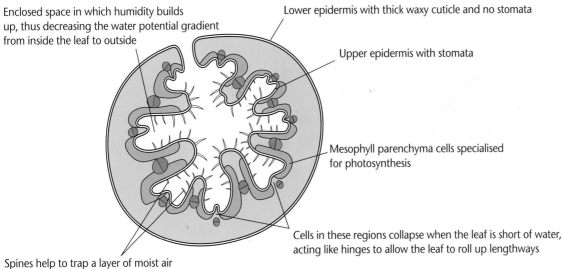

Enclosed space in which humidity builds up, thus decreasing the water potential gradient from inside the leaf to outside

Lower epidermis with thick waxy cuticle and no stomata

Upper epidermis with stomata

Mesophyll parenchyma cells specialised for photosynthesis

Cells in these regions collapse when the leaf is short of water, acting like hinges to allow the leaf to roll up lengthways

Spines help to trap a layer of moist air near the leaf surface

Once stabilised by marram grass, the dunes can become several metres high.

Now a few other species are able to set up home in the shelter that the marram provides. Humus builds up in the dry sand, so that it begins to look rather more like soil. Water retention is improved and mineral nutrients do not drain away so quickly. Other grasses and plants adapted for living in grassland colonise the dunes. Eventually, as conditions begin to improve, the marram grass is out-competed and dies out.

On the oldest dunes, furthest from the sea, other plants such as heath violets can grow.

If left undisturbed for long enough, these old dunes will eventually become woodland. The precise type of woodland depends on the abiotic factors that prevail in that place. If it is very wet, then the climax community may be willow carr – woodland containing willow and other trees, where water often lies between them. In drier places, oak, beech or ash woodland may develop.

Willow carr is often the climax community in ground that floods regularly.

2 List four abiotic factors that make colonisation by plants particularly difficult in the sand just above the high tide line. (Some have been mentioned in the text, and you should be able to think of others.)

3 Explain the adaptations of marram grass that enable it to colonise sand dunes next to the sea.

4 Suggest how grassland plants reach the dunes that they colonise.

how science works

hsw

Succession on Mount St Helens

In May 1980, snow-capped Mount St Helens, in southwest Washington State, USA, exploded. The north side of the mountain collapsed, creating a debris avalanche – a mixture of ice blocks from the shattered glaciers, hot rocks and mud. Two and half square kilometres of material plunged down the mountainside. When it eventually came to rest, it formed a layer with an average depth of 45 m. Temperature measurements in the deposits, taken 10 days after the eruption, varied from 68 °C to 98 °C. Not surprisingly, every living thing beneath the avalanche was killed.

It wasn't long before literally hundreds of researchers turned up, eager to take this opportunity to study how life would recolonise the barren ground. Different teams worked in different areas. One group chose to study a huge area of the debris avalanche. They set out markers surrounding 103 circular plots, each with an area of 250 m², and recorded the species they found there. The same group returned to the site on numerous occasions for 20 years following the eruption.

Some of their results are shown in the graphs. Fig. 4 shows how the percentage cover in the plots changed over time. Fig. 5 shows the mean number of species per plot.

Fig. 4

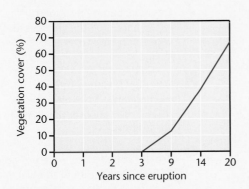

Fig. 5

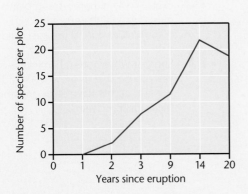

The unexpected and devastating eruption ripped away huge areas of Mount St Helens.

Pioneer plants were able to colonise the lifeless debris.

– and also some western hemlocks – were present very early on in the succession as small seedlings. But most of the western hemlock seedlings soon died, and it took some time before the numbers of alder trees really began to increase, as shown in Fig. 6.

Today, the shattered sides of Mount St Helens are no longer grey and barren. The communities that live there are still changing and it will take many more years before the climax community – western hemlock forest – will become established. Before that happens perhaps there will be another eruption of this active volcano, and the whole process will start over again.

The most common species during the very early stages of the succession were species with light, wind-dispersed seeds, that were fast growing and able to mature quickly. These included a small plant called pearly everlasting, which grows close to the ground where it escapes most of the wind. As other plants, including grasses, colonised, the little pearly everlastings disappeared. Within a few years, lupines had begun to grow. These are much larger plants than the everlastings, and they have nitrogen-fixing bacteria in their roots. Other nitrogen-fixers, including white clover and bird's-foot trefoil also took residence in the area.

The first trees to make a substantial showing were alders, which are also nitrogen fixers. In fact, a few of these

5 Describe the abiotic factors making it difficult for plant life on the debris avalanche soon after the Mount St Helens eruption.

6 Suggest the adaptations of pearly everlasting that make it a successful pioneer plant. (You could look this up on the Internet. The Latin name for this plant is *Anaphalis margaritacea*.)

7 Describe, and then suggest explanations for, the changes in vegetation cover over time shown in Fig. 4.

8 Describe, and then suggest explanations for, the changes in species diversity over time shown in Fig. 5.

9 Explain why lupines were not able to grow on the site in the first year or so but became quite common later on.

10 Explain why nitrogen-fixing plants made up a large part of the early communities during this succession.

11 Suggest why western hemlock is only likely to be able to grow well on this site after alders have been growing there for some years.

Fig. 6

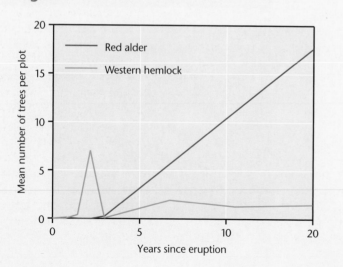

key facts

● Succession is a gradual change in communities over time. Each community changes the environment so that it becomes suitable for new species to live there. The final community is known as the climax community.

● In the early stages of succession, abiotic factors are most important in determining which species can survive. Pioneer species have adaptations that allow them to colonise these hostile environments. As succession proceeds, biotic factors become more significant.

● Species diversity tends to increase as succession proceeds.

6.2 Conservation

As most of us are only too well aware, human activities can do great damage to the environment. We cut down forests to provide us with wood for building and for fuel, or to clear the land so that it can be used for housing. We pollute the land, air and water, making it impossible for some species to live there. The more people on Earth, the more effects we have on the other organisms with which we share our planet.

Conservation aims to maintain the biodiversity around us. Biodiversity is a difficult term to define but it is often taken to mean the variety of different habitats in an area, the variety of different species living there, and the genetic diversity within those species. Conservation attempts to ensure that we do not allow a decline in the number and range of different ecosystems and species, or the genetic diversity within species.

Planning conservation programmes

Before conservation can begin, we need as much information as possible about what we are trying to conserve. Imagine, for example, that a small woodland has been recognised as an important wildlife habitat, and that an environmental organisation wants to conserve it. They would need to sample and collect data about the species present and the sizes of their populations (see Chapter 1), and about the abiotic characteristics in the woodland. Then they would need to identify which species or features are most under threat and identify the threats as precisely as possible. It would also be helpful to monitor the woodland over time to look for any changes that may be taking place in the community. Armed with this information, the environmental organisation could then draw up a management plan stating the aims of the conservation programme, and giving details of how to carry it out, including sources of funding.

You might think that a better way of conserving such a woodland would be to put a strong fence around it to keep everyone out, and then just to leave it alone. Cases where such an approach is useful are relatively rare, however. The 'leave it alone' method is most likely to maintain biodiversity if the area is truly wild and untouched by humans. Such wild lands still do exist in places; for example, deep in the Amazonian rainforests in some parts of Brazil, some of the rainforests in Papua New Guinea, the hearts of deserts such as the Sahara and the icy lands of Antarctica. But even there, pressures from forestry and mining make it difficult to be sure that these special places will remain unspoilt in the future.

In many cases, what we want to conserve is something that is the result of previous human activity. Many of the habitats that we think of as special only exist because human activity has prevented them from undergoing succession. If we stop grazing sheep on chalk downland, or stop clearing silt from freshwater ponds, they will undergo succession and end up as woodland. The woodland in our example may contain species that would be lost if we allowed natural succession to occur. Just leaving things alone will not be enough to maintain biodiversity in this area – we need to intervene actively.

Until recently, sheep had been grazing on this chalk downland.

Volunteers clearing scrub from chalk downland.

key facts

- Conservation attempts to maintain biodiversity. This involves maintaining a variety of different habitats, which between them can support a wider range of species than a single type of habitat could do.

- Conservation frequently involves preventing succession from occurring, or the careful management of succession. Without intervention, many habitats in Britain would gradually turn into woodland.

1 The diagram shows the stages in a succession from colonisation of bare soil to the formation of woodland.

| Mosses and lichens | → | Grasses | → | Broad-leaved herbaceous plants | → | Shrubs | → | Trees |

a What name is used to describe the final stage in a succession? (1)

b Explain **one** way in which farming practices prevent the formation of woodland. (1)

c Clover plants are able to reproduce by vegetative propagation. Suggest **three** advantages of this form of reproduction when clover colonises a new habitat. (3)

Total 5

AQA, January 2007, Unit 5, Question 1

2

a Give **two** aims of biological conservation. (2)

b The graph shows the rate of extinction of species of birds on islands of different size.

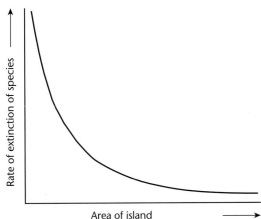

Area of island →

i Describe the relationship between the rate of extinction of species and the areas of the islands. (2)

ii Suggest one cause of this relationship. (1)

Total 5

AQA, June 2005, Unit 6, Question 2

3 A study was made of a transect through sand dunes, from dunes near the sea-shore to woodland. Samples of quadrats at five positions along the transect were analysed. The results are shown in the table.

	Dunes near sea-shore	Mobile dunes	Fixed dunes	Heath and scrub	Wood-land
Mean percentage plant cover	2	25	90	100	100
Number of plant species per unit area	12	36	95	140	92

a i Woodland is the final stage in this ecological succession. Give the term used to describe the final stage in an ecological succession. (1)

ii The number of plant species per unit area in the woodland is less than that in the heath and scrub. Suggest an explanation for this. (2)

b Several of the species of plants living on the dunes have small leaves and their stomata are located in grooves on the underside of the leaves. What do these features suggest about the soil conditions where they live? Explain your answer. (3)

Total 6

AQA, June 2003, Unit 5, Question 3

4

a Explain what is meant by

i succession; (2)

ii a climax community. (1)

Heather plants are small shrubs. Heather plants are the dominant species in the climax community of some moorlands. The structure and shape of a heather plant changes as it ages. This results in changes in the species composition of the community. A large area of moorland was burnt leaving bare ground. The table shows four stages of succession in this area.

Time after burning (years)	Appearance of heather plant	Mean percentage cover of heather	Other plant species present
4		10	Many
12		90	Few
19		75	Several
24		30	Many

b Explain why the number of other plant species decreases between 4 and 12 years after burning. (2)

c The rate at which a heather plant produced new biomass was measured in g per kg of heather plant per year. This rate decreased as the plant aged. Use the information in the table to explain why. (3)

Total 8

AQA, June 2006, Unit 5, Question 6

5 When coal is mined by open-cast mining, the top layer of soil is first scraped off and stored in a large heap. Once mining has finished, the area can be reclaimed. Soil from this store is then spread back over the surface. Some of the bacteria living in the soil store respire aerobically and some respire anaerobically.

Table 1 shows the numbers of aerobic and anaerobic bacteria found at different depths in a soil store.

Table 1

Depth (cm)	Mean number of bacteria per gram of soil ($\times 10^7$)			
	Aerobic bacteria		Anaerobic bacteria	
	after 1 month	after 6 months	after 1 month	after 6 months
0	12.0	12.1	0.6	0.8
50	10.4	8.6	0.8	1.3
100	10.1	6.1	0.7	4.1
150	10.0	3.2	0.7	7.9
200	11.6	0.8	0.7	8.4
250	11.9	0.7	0.8	8.8
300	11.0	0.8	0.6	9.1

a Some of the soil used to determine bacterial numbers was collected from the surface of the soil store. Describe how you would ensure that this soil was collected at random. (2)

b i Describe how the numbers of aerobic bacteria after 6 months change with depth. (2)

ii Explain the difference in the numbers of aerobic bacteria at a depth of 300 cm between 1 and 6 months. (2)

c Explain how the changes in bacterial numbers which take place at 150 cm illustrate the process of succession. (3)

Dehydrogenase is an enzyme involved in aerobic respiration. Dehydrogenase activity in a soil sample can be used as a measure of the activity of aerobic bacteria. The graph shows the mean dehydrogenase activity of soil samples taken from the same depth in a soil store at different times. The bars on the graph represent two standard errors above and below the mean.

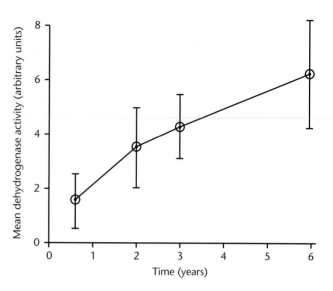

d i From what depth in the soil store would you expect these soil samples to have been taken? Use information from **Table 1** to explain your answer. (2)

ii How would you expect dehydrogenase activity to vary with depth after 6 months? Use information from **Table 1** to explain your answer. (3)

e What do the error bars tell you about the difference between the mean dehydrogenase activity at 6 months and 3 years? Explain your answer in terms of probability and chance. (3)

f **Table 2** shows the dehydrogenase activity and the number of aerobic bacteria present in some soil samples.

Table 2

Dehydrogenase activity (arbitrary units)	Number of aerobic bacteria per gram of soil ($\times 10^7$)
13.1	12.7
9.2	8.7
5.5	6.5
3.0	4.6
2.2	2.7
0.4	0.6

A sample of soil was found to have dehydrogenase activity of 8.7 arbitrary units. Explain how you would use the data in **Table 2** to predict the likely number of aerobic bacteria in 1 g of this soil sample. (3)

Total 20

(AQA, January 2004, Unit 9, Question 1)

how science works **assignment**

hsw

Blue butterflies run out of thyme

The large blue butterfly, *Maculinea arion*, became extinct in Britain in 1979 (but is now slowly making a comeback). It lives on chalk grasslands where there are colonies of a particular species of red ant, *Myrmica sabuleti*, and plenty of the low-growing herb, thyme.

The caterpillars of this beautiful butterfly feed exclusively on wild thyme, burrowing into the flower heads to feed on the flowers and then the developing seeds. When the larvae reach about 4 mm in length, they leave the flower and drop to the ground – and wait.

If they are in a suitable place, the larvae are found by worker ants. The caterpillars have a 'honey gland', which attracts the ants to them. The ants pick up the caterpillars and take them into their underground nests.

The caterpillars are not good guests. They eat the ant grubs, growing fat themselves and eventually pupating. They hatch into adult butterflies the following summer. It is a strange sight to see a crumpled blue butterfly crawling up out of the ground, before spreading its wings in the sun and preparing for its first flight.

Wild thyme has short stems and grows as mat-like clumps in short grass. The chalk downs used to be heavily grazed by sheep and rabbits, but the numbers of sheep have been reduced as a result of changes in farming practices. Then the viral disease myxomatosis spread through the rabbit population, reducing their numbers. As a result, the grass was no longer cropped short. Thyme was unable to compete with the tall grasses for light.

With the loss of the short grass and thyme came the loss of the butterflies. However, conservationists are determined not to let this butterfly become permanently extinct in Britain. In several areas, including some in Cornwall, areas of downland are being managed to provide perfect conditions for thyme, red ants and the large blue butterfly. The aim is to produce a turf 2–5 cm tall, well sheltered from wind and on well-drained soil – enabling red ants to make thriving underground nests. Sheep or cattle are being grazed on the sites. As bushes start to grow, threatening to turn the grassland into scrub, they are cut down or dug up. However, some bushes, such as gorse, are left so that they can provide shelter. Thyme seeds have been sown, taking care to use seed harvested from plants growing locally.

Large blue butterflies have now been successfully re-introduced at several different sites. With constant attention and careful management, it is hoped that the populations of this globally endangered species will once again increase.

The large blue butterfly, *Maculinea arion*.

Wild thyme, *Thymus serpyllum*.

A1 Explain how the large blue butterfly became extinct in Britain.

A2 Suggest how the specialised niche of this butterfly has contributed to its status as a globally endangered species.

A3 With reference to succession, explain how and why the sites for re-introduction of this butterfly are being managed.

A4 Suggest why the thyme seeds being sown are collected from locally growing plants.

7 Inheritance

In 1968, the mother of geneticist Nancy Wexler was diagnosed as having Huntington's disease. Nancy's three uncles, all brothers of her mother, had already died of the distressing condition, the symptoms of which only appear in middle age and include relentless loss of coordination and mental deterioration. Nancy realised that she had a 50:50 chance of developing the disease in later life. However, the Wexler family were not prone to giving in easily.

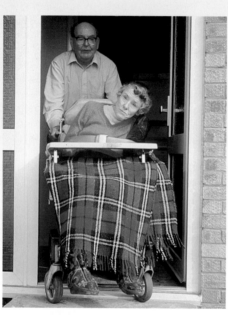

This woman has Huntington's disease, a genetic disorder that results in degeneration of the basal ganglia in the brain.

Nancy's father, Milton, resolved to set up a research foundation to find a cure, if possible in time to save his wife. Sadly, the search was to take much longer than their optimistic hopes. Nancy's mother died 10 years later, but Nancy remained a driving force in the hunt for the elusive gene that causes the disease.

The first breakthrough came when a Venezuelan doctor reported a family in a remote part of his country with large numbers of cases of the disease. Nancy Wexler immediately organised a research team to collect blood samples and eventually traced 11 000 descendants of one 19th century woman who had carried the fatal gene. Back in the USA, the team analysed the blood samples and discovered that the gene was situated somewhere on chromosome 4.

The gene was much more tricky to find than those of several other genetic disorders, but in 1993, after searching through some five million nucleotide bases on chromosome 4, its position was located. The mutant form of the gene responsible for the disease is nicknamed the 'stuttering gene' because one DNA triplet was found to be repeated up to 120 times in people suffering from the disease. Identifying the gene makes it much easier to test, before any symptoms appear, whether individuals in affected families have inherited the disease, and to identify affected children before birth. The next stage is to understand how the mutant form of the gene affects the brain and wreaks its terrible havoc. Then treatment and a cure can become a real possibility.

By 2008 there was no still no cure for this invariably fatal disease. Nevertheless, research continues to progress. We now know that the disease is caused by a mutation in the gene that encodes a protein called huntingtin. In people with the normal form of the gene, there are fewer than 35 glutamines (one of the 20 types of amino acid) in this protein, but in people with Huntington's disease the protein has 36 or more glutamines. This appears to make the protein molecules clump together. Some results suggest that the damage to neurones may be caused by the person's own immune system attacking them, as a result of recognising this protein and behaving as though it was foreign.

Source: adapted from an article by Susan Katz Miller, New Scientist, 24 April 1993.

7.1 Genes and phenotypes

In this chapter, we will look at how characteristics can be passed on from parents to their offspring. This branch of biology is called genetics, and it has a weird and wonderful vocabulary all of its own. It is important that you become familiar with these technical terms and are able to use them with confidence. They will be explained one by one as you work through this chapter, and the questions will provide you with opportunities to practise using them.

What is a gene?

Let's start with one of the most obvious – the term **gene**. You probably remember from AS that a gene is a length of DNA that codes for a particular protein. Despite all the discoveries made by the Human Genome Project, we still do not know exactly how many genes there are in a human cell, but we think it is probably about 20 000. These genes are found on **chromosomes**, of which we have 46. Each chromosome is a very long molecule of DNA. Just before a cell divides, each DNA molecule is replicated, and the two identical DNA molecules lie side by side to form two **chromatids** joined at the **centromere**.

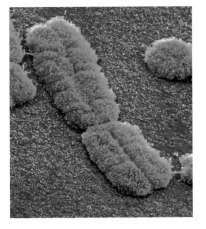

False-colour scanning electron microscope image of a chromosome just before cell division, showing the two chromatids joined at the centromere. Magnification ×8500.

Each human body cell has two complete sets of chromosomes. A cell with two sets of chromosomes is said to be **diploid**. We have 23 chromosomes in each set, making 46 in all. One set came from the father and one set from the mother. Different species have different numbers of chromosomes. Guinea pigs have 64. Christmas trees, or conifers, have 24.

Genotypes

The photograph below shows all the chromosomes in a human cell arranged in order. The 23 chromosomes of one set have been arranged next to the 23 matching ones of the other set. Any two matching chromosomes are said to be **homologous**.

Each chromosome of a homologous pair carries genes for the same characteristics in the same position. The particular place on a chromosome at which a gene is found is known as its **locus**. For example, on chromosome 1 there is a gene that codes for the enzyme acetylcoenzyme A dehydrogenase, and another for the structural protein collagen. So we have two copies of each of these genes – and of almost every other gene too.

Almost all genes occur in more than one form. These different forms of the same gene are called **alleles**. For example, on chromosome 4 there is a gene for the protein huntingtin, an important component of neurones. The normal form of this gene has one stretch that contains between 10 and 35 repeats of the base sequence CAG. Other forms, or alleles, have many more repeats of the CAG sequence, sometimes as many as 120.

It makes things easier if we use shorthand symbols to represent different alleles of a gene. The convention is to use one letter for the two alleles. The uppercase letter is used for the dominant allele, and the lowercase letter for the recessive allele. So, here, we could use h to represent the normal gene for huntingtin (which is recessive), and H to represent one of the abnormal alleles.

As there are two copies of each gene in each cell, there are three possible combinations of alleles.

They are: HH   Hh   hh

We do not need to write down hH as well, because this is no different from Hh.

The combination of alleles is called the **genotype** of the cell or the organism.

Phenotypes

The combinations of alleles in an organism affect many of its characteristics. These observable or measurable features are the organism's **phenotype**. (Phenotype can also be affected by the environment, but here we will just concentrate on genes.)

The full complement of human chromosomes (male) arranged in numbered homologous pairs. This full set of chromosomes is called the karyotype.

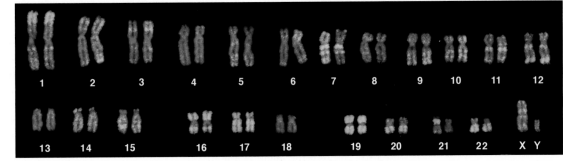

For the huntingtin alleles, allele h is the 'normal' allele, and it confers a 'normal' phenotype on its owner. Allele H causes abnormal huntingtin protein to be made, and it causes Huntington's disease. If a person has the genotype HH, they will have the disease. If they have the genotype hh, they will not have the disease. If they have the genotype Hh, they will have the disease. This happens because the H allele is **dominant** – it has its effect even when another different allele is present. Allele h is said to be **recessive** – it has an effect only when the dominant allele is not present.

You'll often find it helpful when you are answering genetics questions on examination papers to start by quickly jotting down all the different possible genotypes and their corresponding phenotypes, like this:

genotype	phenotype
HH	Huntington's disease
Hh	Huntington's disease
hh	normal

There are two more terms to become familiar with. When the two copies of a gene are the same allele, for example, HH or hh, the genotype is said to be **homozygous**. When they are different, for example, Hh, it is **heterozygous**.

Not all alleles behave like this. For example, there is a gene that determines our blood group – whether we are group A, group B, group AB or group O. There are three alleles of this gene. The correct symbols for them are:

I^A	The allele that gives blood group A
I^B	The allele that gives blood group B
I^o	The allele that gives blood group O

This is a seemingly strange set of symbols to use, but there is good reason for it. When the two alleles I^A and I^B are present together, *both* of them affect the phenotype. The person has blood group AB. These alleles are therefore said to be **codominant**. This explains why they are written using the same letter with a different superscript. If we used just the letters A and B, this would imply that they were different *genes*, not just different alleles of the same gene. If we used A to represent the allele giving blood group A, and a to represent the allele giving blood group B, this would imply that allele A is dominant and allele a is recessive, which is

not true. So, when choosing symbols for codominant alleles, use one symbol to represent the gene, and then different superscripts to code for its different alleles.

If you look very carefully, you may be able to spot that the superscript in I^o is actually a small o, not a capital O. This is done because this allele is recessive to both I^A and I^B.

Armed with this information, we can write down all the possible genotypes and phenotypes:

Blood group genotype	Blood group phenotype
$I^A I^A$	group A
$I^A I^B$	group AB
$I^A I^o$	group A
$I^B I^B$	group B
$I^B I^o$	group B
$I^o I^o$	group O

The term used to describe a situation where there are three or more different alleles of a gene is **multiple alleles**.

1 In Dalmatian dogs, the gene that determines the colour of the spots has two alleles. One produces black spots and the other produces liver (brown) spots. Heterozygous dogs have black spots.

Which allele is dominant? Explain your answer.

2 A faulty allele of the gene that codes for haemoglobin produces a form of this protein that cannot transport oxygen when oxygen concentrations are low. A person with two of these faulty alleles has a disease called sickle cell anaemia. A person with one copy of the faulty allele and one normal allele has a mild form of the disease called sickle cell trait.

a Do these alleles show dominance or codominance? Explain your answer.

b Choose suitable symbols for the sickle cell allele and the normal allele.

c Write down the possible genotypes and the phenotypes that they produce.

key facts

- A gene is a length of DNA that codes for the production of a particular protein.

- Genes affect the characteristics of an organism, known as its phenotype. Environment can also affect phenotype.

- A diploid cell contains two sets of chromosomes. There are therefore two copies of each type of chromosome, and two copies of the genes that they carry.

- Homologous chromosomes carry the same genes at the same loci.

- Genes often come in different forms, called alleles.

- The alleles of a gene that an organism has are known as its genotype. If the two alleles are the same, it is homozygous. If they are different, it is heterozygous.

- An allele that has its full effect on the phenotype even when a different allele of the same gene is present is said to be dominant. An allele that only has an effect when no other allele is present is said to be recessive. If both alleles have an effect in a heterozygote, they are said to be codominant.

- Some genes have three or more different alleles, known as multiple alleles.

7.2 Inheriting genes

As you know, your genes came from your parents. Half of your genes came from your mother, and half from your father. They were passed on to you when the nucleus of a sperm and the nucleus of an egg fused. The two sets of genes became the genes of the zygote, and were copied over and over again as the zygote divided by mitosis to form all the millions of cells in your body.

To understand how genes are inherited, you need to think back to what you know about how gametes are produced. You'll probably remember that gametes are made by **meiosis**, a special kind of cell division in which the normal chromosome number is halved. Meiosis of diploid cells in a woman's ovary or a man's testis produces cells with only a single set of chromosomes. They are said to be **haploid**. Each gamete has only 23 chromosomes, rather than 46.

Because of this, each gamete has only one copy of each gene, not two. So, if a man has the genotype HH, each of his sperm will contain one copy of the H allele. If he has the genotype Hh, approximately half of his sperm will contain an H allele, and the other half will have an h allele. Understanding this enables us to predict the genotypes – and therefore the phenotypes – of the offspring of a couple. Let's take an example of a woman with the genotype hh and a man with the genotype Hh.

Each of the woman's eggs will contain an h allele. As we have seen, half of the man's sperm will have genotype H, and half will be h. When a sperm fertilises an egg in the woman's oviduct, there is an equal chance that it will be an H sperm or an h sperm. This means that there is an equal chance that the child will have the genotype Hh or hh.

We can show all this in a particular format known as a **genetic diagram**. These may seem a nuisance to write out, especially if you are quick at seeing how things work, but it is important to do it correctly if you are going to get all the available marks in an examination question. The genetic diagram for this cross is shown in Fig. 1.

Fig. 1   Genetic diagram for Huntington's disease

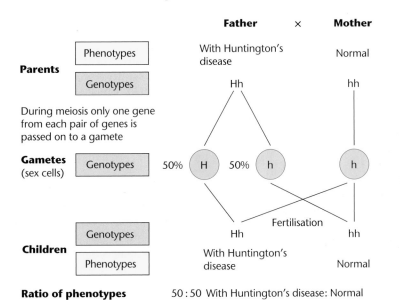

Ratio of phenotypes   50 : 50 With Huntington's disease : Normal

Notice that the genotypes of the gametes have been written inside circles. This is good practice, and you are strongly recommended to do this. It is helpful in keeping track of all the different symbols, particularly in a complicated cross.

It is easy to get muddled when drawing the lines between the gametes to show the possible ways in which they might fuse. A different solution is to use a Punnett square within your genetic diagram, like this:

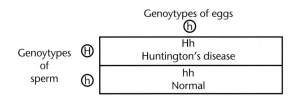

This Punnett square goes *after* you have written down the genotypes of the gametes, so you write them down twice – once underneath the genotypes of the parents, and then again at the top and side of your Punnett square.

Inheritance of cystic fibrosis

Cystic fibrosis is a genetic disease caused by a faulty allele of a gene that controls the formation of an important membrane protein. This protein is called CFTR. Its function is to control the movement of chloride ions out of cells. This is especially important in the cells in the lungs. Normally, chloride ions are moved out of the cells through this protein, and water molecules follow. If the protein is not working, chloride ions cannot move out. This means there is less water outside the cells. This results in sticky mucus, which causes difficulty in getting air in and out of the lungs, as well as problems with gas exchange across the walls of the alveoli. The mucus builds up and is prone to becoming infected with bacteria.

Unlike the allele that causes Huntington's disease, the allele that causes cystic fibrosis is recessive. We can use the symbol F for the normal (dominant) allele, and f for the faulty one. (C and c are not a good choice, because you won't be able to tell the difference between them if you are writing quickly.) The genetic diagram in Fig. 2 shows how two parents who do not have cystic fibrosis can have a child who does.

Probabilities in genetics

In the genetic diagram showing the inheritance of cystic fibrosis (Fig. 2), we have a list of four genotypes for the offspring. This does not mean the couple will have four children, or that they will have these genotypes. It simply shows the possible results of the different gametes fusing together. It indicates the *probability* of each of these genotypes occurring. It shows us that, each time they have a child, there is a one in four chance that the child will get one copy of the cystic fibrosis allele from each parent, and therefore have cystic fibrosis.

There are several ways in which we can write down these probabilities. We can say that:

- The ratio of normal to offspring with cystic fibrosis is likely to be 3 : 1.
- The probability of the child having cystic fibrosis is 0.25.
- Of the offspring, 25% are likely to have cystic fibrosis.

These all mean the same thing.

If you toss a coin 50 times, you'd expect it to fall heads down 25 times and tails down 25 times. But it might not, as we are dealing with chance and probability. In fact, the more often you toss the coin, the closer you are likely to get to a 1 : 1 ratio of heads to tails. If you toss the coin six times, you might get three heads and three tails, but you could also get four of one and two of the other, or even five of one and one of the other. Or, you could get six heads or six tails.

It's the same with genetics. When there are large numbers of offspring, we tend to see the ratios that we would expect. When there are small numbers, we should not be surprised if these ratios are not achieved. If the couple in Fig. 2 had four children, we would be correct in predicting that they would probably have three who were normal and one with cystic fibrosis.

Fig. 2   Inheriting cystic fibrosis

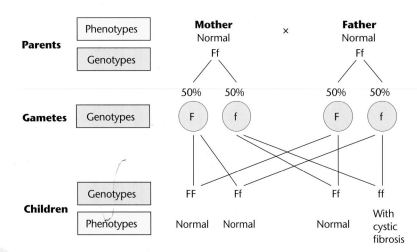

But they might have four normal children, or two with cystic fibrosis.

And, of course, they might not have four children at all. They might have none, or one, or three, or seven. All we can say, is that each time they have a child, there is a one in four chance that the child will have cystic fibrosis.

Another important thing to remember is that whenever the couple has a child, the chance of that child having cystic fibrosis is *not affected* by whether or not their other children have it. Each time, the probability of the child having cystic fibrosis is 0.25, no matter what the genotype of any brothers or sisters.

3 Construct a genetic diagram to show the possible genotypes and phenotypes of a man with the genotype $I^A I^A$ and a woman with the genotype $I^B I^O$. Complete your genetic diagram with a statement about the expected ratios of each genotype and phenotype.

4 A woman with blood group A has a child with blood group O. The paternity of the child is not known. One possible father has blood group AB and another has blood group A.

Construct genetic diagrams to show which of these men could definitely *not* be the father of the child.

hsw

how science works

Huntington's disease

The pedigree diagram in Fig. 3 shows the incidence of Huntington's disease in three generations of a family.

Fig. 3 Pedigree diagram for Huntington disease

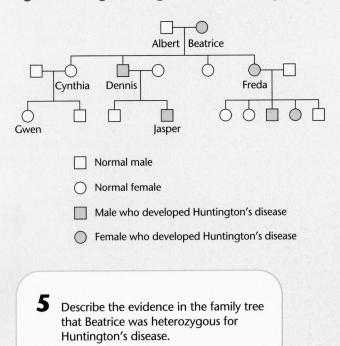

☐ Normal male

◯ Normal female

▨ Male who developed Huntington's disease

◯ Female who developed Huntington's disease

5 Describe the evidence in the family tree that Beatrice was heterozygous for Huntington's disease.

6 Gwen has three children. She is worried that they might develop Huntington's disease. What advice would you give her? Construct a genetic diagram as part of your explanation.

7 What are the chances that a child of Jasper will develop Huntington's disease?

8 It would be possible to test each of Jasper's children when they are young, or even before birth, to find out if they have the allele for Huntington's disease. If you were Jasper or his wife, would you want to have this test done? Do you think it would be right to offer this test to parents? Discuss your reasons – both scientific and ethical – and explain the possible problems that might result from the use of such a test.

9 Although Huntington's disease is always fatal, it continues to be passed on from parents to offspring. Look back at the introduction to this chapter, and explain why this is so.

Good breeding

The study of genetics has helped us to understand human genetic disorders such as Huntington's disease and cystic fibrosis. It is also extremely useful when it comes to breeding domestic animals and improving the quality of crop plants. Most of the animals that we commonly use for food or have as pets are very different from their wild ancestors.

A Siamese cat.

A Burmese cat.

In cats a gene, C, is concerned with the colour of the coat. The normal allele, C^C, makes the cat's coat blackish and is dominant. Siamese and Burmese cats both have particular alleles of this gene. Siamese cats have a recessive allele, C^s, and they must be homozygous for this allele. This mutant allele codes for an enzyme that synthesises black pigment, but only when below body temperature. This is because the mutant allele codes for an enzyme with a tertiary structure that is slightly different from that of the normal enzyme; molecules of this enzyme happen to unfold (denature) at about 37 °C. The Siamese cat is therefore only black in the cooler parts of the body, such as the tail, ears and lower legs. The rest of the coat is pale-cream coloured. This shows how both the genotype and the environment can affect the phenotype.

Burmese cats have a different recessive allele of the gene, C^b, which makes the coat colour dark brown instead of pale cream. The extremities are black, as in Siamese cats.

10

a What are the genotypes of a Siamese cat and of a Burmese cat?

b Most cats have the normal allele, C^C. What will the kittens be like if a Siamese cat mates with an ordinary homozygous black cat? Explain your answer with a genetic diagram.

11
When a cat breeder mated a Siamese with a Burmese cat, all the kittens in the litter had pale brown coats.

a What is the genotype of this breeder's pale brown kittens?

b Are the alleles in this genotype codominant? Explain your answer.

c What results would you expect if two of these pale brown cats interbreed, producing large litters?

d The breeder wants to sell kittens with the pale brown coat colour. How should she obtain litters of kittens with this coat colour?

12

a When the two alleles of a gene are dominant and recessive, crossing heterozygotes gives a 3 : 1 ratio of phenotypes in the offspring. What is the ratio of phenotypes when the two alleles are codominant?

b Manx cats are tailless. When bred together they produce an apparently odd ratio of 2 tailless : 1 normal tailed. This is because embryos that are homozygous for the Manx allele fail to develop in the womb and are never born. Fig. 4 shows how the ratio arises. What proportion of kittens will have tails if a Manx cat mates with a tailed cat? Explain your answer.

Fig. 4 Genetic diagram showing breeding of Manx cats

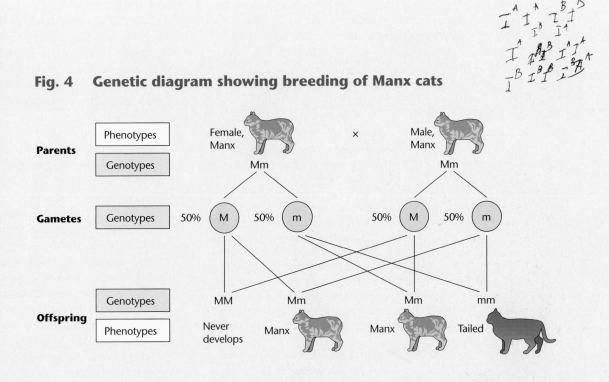

Sex inheritance

Whether a baby develops as a girl or as a boy depends not on a single gene but on a pair of chromosomes – which you can see in the karyotype in the photograph on page 125. The karyotype, taken from a male, has one X chromosome and a much smaller Y chromosome. A female has two X chromosomes.

Fig. 5 shows how the sex of a baby is determined. The proportion of males and females in the population is kept more or less equal.

The full details that explain how the chromosomes actually determine the sex of the child are not yet known. The Y chromosome is much smaller than the X, so many of the genes that are present on the X chromosome are missing from the Y. However, there is one gene, the SRY gene, which is thought to be the key to maleness. It is situated on the Y chromosome but is absent from the X chromosome. This gene causes the developing reproductive organs to become testes rather than ovaries. It is probable

Fig. 5 How sex is determined

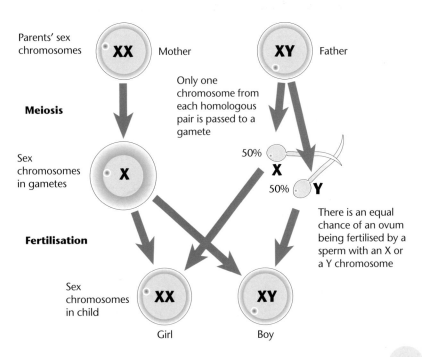

that the hormones that the testes produce then stimulate other male features to develop. However, other genes may well be involved.

13

a   Does the egg or the sperm determine whether a baby is a boy or a girl? Explain your answer.

b   Dairy farmers may want all their calves to develop into cows for milk production. The cows in the herd are artificially inseminated, that is sperm is collected from a bull and injected into each cow's vagina. Suggest how, in principle, it would be possible to ensure that only cows and not bulls were produced.

c   What problems do you think could arise if parents were able to choose the sex of their child?

Sex linkage

The X and Y chromosomes are not just concerned with sex determination – they also have genes for other characteristics. As the X chromosome is much larger than the Y, many of these genes occur on the X chromosome but not on the Y. This means that the pattern of inheritance of affected genes is linked to the sex of the individual.

One gene that occurs only on the X chromosome affects a pigment in the retina of the eye. A mutant allele of this gene causes red/green colour blindness. A person with this allele cannot distinguish between red and green. Since males have only one X chromosome, they have only one copy of this gene. If this gene is the mutant allele, they will be red/green colour blind. Females will only be red/green colour blind if they inherit the allele on both chromosomes. This explains why this type of colour blindness is much more common in men than women.

Males have a one in two chance of inheriting the condition if their mother carries the mutant allele, as shown in Fig. 6.

Notice how the alleles are written. For sex-linked genes, you need to show that they are sex-linked by writing them as X plus a superscript to indicate the particular allele of the gene. The Y indicates the Y chromosome, which of course has no superscript because it does not contain that gene.

Fig. 6   **The inheritance of red/green colour blindness**

In genetic diagrams, a sex-linked gene is shown alongside the sex chromosome. In red/green colour blindness, the allele for normal colour vision is C, and the mutant is c. A chromosome with the normal allele is written X^C, and with a mutant allele, X^c. The Y chromosome has no gene for this characteristic.

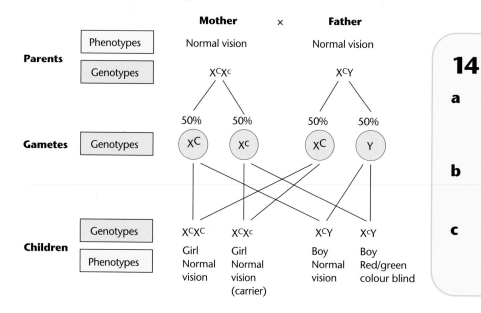

14

a   What percentage of children from the cross shown in Fig. 6 are colour blind? What percentage of girls and what percentage of boys?

b   What parental genotypes could produce a red/green colour-blind girl?

c   Explain why a son never inherits red/green colour blindness from his father.

Sex can be confusing

We can often tell whether a cell is from a man or a woman by looking for a tiny structure called a Barr body. Although a woman has two X chromosomes, in each cell one of these is inactivated early in development. It forms a distinctive dark blob, called the Barr body (see the photograph below). If we can see a Barr body, then the cell is from a woman. This is quicker and easier to do than looking to see whether the person has two X chromosomes or an X and a Y, because the chromosomes only become visible just before a cell is about to divide. Barr bodies are visible – if the cell is suitably stained – at all stages of the cell cycle.

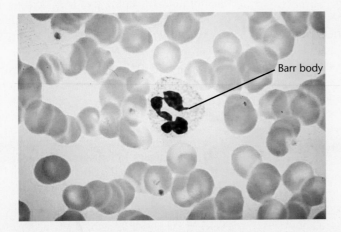

Barr body

A light micrograph of a white blood cell, surrounded by red blood cells. The white blood cell has a large nucleus with five lobes. On the right of the top lobe, you can see a drumstick sticking out. This is the condensed X chromosome, or Barr body.

A variety of chromosomal abnormalities can obscure the distinction between the sexes. For example, about one girl in 3 000 has only one X chromosome in each body cell. This is called Turner's syndrome. The condition often results in late or non-development of secondary sexual characteristics and infertility. It is also quite common for individuals to have extra sex chromosomes. In Klinefelter's syndrome, an individual has a chromosome complement of XXY. People with XXX, XXXY, XYY and various other combinations are also not unusual.

It is possible to have females with XX chromosomes who have abnormally masculine bodies and muscle strength because their adrenal glands do not respond to female hormones in the usual way. It is also possible for people with female body shape and muscle strength to have XY chromosomes, because their cells do not respond to male hormones as normal. This is called Androgen insensitivity syndrome (AIS). The person has all the characteristics of a woman, and will always think of herself as female.

Santhi Soundarajan won a silver medal in the 800 m race at the 2006 Asian Games but later failed a gender test. It is thought that she has Androgen insensitivity syndrome (AIS).

15 How many chromosomes would a person with Turner's syndrome have in her body cells? Explain your answer.

16 For each of the following chromosome complements, would you expect the person to show male or female characteristics?

  i   XXY
  ii   XXX
  iii XYY
  iv XXXY.

Explain your answers.

17 Women competing at events such as the Olympic Games may be checked to ensure that they really are female. Do you think that someone with the genotype XY, but who has Androgen insensitivity syndrome (AIS), should be allowed to compete as a woman? Discuss your reasons.

Haemophilia as an example of a sex-linked disorder

Haemophilia is an inherited disease in which the blood fails to clot easily. This causes internal bleeding, especially in the joints. The most common reason for haemophilia is shortage of a blood-clotting protein known as Factor VIII. The gene for Factor VIII synthesis occurs only on the X chromosome, so it is sex-linked. Males are much more likely to inherit haemophilia because they have only one copy of the X chromosome. Females can carry the faulty allele, and pass it on to their sons, but they themselves show no signs of haemophilia. A zygote that is homozygous for this allele usually dies soon after fertilisation, so women with haemophilia are rare.

The normal dominant allele that produces Factor VIII is written as X^H, and the recessive allele that results in failure to produce Factor VIII is X^h.

Genetic counselling can help people to decide whether or not to have children. Couples whose babies are at risk from a serious genetic disorder such as haemophilia may choose to have tests during the early stages of pregnancy and perhaps opt for a termination if the embryo is affected. When a haemophiliac man marries a woman who does not have haemophilia, the couple may opt for genetic testing to find out if the woman is carrying the allele for haemophilia. This would be very unlikely, but if it were the case, all boy children born to the couple would have haemophilia and all girl children would be carriers. Under these circumstances, the couple may make an informed decision not to have children at all.

If the woman is not carrying the allele, none of their children will have the disorder, because sons will inherit an X chromosome with a normal allele from their mother. None of the boys will carry the gene for haemophilia so they cannot pass it on to their children. However, the girls will carry the haemophilia allele, because they inherit one of their X chromosomes from their father and this must carry the recessive allele. Later in life, they should be aware that they have a 50% chance of their sons having haemophilia, assuming that they do not themselves marry a man with haemophilia.

18

a What is the genotype of a male with haemophilia?

b Females with haemophilia are rare. What would be the most likely genotypes of the parents of this girl?

---

key facts

- Gametes are haploid, and therefore contain only one copy of each gene. If a person is homozygous, then the genotype of all the gametes is the same. If they are heterozygous, then half of their gametes contain one of the alleles and the other half of their gametes contain the other allele.

- There is an equal chance of any gamete from one parent fusing with any gamete from the other. We can show these chances in a genetic diagram. The genetic diagram shows the different genotypes that can arise in the offspring, and the relative chances of each genotype occurring.

- It is important to remember that the offspring ratios predicted by genetic diagrams are only probabilities, and the actual results may not be exactly the same, especially if small numbers are involved.

- In a situation involving dominance, if two heterozygous organisms are crossed we would expect a 3 : 1 ratio of offspring showing the dominant characteristic to those showing the recessive characteristic in their phenotype. If a heterozygous organism were crossed with a homozygous recessive one, then we would expect a 1 : 1 ratio in the offspring.

- In humans, sex is determined by the X and Y chromosomes. Genotype XX produces females, and XY produces males.

- Numerous genes that are found on the X chromosome are not found on the Y chromosome. These are called sex-linked genes. A man has only one copy of each gene. A recessive allele will therefore always show up in a man, whereas it will often be masked by a dominant allele in a woman. Examples of sex-linked conditions are haemophilia and red/green colour blindness.

7.3 Genes in populations

In this chapter, we have been looking at just one gene at a time. But each of us has about 20 000 genes, many of them with different alleles. All the genes in a population are known as the **gene pool** of that population.

By definition, members of the same population are able to breed with each other. This means that all the different alleles of all the different genes in the gene pool are 'available' to any one individual. The different alleles of all the genes in the gene pool are constantly being passed from parents to offspring within the population, theoretically in any possible combination.

Within a species, there may be different populations. For example, meadow brown butterflies tend not to move very far from where they were born. The meadow brown butterflies in one meadow may never meet the meadow brown butterflies from a nearby meadow. They are therefore two different populations, because they don't interbreed with each other. Each population has its own gene pool. There may be particular alleles of genes that are present in one population's gene pool, but not in the other. As we shall see in Chapter 8, this is a very important feature in the development of new species.

The Hardy–Weinberg equation

It can be helpful to know something about how common a particular allele is in a population. We call this the **frequency** of that allele. For

Meadow brown butterflies may spend their whole lives in one field.

any one gene, the frequencies of all its different alleles add up to 1. If all the alleles are the same, then the frequency of that allele is 1. If two different alleles are equally common, then the frequency of each allele is 0.5.

About one child in every 3 300 in Britain is born with cystic fibrosis. We can use this fact to work out the frequency of the cystic fibrosis allele in the whole population of Britain. To do this, we use an equation called the **Hardy–Weinberg equation**. This is:

$$p^2 + 2pq + q^2 = 1$$

In this example for cystic fibrosis:

p is the frequency of the dominant allele, F
q is the frequency of the recessive allele, f
p^2 is the frequency of genotype, FF
$2pq$ is the frequency of genotype, Ff
q^2 is the frequency of genotype, ff.

It is also useful to remember that:

$$p + q = 1$$

In other words, between them the frequencies of the f and the F allele add up to 1.

Let's see how this works.

A child with cystic fibrosis is homozygous for the recessive allele, f. We know that one in 3 300 children have cystic fibrosis. So immediately we know that:

$$q^2 = \frac{1}{3\,300}$$
$$= 0.000\,3$$

Now we can work out q, the frequency of the f allele in the population.

$$q = \sqrt{0.000\,3}$$
$$= 0.017$$

We can go on to work out the frequency of the F allele, p. We know that $p + q = 1$, so:

$$p + 0.017 = 1$$
$$p = 1 - 0.017$$
$$\text{so } p = 0.983$$

And now we are in a position to be able to work out what proportion of the population is heterozygous, Ff.

The proportion of heterozygous people in the population = $2pq$

$$= 2 \times 0.983 \times 0.017 = 0.033\,4$$

So now we know that, out of every 100 people in the population of Britain, 3.3 people are carriers for cystic fibrosis.

---

19 PKU stands for phenylketonuria. This is a genetic disease caused by a recessive allele, p. About one in every 15 000 babies born in the UK has PKU.

Calculate the frequency of carriers of PKU in the UK population.

How reliable are these calculations? On the whole, the values that we calculate do seem to match other data that we can collect. But they only work if the allele frequencies do not change from generation to generation. They assume, for example, that a person with cystic fibrosis, genotype ff, is just as likely to have as many children as a person with genotype Ff or FF. For some alleles, this may not be true. Being homozygous for some alleles may reduce the chances of having children. The calculations also assume that the different alleles really do mix randomly in the population – in other words, that an individual with any genotype will mate with an individual of any other genotype, at random. But this may not always be true. In a population of one species of gull, for example, some have blue eyes and some have brown eyes. It has been found that a gull is more likely to mate with one who has the same eye colour that it has. (No one has quite worked out how a gull knows what colour eyes it has. Perhaps it looks at its parents' eyes.) This could affect the proportions of the alleles in the next generation.

Another thing that can throw out our calculations is mutation. A mutation is a change in the sequence of bases in a DNA molecule. This is how new alleles of a gene originally come into being. If there is a lot of mutation in the population we are studying, then there may be *more* of a particular allele in the next generation than we have calculated, because new occurrences of the allele have arisen through mutation.

key facts

- The gene pool is all the alleles of all the genes that are present in a population.

- A population is a group of organisms of the same species that can interbreed. A species may contain many different populations, which interbreed within their population but not with other populations.

- The Hardy–Weinberg equation allows us to calculate the frequency of a particular allele in a population, assuming that allele frequencies do not change from generation to generation. It is:

  $$p^2 + 2pq + q^2 = 1$$

  where p^2 is the frequency of the homozygous dominant genotype (for example, FF), $2pq$ is the frequency of the heterozygous genotype (Ff) and q^2 is the frequency of the homozygous recessive genotype (ff).

1 IQ test scores have been used as a measure of intelligence. Genetic and environmental factors may both be involved in determining intelligence. In an investigation of families with adopted children, the mean IQ scores of the adopted children was closer to the mean IQ scores of their adoptive parents than to that of their biological parents.

a Explain what the results of this investigation suggest about the importance of genetic and environmental factors in determining intelligence. (1)

b Explain how data from studies of identical twins and non-identical twins could provide further evidence about the genetic control of intelligence. (4)

Total 5

AQA, B, June 2006, Unit 4, Question 2

2

a ABO blood groups in humans are an example of discontinuous variation, whereas height in humans is an example of continuous variation. Describe how discontinuous variation differs from continuous variation in terms of

i genetic control;

ii the effect of the environment;

iii the range of phenotypes. (3)

b Genetically identical twins often show slight differences in their appearance at birth.

Suggest **one** way in which these differences may have been caused. (1)

Total 4

AQA, B, January 2006, Unit 4, Question 4

3 Yarrow is a herbaceous plant which grows in California at altitudes from 1500 m to 3000 m. The mean height of the stems of plants growing at 3000 m is smaller than that of plants growing at 1500 m.

a The higher the altitude, the lower the mean temperature. Explain how the lower temperature at high altitude reduces the growth of plants. (4)

b The relative contribution of environmental and genetic factors on the growth of the plants was investigated. Samples of young plants were taken and grown outdoors in prepared plots at altitudes of 1500 m and 3000 m.

Altitude at which young plants were collected/m	Mean maximum height of stems of plants/cm	
	Grown at 1500 m	Grown at 3000 m
1500	80.4	35.3
3000	31.5	24.7

Describe the evidence from the table that the variation in height is

i partly genetically determined; (1)

ii partly environmentally determined. (1)

Total 6

AQA, B, January 2006, Unit 5, Question 5

4

a A protein found on red blood cells, called antigen G, is coded for by a dominant allele of a gene found on the X chromosome. There is no corresponding gene on the Y chromosome.

The members of one family were tested for the presence of antigen G in the blood. The antigen was found in the daughter, her father and her father's mother, as shown in the genetic diagram below. No other members had the antigen.

	Grandmother (has antigen G)	Grandfather	Grandmother	Grandfather
Genotypes	.... or	....	....	....
Gamete genotypes	.... or	....	....	....
	Father (has antigen G)		Mother	
Genotypes	....		....	
Gamete genotypes	....		....	
		Daughter (has antigen G)		
Genotype		....		

i One of the grandmothers has two possible genotypes. Write these on the genetic diagram, using the symbol **XG** to show the presence of the allele for antigen G on the X chromosome, and **Xg** for its absence. (1)

ii Complete the rest of the diagram. (3)

iii The mother and father have a son. What is the probability of this son inheriting antigen G? Explain your answer. (2)

b During meiosis, when the X and Y chromosomes pair up, they do not form a typical bivalent as do other chromosomes. Explain why. (2)

Total 8

AQA, B, June 2004, Unit 4, Question 5

5 In a species of fruitfly, females have two X chromosomes, and males have an X and a Y chromosome. A gene controlling eye shape in fruitflies is sex-linked, and found only on the X chromosome. This gene has two alleles, **R** for round eyes and **B** for bar eyes. A homozygous, round-eyed female (**X^R X^R**) was crossed with a bar-eyed male. In the offspring (Offspring 1), all the female offspring had wide bar eyes (intermediate in size) and all the males had round eyes.

The figure shows the heads of three fruitflies.

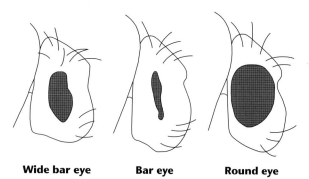

Wide bar eye   **Bar eye**   **Round eye**

a Name the relationship between the two alleles that control eye shape. (1)

b Give the genotype of the male parent. (1)

c Offspring 1 were allowed to interbreed. Copy and complete the genetic diagram to show the phenotypic ratio you would expect in the resulting Offspring 2.

Parental phenotypes	Round-eyed female	Bar-eyed male
Parental genotypes	**X^R X^R**	
Offspring 1 phenotypes	Wide bar-eyed female	Round-eyed male
Offspring 1 genotypes		
Gametes		
Offspring 2 genotypes		
Offspring 2 phenotypes and ratio		(3)

Total 5

AQA, B, January 2002, Unit 4, Question 4

6 A sex-linked gene controls fur colour in cats. Ginger-coloured fur is controlled by the allele **G**, and black-coloured fur is controlled by the allele **g**. Some female cats have ginger and black patches of fur. They are described as tortoiseshell. Male cats cannot be tortoiseshell.

a What is meant by a *sex-linked* gene? (1)

b A male cat with the genotype **X^g Y** mates with a tortoiseshell female.

i Give the phenotype of the male. (1)

ii Give the genotype of the tortoiseshell female. (1)

iii Complete the genetic diagram to show the genotypes and the ratio of phenotypes expected in the offspring of this cross.

Parents	Male	Tortoiseshell female
Parental genotypes	**X^g Y**	....
Parental gametes		
Offspring genotypes		
Offspring phenotypes		
Ratio		(3)

Total 6

AQA, B, January 2004, Unit 4, Question 5 (a–b)

hsw

how science works **assignment**

Genetics and human history

By studying the distribution of allele frequencies in different human populations, it is possible to build up a picture of the complex patterns of migration that have taken place in human history. It seems that the first humans originated in Africa and then spread around the world. Study of the genes of the native (original) inhabitants of North and South America suggests that a small group of people, with a limited range of alleles of some genes, reached North America from Siberia and then gradually populated the continent.

The map in Fig. 7 shows the different percentages of people with blood group B in some populations in different parts of the world.

Fig. 7   Map showing different percentages of people with blood group B

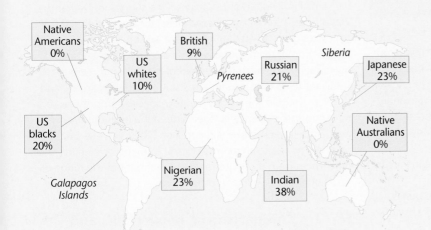

A1 It is suggested that none of the group of people that reached North America from Siberia carried the blood group allele I^B.

a What evidence from the map supports this suggestion?

b Which two blood groups would you expect to be absent in these people, and which genotypes would you expect to be present?

A2

a Is the occurrence of blood group B correlated with skin colour? Use evidence from the map to explain your answer.

b How might the different percentages of group B in US blacks and US whites be explained?

A3 Most West European populations have 9%–10% with blood group B. However, the Basque people who live in the Pyrenees have a much lower percentage.

a What does this suggest about the origins of the Basque people?

b What additional information might now be available to help confirm relationships between peoples from different parts of the world?

8 Selection and speciation

In October 2004, a report in the journal *Nature* described the finding of human-like remains on a small island called Flores, in Indonesia. The group of Indonesian scientists who made the discovery were searching a rock shelter for possible remains of humans and other mammals.

The bones were not fossilised, just very old. The first find was a jawbone. It was tiny. Its size suggested that its owner had been only one metre tall, with a brain only one-third the size of a modern human brain. More finds followed. By 2004, a variety of bones from eight individuals had been unearthed. Dating of the bones indicated that the people had lived about 18 000 years ago.

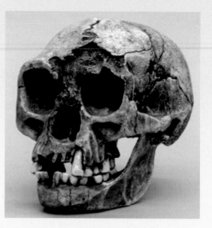

A Flores skull; refer to Fig. 1 for a comparison of *Homo* skulls and skeleton size.

The cave where human remains were found on Flores, Indonesia.

There were arguments among different groups of scientists from the start. The Indonesians who had found the first bones had been working with a team, including Australian scientists, who did not happen to be on the island at the time. The Australians were angry that the bones had been taken back to an Indonesian laboratory, and also that they were not given the permits they needed to return to continue research in the caves. Eventually these disagreements were sorted out, and the research teams continue to be made up of both Indonesian and Australian scientists.

There were more arguments about exactly what the bones represented, and it looks as though there will be no agreement

about this for a long time. The team who discovered the bones believe that they belong to a newly discovered species of human, which they have named *Homo floresiensis*. These tiny people were quickly nicknamed 'hobbits'. They would have lived at the same time as our ancestors, members of our own species *Homo sapiens*. They may have known about these little people, so perhaps this is where the dwarf and fairy legends arose.

Others, though, don't think this is a new species at all, but just very small individuals of *H. sapiens*. One suggestion is that this was a population of *H. sapiens* where individuals suffered from microcephaly, a condition in which a person has a very small head. Another team suggested that they suffered from myxodoema, in which the thyroid gland does not secrete sufficient thyroxine, resulting in a sluggish metabolic rate and slow growth. On the whole, the new-species camp seems to have stronger evidence on their side, but a lot more data is needed before everyone will agree on whether these small people really were another species of humans, or whether their differences from us are just extreme examples of the range of variation that can occur in our species.

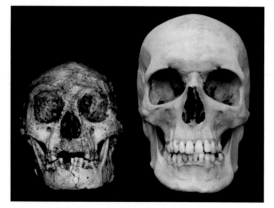

A Flores skull shown beside a human skull.

Fig. 1 Comparison of *Homo* skulls and skeleton size

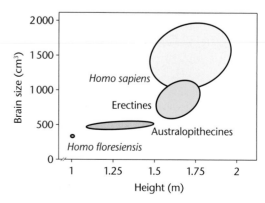

8.1   Natural selection

It is rare to find two living organisms that are identical. Even if they have exactly the same genotype – such as identical twins – we can usually pick out some differences between them. These differences between organisms of the same species are known as **variation**.

These twins have identical genotypes. Most of us have collections of alleles that are unique.

In your AS course, you looked at the different causes of variation. In Chapter 7 (Inheritance), we have seen that a great deal of variation is caused by organisms having different alleles of particular genes. Variation is also caused by effects of the environment. For example, a person who trains hard to be an athlete will develop much larger and stronger muscles than his or her identical twin who isn't interested in sport at all.

The advantages of variation

Despite the wide range of variation within our species, *Homo sapiens*, people of all nationalities are quite clearly human beings. Human beings always give birth to human beings; cats always produce cats; earthworms produce earthworms; and so on.

Why, then, do so many species reproduce sexually and thereby boost the amount of variation? What are the advantages of individuals of the same species being different? Why don't members of the same species become more and more alike until they have produced the perfect specimen? Alternatively, why do members of a species not become more and more varied until they are no longer remotely similar to one another?

Fig. 2 shows a theoretical model of how a group of rodents on an isolated island might change as a result of variation over a period of time. This model is simple compared with what happens in a real ecosystem. However, you can see that even quite a small selective advantage or disadvantage can, over just a few generations, have a significant effect on a particular phenotype.

Despite the broad range of variation, you might predict that, in time, all the unfavourable alleles would disappear from the population and that the rodents would all be homozygous for the favourable alleles. In practice, however, things change.

In the example in Fig. 1 the climate of the island might become even colder or it might warm up. Some new predators might reach the island. As a result of the rodents feeding on the plants, there might be a change in the plant populations; for example, it might be that only very prickly shrubs or shrubs that produced fruits with a very hard shell, or poisonous flesh, were able to survive.

This might mean that alleles that were disadvantageous in one situation now become advantageous. If a population has many different alleles, then there is a good chance that some of them may help some organisms to survive even if environmental conditions change. Having a large gene pool, with lots of different alleles, is a kind of insurance policy for the future, no matter what it may bring.

1

a   In the theoretical model in Fig. 2 overleaf, what factors could limit the size of the population of rodents that can live on the island?

b   Suppose that each pair of rodents produces about 20 young in a year. What would happen to the population if all of these young survived and bred?

c   If the adults live on average for only a year, how many of these young must survive if the size of the population is to stay roughly constant?

Fig. 2   Change over time

❶

Imagine a small island on which just shrubs and grasses grow. A group of small rodents reaches the island, perhaps floating to it on tree trunks blown out to sea by a storm. The rodents feed on the seeds of the grasses and shrubs. The climate of the island is somewhat colder than the mainland from which the rodents came. There are no predators on the island.

What would you expect to happen to the size of the population of rodents during the first few years?

❷

With no predators and plenty of food, the rodents do very well. There is, however, a limit to the number of rodents that can survive on the island. Many of the young rodents will die before they can breed.

What might cause the rodents to die?

❸

Some might die from sheer bad luck, such as a rock falling on them. Some may inherit a pair of recessive genes that cause a fatal disorder. As a result of continuous variation caused by different combinations of alleles, some will be smaller than others, some will have thicker layers of fat under the skin, some will have longer fur, some may be better at finding seeds, some may be able to stretch higher to reach seeds on the shrubs, some may produce enzymes that help them to digest other parts of the plants, and so on.

Which of these features will help them to compete more successfully for food? Which will help the rodents survive the colder climate?

❹

The gene pool of the rodents includes two alleles for a hair length gene: long, H, and short, h. On the mainland, from where the rodents originally came, the two phenotypes happened to be present in exactly equal numbers. The longer hair length gives better insulation against the cold, so on the island the short phenotype becomes less favourable (Table 1).

What will happen to the rodents with genes for features that favour survival? What will happen in the population to the frequency of the alleles that favour survival?

❺

It seems a reasonable prediction that the rodents with alleles that help them to survive are more likely to be successful. They are therefore more likely to breed and pass on these alleles to their offspring. Alleles of genes for unfavourable features can be expected to disappear from the population as fewer and fewer rodents with these alleles are successful in breeding.

2

a Table 1 shows the percentage of short-haired rodents in the population on the island when different percentages fail to breed in each generation. Draw a graph to show the data in Table 1.

b Assuming that the population stayed constant at 10 000, how many individuals would show the phenotype of the recessive allele after 20 generations at each of the three selection pressures?

c What would happen to the frequency of the h allele in the population in each case?

Table 1   Effects of differential breeding ability

Generation	Percentage of genotype hh in the population when:		
	2% fail to breed	10% fail to breed	50% fail to breed
0	50	50	50
5	48	40	11
10	46	31	4
15	44	24	2
20	42	18	1

hsw

how science works

Selection and the sickle cell allele

Red blood cells contain the protein haemoglobin (see Fig. 3). This is a soluble, globular protein, made up of four interlinked polypeptide chains. There are two alpha chains and two beta chains. Each one contains a haem group, which is capable of combining with oxygen when oxygen concentrations are high, and releasing it when oxygen concentrations are low. This allows red blood cells to pick up oxygen in the lungs, carry it in the blood to the tissues, and release it where oxygen concentrations are low, such as in respiring muscles.

There is an allele of the gene that codes for the beta chain of haemoglobin that has an incorrect base in it. This results in a different amino acid being present in the chain. This amino acid makes the haemoglobin molecules less soluble, and also causes them to stick together. This happens especially when oxygen concentrations are low. The haemoglobin molecules inside the red blood cells clump together, pulling the cell into a kind of sickle shape. In this condition, the haemoglobin cannot carry oxygen and the red blood cell cannot get through capillaries. The person's tissues are starved of oxygen, and he or she also experiences pain caused by the blood cells stuck in their capillaries. The person is said to be having a sickle cell crisis. Without treatment this is often fatal.

The allele that causes this is known as the sickle cell allele, and we can use the symbol H^S to represent it. The normal allele is H^A. A person with genotype H^SH^S has sickle cell anaemia, and suffers crises as described above. A person with genotype H^SH^A has a mixture of two sorts of haemoglobin in the blood. Usually the person is fine, but problems may arise if he or she is doing something very strenuous, such as climbing at high altitude where oxygen

Fig. 3   Haemoglobin

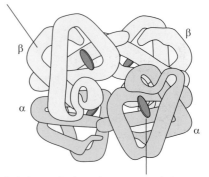

Four polypeptide chains make up the haemoglobin molecule. Each chain contains 574 amino acids.

Each chain is attached to a haem group that can combine with oxygen.

levels are low. The person is said to have sickle cell trait.

Why hasn't this harmful allele been eliminated from the human population? The reason is that, in the homozygous state, it confers a selective advantage to the affected person. For example, in many parts of the world, malaria is a common and frequently fatal disease. Many children die from this disease in sub-Saharan Africa and in tropical Asia (see Fig. 4). And it is here that we find the highest frequencies of the sickle cell allele. There is a direct connection between them. If these populations did not have a high incidence of the sickle cell allele, even more people would die. So, the sickle cell allele gives a survival advantage when malaria is present.

hsw

how science works

Fig. 4   The global distribution of malaria and sickle cell anaemia

Distribution of the most dangerous form of malaria in Africa and Asia.

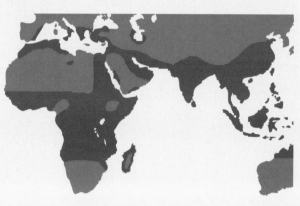

■ Areas with endemic *P. falciparum* malaria

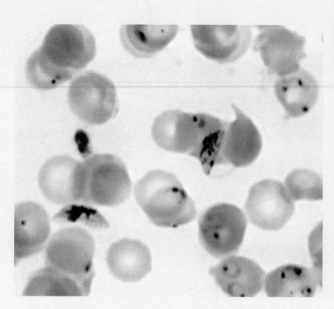

A light micrograph that shows malarial parasites growing in human blood cells.

Distribution of the sickle cell allele in the human populations of Africa and the Indian subcontinent.

Percentage of population that has the sickle cell allele (haemoglobin S)

■ 14+	■ 6–8
■ 12–14	■ 4–6
■ 10–12	■ 2–4
■ 8–10	■ 0–2

Malaria is caused by a protozoan, called *Plasmodium*, that gets inside red blood cells where it feeds and breeds. The most dangerous kind is *P. falciparum* malaria, which can affect the brain and frequently kills. But the malarial parasite doesn't seem to be able to thrive if the haemoglobin in the cells is the sickle variety. Humans with the genotype H^{S}H^S almost never get malaria. Those with H^{S}H^A may get malaria, but it is a much milder form and rarely fatal. Those with H^{A}H^A are much more prone to contracting malaria, and it is often in a more serious form. Many people die from it.

3 In one area of Africa, 2% of children have sickle cell anaemia.

a Use the Hardy–Weinberg equation (see page 135) to calculate:

   i the frequency of the sickle cell allele in the population
   ii the percentage of people in the population who have sickle cell trait.

b What assumptions must we make if we are using the Hardy–Weinberg equations in this way?

4 A couple both have sickle cell trait. Draw a genetic diagram to show the probability that their first child will have sickle cell anaemia.

5 Many black people in the USA are descended from people who lived in Western Africa. The sickle cell allele is still relatively common in the black population in the USA, but is becoming rarer. Suggest why:

   i the allele is relatively common
   ii the allele is becoming rarer.

How natural selection works

The situation described in the example in Fig. 5 shows the main steps in the theory of **natural selection**.

Fig. 5 Natural selection

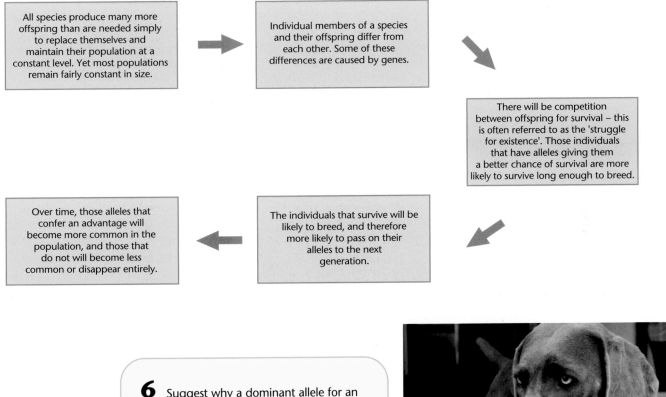

All species produce many more offspring than are needed simply to replace themselves and maintain their population at a constant level. Yet most populations remain fairly constant in size.

→

Individual members of a species and their offspring differ from each other. Some of these differences are caused by genes.

There will be competition between offspring for survival – this is often referred to as the 'struggle for existence'. Those individuals that have alleles giving them a better chance of survival are more likely to survive long enough to breed.

The individuals that survive will be likely to breed, and therefore more likely to pass on their alleles to the next generation.

Over time, those alleles that confer an advantage will become more common in the population, and those that do not will become less common or disappear entirely.

6 Suggest why a dominant allele for an unfavourable condition is likely to disappear from a population by selection more rapidly than a recessive allele for an unfavourable condition.

All dogs belong to the same species, but there are many different breeds, each with their own gene pools. Breeders have selected dogs with particular features to breed, causing huge changes in phenotype.

The ideas that you have just worked through form the basis of Charles Darwin's theory of natural selection. Alfred Russel Wallace independently came up with much the same hypothesis, and it was his pressure that persuaded Darwin to publish his ideas in 1858. The idea of evolution had been around for many years; the ancient Greeks had proposed it well over 2 000 years before Darwin. Darwin's theory was put forward as an explanation of how evolution could have occurred.

Other ideas had been suggested before, but Darwin was the first scientist to supply detailed supporting evidence for his theory. He pointed out that artificial selection of features when breeding animals can produce startling changes within a few generations. Today, the breeds of dogs that have been produced in the last 100 years or so are a good example of this. At the time, however, Darwin could not provide direct evidence that natural selection had produced change in a particular species, and many people challenged his explanations.

Since Darwin's time, evidence has been accumulated in support of his theory. One well-documented example of change involves the peppered moth, *Biston betularia*. The typical form of peppered moth has white wings speckled with black scales in an irregular pattern. Collections of moths from about 1850 show that at this time almost all peppered moths in Britain were this speckled form.

A major cause of mortality in these moths is predation by birds. The moths rest on tree trunks during the day, and rely on camouflage for protection. Speckled moths are superbly camouflaged on tree bark covered by lichens, whereas black moths are very visible. They are better camouflaged on trees with few lichens, especially if the tree trunks have soot deposits on them.

The melanic form of peppered moth has wings that are almost entirely black. A small number of specimens first appear in moth collections made after 1850, just after the industrial revolution had

begun. By the end of the century nearly all the peppered moths in some areas, such as around Manchester, were melanic. Over the same period of time, sulfur dioxide killed most of the lichens (compound organisms consisting of an association of fungi and algae) living on tree trunks in industrial areas, and deposits of soot particles blackened the bark.

It is important to realise that the melanic form of the moth developed as a result of a chance mutation. There were probably always a few melanic moths around, but in times before pollution darkened the trees they would mostly have been eaten before they bred. It was because the environment became black and sooty that black moths could not be seen by predators, and so survived more successfully than speckled moths. The mutation was entirely random; the moths did not purposely change their colour (any more than you would be able to do), nor did they become black because they were covered by soot. The change in the environment simply shifted the selection pressure – it was now an advantage to be black rather than speckled.

Wing colour in the peppered moth is controlled by a single gene, the melanic allele being dominant to the speckled allele. Intermediate phenotypes also occur in which the blackness of the wings is modified by genes at other loci. Studies of the distribution of the typical and melanic forms showed that by the 1950s the melanic form was by far the more common in the heavily polluted parts of Britain, whereas in western areas with very little pollution almost 100% of the peppered moths were still the typical speckled form.

These two moths belong to the same species, the peppered moth *Biston betularia*. The dark, melanic, moth has a dominant allele that causes the dark colour to be produced.

7 Suggest how the theory of natural selection could explain the change from speckled to melanic forms of peppered moth in industrial areas of Britain.

8 The melanic moths spread very rapidly in industrial areas. How would the genetics of wing colour explain this rapid spread?

Although the increase in the proportion of melanic moths correlated with the increase in pollution, this did not prove that natural selection had occurred. It was just possible that there was some other explanation, such as that a pollutant was causing black pigment to be produced in the moths. The first step was to check that the melanic moths really did have a selective advantage in polluted areas. A biologist called Henry Kettlewell released equal numbers of marked moths of each type – melanic and speckled – in an unpolluted wood in Dorset and in a polluted wood near Birmingham. He then compared the proportions of each type recaptured after a few days. The assumption was that the more moths had survived, the more he would recapture. The results are listed in Table 2.

Table 2   Results of peppered moth experiment

Site where moths were released	Percentage of released moths that were recaptured	
	Speckled	Melanic
Dorset (unpolluted)	12.5	6.3
Birmingham (polluted)	15.9	34.1

9 Describe how the results listed in Table 2 support the hypothesis that natural selection occurred.

Next, Kettlewell attempted to establish whether the difference in survival rates really was due to differences in predation. The fact that *people* can see the speckled form more easily on polluted tree trunks does not prove that birds find the moths more easily. So Kettlewell placed moths of each type on tree trunks and then filmed them. He found that birds did indeed catch more of the speckled form on blackened trunks, and vice versa. More recent researchers have criticised this experiment because the moths rarely rest on exposed trunks. Usually the moths select sites such as the underside of smaller branches. Increasingly detailed knowledge of the behaviour and habits of the moth have shown that the story is more complex than was first thought. However, repeats of the experiment, in which moths were placed in places where they would naturally be found, have confirmed Kettlewell's original conclusions. The evidence strongly suggests that the melanic form has been selected in industrial areas mainly because it survives predation more successfully than the speckled form.

10 As sulfur dioxide and soot pollution are reduced, in many areas the lichens are returning and tree trunks are becoming less black. What do you predict will be the effect of natural selection on the peppered moth populations in these areas?

Darwin's finches

One hallmark of a good scientific theory is that it allows us to make predictions. One prediction that follows from the theory of natural selection is that: a species will adapt to changing conditions within the limits of its range of variation. Is there any evidence to support this prediction?

The medium ground finch lives in the Galapagos Islands (see Fig. 6), an isolated group of volcanic islands in the Pacific Ocean, about 600 miles west of South America. Darwin visited these islands in 1835 and was impressed by the fact that on each island the animals and plants were slightly different. Thirteen species of finches inhabit the islands, even though few other species of small birds live there. Darwin's observations on the islands were a major influence in the development of his ideas about natural selection.

The finches of the Galapagos Islands have been closely studied for many years. Many of them feed on seeds. Birds with large beaks tend to feed on hard seeds that need to be cracked before they are eaten. Birds with small beaks tend to feed on smaller, softer seeds.

In 1983 a significant climate change occurred when a warm ocean current brought prolonged rainfall to the normally dry islands. Many of the cacti died in the wet conditions. This reduced the supply of large, hard seeds. On the other hand, plants that produced small, soft seeds flourished. So, the range of food available

Fig. 6   Medium ground finches in the Galapagos Islands

to the seed-eating finches changed considerably. There was a remarkably rapid response in the population. Whereas birds with large beaks had been particularly successful in dealing with cactus seeds, they could not pick up the smaller seeds of other plants very easily.

The medium ground finches with small beaks prospered; within a few generations the mean size of beak in the population had decreased appreciably. Mathematicians predicted the change in beak size on the basis of the estimated selective advantage. The actual results closely matched their predictions. Moreover, as the climate became drier again in the following years, the trend was reversed in precisely the expected way. This adaptation to changing conditions was only possible because there was continuous variation in beak size in the population, and this variation could be inherited.

Directional and stabilising selection

The increase in the frequency of the melanic allele in the peppered moth population during the industrial revolution, is an example of **directional selection**. A change in the environment changed the selection pressures on the population. Instead of the speckled form having a better chance of survival, the melanic form was the better adapted. Directional selection shifts the selection pressure, resulting in a change in allele frequency.

Most of the time, however, natural selection does not do this. If the environment does not change, and if the population is already well

adapted to its environment, then selection tends to keep things as they are. This is called **stabilising selection**. In Devon and Cornwall, for example, where there was no air pollution

Fig. 7   Stabilising and directional selection

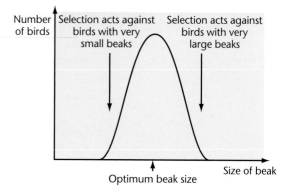

Stabilising selection

Number of birds | Selection acts against birds with very small beaks | Selection acts against birds with very large beaks

Optimum beak size

Size of beak

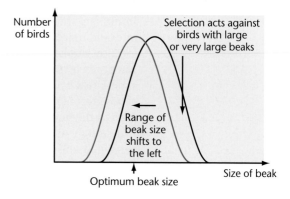

Directional selection

Number of birds

Selection acts against birds with large or very large beaks

Range of beak size shifts to the left

Optimum beak size

Size of beak

during the industrial revolution, the peppered moth population remained almost entirely of the speckled type.

Fig. 7 shows how stabilising and directional selection occurred amongst the medium ground finches in the Galapagos islands at different times. Beak size in this population shows a normal distribution, with the middle of the range representing the beak size that has the greatest selective advantage. Birds with beaks that are similar in size usually have the best chance of getting plenty to eat, and therefore of surviving and reproducing. Generation after generation, this beak size is selected for. The frequency of the different alleles that produce different beak sizes stays the same. The position and shape of the graph stay the same. This is stabilising selection.

But during the very wet year described on page 147, the birds with smaller beaks had the advantage. The selection pressure shifted. Now the birds towards the lower end of the range had the best chance of survival. They reproduced more than other members of the population with larger beaks, so the next generation had a range of beak sizes that shifted to the left. The frequency of the alleles that produce smaller beak sizes increased, while the frequency of the alleles that produce larger beaks decreased. This is an example of directional selection.

11 Another species of ground finch on the islands has a much larger beak, specialised to eat cactus seeds. Numbers of this species fell sharply during the very wet years. There was no adaptation of beak size such as occurred in the medium ground finches. Suggest why this species was unable to adapt to the changing conditions.

key facts

● All species show variation, some of which is caused by their genes. Individuals with combinations of alleles that give them a better chance of survival than others with different combinations of alleles have a selective advantage. They are more likely to survive and breed, so their alleles are more likely to be passed on to the next generation. The frequency of the advantageous alleles in the gene pool increases. This is known as natural selection.

● Normally, when a species is already well adapted to its environment, and that environment is fairly stable, natural selection keeps things as they are. The frequencies of alleles in the population stay very much the same from generation to generation. This is known as stabilising selection.

● If the environment and therefore selection pressures change, or if a new allele arises by mutation, then there may be a shift in the allele frequencies in subsequent generations. This is known as directional selection, and it can bring about evolution.

8.2 Speciation

In your AS course, you learnt that a species is a group of organisms that share similar features, and that can interbreed to produce fertile offspring. They do not normally breed with other species, and – if they do – then any offspring are not fertile.

How do new species arise? We have seen how natural selection can cause a shift in the frequency of alleles in a population. But our examples so far haven't actually produced a new species. The melanic and speckled forms of the peppered moth are still peppered moths and can interbreed successfully. The medium ground finches with small beaks and large beaks are still medium ground finches.

Looking at the definition of a species, we can see that the crucial feature of a species is that it can breed within itself, but not with other species. So, in order to create a new species from an existing species, a group of organisms must somehow become unable to breed with the other members of the original species. They must become **reproductively isolated** from all the members of the original species.

Geographical isolation

One way in which this can happen is if a population of the original species becomes geographically separated from all the rest.

Imagine that you could watch what happens to a species of animal over many years. The species is spread over a large area. Part of a population becomes separated from the rest by a river in flood, or by a volcanic eruption, or because part of the land dries out and turns to desert. Over the thousands of years that you keep watch, major geological upheavals cause areas of land to split away and volcanic activity produces new islands. The separated populations do not experience the same conditions – the climate, the food supply, the competition, the physical environment could all be significantly different. Some adaptations would be successful on one side of a mountain range, others would be favoured on the opposite side and so the separated populations would evolve in different ways.

Because the selection pressures are different, the processes of natural selection go in slightly different directions on opposite sides of the mountain. In the separated populations there will be a range of variation. Individuals will have different combinations of alleles, and gene mutation may produce new alleles. The animals with alleles that favour survival in their local environment will be more likely to pass on these alleles to their offspring. As the allele frequency changes, the phenotypes of the two populations will become more and more different.

After a while, the animals in the two locations might be quite dissimilar, both from each other and from their ancestors. They may be so different that they can no longer interbreed successfully, even if brought together again. If this happens, then they have become new species.

How long does this process take? It could take hundreds, thousands or even millions of years. It's very rare to be able to see a new species developing within a person's lifetime, so we usually have to work out how it is happening by looking at what we can see at one moment in time. For example, in the Galapagos Islands, we can see that there are different species of finches on the different islands. They look very similar to one another. This could be explained if one species of finch arrived on one of the islands from the mainland. Some spread to different islands, where the selection pressures were different. They were separated from other members of their species by the water between the islands, so each evolved along its own path, eventually becoming completely reproductively isolated from each other.

For new species to develop, separated populations must be genetically isolated. If they live close together and can still interbreed regularly, the groups will continue to exchange alleles. Then they will not separate into populations with distinct sets of alleles. Their gene pools will remain as one, rather than each having its own gene pool.

Other methods of reproductive isolation

Geographical isolation isn't the only way that two populations can become reproductively separated from each other.

Imagine a field with a species of plant that flowers all day. A new mutation produces an allele that causes some plants to flower only in the morning. These may attract a species of insect that gathers pollen only in the early part of the day. Another mutation in other plants may produce an allele that results in plants that flower only in the evening or at night. These attract a different species of insect for pollination. Both alleles are successful, but the plants form separate populations that are pollinated by one species of insect only. They no longer exchange alleles. Their gene pools are separated.

Further changes in phenotype can occur in both groups – slightly altered petals or stigmas could make pollination by the insect concerned even more efficient. There would then be selection for these alleles and, gradually, the two sets of plants could develop into two distinct species.

So, to summarise, new species arise as a result of:

- isolation – two populations are separated, for example, geographically
- genetic variation – each population contains a wide variety of alleles and therefore has a range of phenotypes among its members

- natural selection – in each population the alleles that help particular phenotypes to survive in the local conditions are selected
- speciation – the populations become so different that successful interbreeding is no longer possible.

key facts

● A population of organisms may become geographically isolated from the rest of the species. The selection pressures on this population may be different from those on the other populations of the species, and so the allele frequencies in the two groups begin to diverge.

● Eventually, the allele frequencies may be so different in the new population that the individuals are unable to breed successfully with the rest of the species, even if the geographical barrier is removed. A new species has been formed.

1 Great tits are small birds. The graph shows the relationship between the number of breeding pairs in the population and the mean number of eggs per nest in different years in a wood.

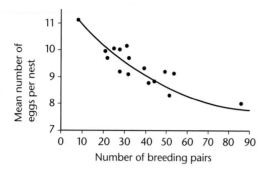

a Explain the relationship shown by the graph. (2)

b Female great tits usually lay between three and 14 eggs in a nest.

  i In the same year, the birds do not all lay the same number of eggs. Explain how **one** factor, other than the number of breeding pairs, could influence the number of eggs laid by a great tit. (1)

  ii Natural selection influences the number of eggs laid. Explain why great tits that lay fewer than three eggs per nest or more than 14 eggs per nest are at a selective disadvantage. (3)

Total 6

AQA, June 2006, Unit 5, Question 5

2 In an investigation, the tolerance to copper ions of the grass *Agrostis tenuis* was determined. Samples were taken of plants growing in waste from a copper mine and from nearby areas just outside the mine. The mean copper tolerance of plants from the mine waste was found to be four times higher than that of plants in the surrounding area.

a Explain how natural selection could produce a copper-tolerant population in the mine waste. (4)

b Copper-tolerant *Agrostis tenuis* plants flower at a different time from those which are not copper-tolerant. Explain how this might eventually lead to the production of a new species of *Agrostis*. (4)

Total 8

AQA, January 2005, Unit 4, Question 7

3 Lake Malawi in East Africa contains around 400 different species of cichlids which are small, brightly coloured fish. All these species have evolved from a common ancestor.

a Describe **one** way in which scientists could find out whether cichlids from two different populations belong to the same species. (2)

b During the last 700 000 years there have been long periods when the water level was much lower and Lake Malawi split up into many smaller lakes. Explain how speciation of the cichlids may have occurred following the formation of separate, smaller lakes. (4)

c Many species of cichlids are similar in size and, apart from their colour, in appearance. Suggest how the variety of colour patterns displayed by these cichlids may help to maintain the fish as separate species. (2)

Total 8

AQA, January 2006, Unit 4, Question 5

4 The land snail, *Cepaea nemoralis*, is found in a number of different habitats. It is prey to birds such as thrushes. The shells of the snail show variation in colour and in the number of dark bands around them. They may be brown, pink or yellow, and they may have one, three or five bands or none at all.

Shell with no bands   Shell with five bands

a What type of variation is shown by the banding of the shells? Explain your answer. (1)

b The graph on the next page shows the frequency of yellow, unbanded snails in three habitats. The frequencies were found to be consistent over a period of time.

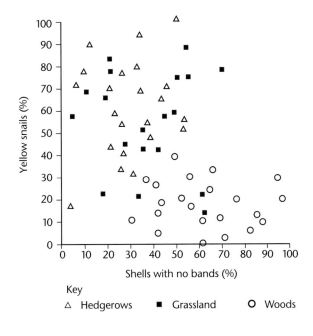

Key

△ Hedgerows   ■ Grassland   ○ Woods

i Describe what the graph shows about the relationship between the habitats and the phenotypes of the snails. (2)

ii Suggest an explanation for this relationship. (4)

Total 7

AQA, June 2005, Unit 4, Question 6

5

a Maize seeds were an important food crop for the people who lived in Peru. The seeds could be kept for long periods. Each year, some were sown to grow the next crop. Archaeologists have found well-preserved stores. The graph shows the lengths of seeds collected from three stores of different ages.

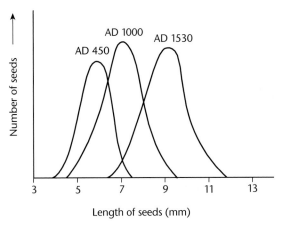

i Within each store the maize seeds showed a range of different lengths. Explain **one** cause of this variation. (2)

ii Use your knowledge of genetics and selection to explain the changes in the mean length of the seeds between AD 450 and AD 1530. (4)

b The Galapagos Islands are an isolated group about 900 km from South America.

Thirteen species of small birds called finches live on the islands. All species are thought to have evolved from a single species which reached the islands from South America. This species feed only on seeds, but the finches on the islands include species which specialise in feeding on buds, nectar and insects, as well as on different sizes of seed.

Explain how evolutionary change could have resulted in this diversity of finch species on the Galapagos Islands. (6)

Total 12

AQA, June 2004, Unit 4, Question 4 (b–c)

Speciation in *Anolis* lizards

The Caribbean islands and Florida in the USA, have a very large number of different species of *Anolis* lizards. These lizards usually have green or brown bodies, and the males have large, colourful, inflatable dewlaps, which they use in display. These displays help the males to defend their territories, and are also used in courtship.

Researchers are interested in how so many different species of *Anolis* lizards have arisen (see photos below). Several teams have been working on the DNA sequences of the different species. The species *Anolis carolinensis* has the distinction of being the very first reptile whose DNA has been completely sequenced.

One investigation looked at the base sequence of the DNA in the mitochondria in five *Anolis* species.

Earlier research had suggested that the original home of the *Anolis* lizards was in Cuba, and that they had spread from there. Data collected in previous studies suggested that each of the different lizard species on the various islands had developed following separate colonisations from Cuba.

The researchers made a prediction that – if this hypothesis were true – the relationships between the different species of lizards would be closer to the Cuban species than to each other. They tested this hypothesis by analysing the base sequence of the mitochondrial DNA in each of the five species. Table 3 shows the researchers' results.

Anolis oculatus.

Anolis sagrei.

Anolis sagrei sagrei.

Anolis carolinensis is the first reptile whose DNA has been completely sequenced.

Anolis equestris.

Table 3   Mean pairwise divergence for the mitochondrial DNA

The larger the number, the greater the differences between the two species. The smaller the number, the closer the relationship between the two species.

Species	Species name	1	2	3	4	5
1	A. longiceps		0.125	0.137	0.175	0.119
2	A. maynardi	0.125		0.119	0.168	0.114
3	A. brunneus	0.137	0.119		0.167	0.113
4	A. carolinensis	0.175	0.168	0.167		0.152
5	A. porcatus	0.119	0.114	0.113	0.152	

A1

a Use the data in Table 3 to work out the species to which each of the following is most closely related:

   i   A. longiceps
   ii A. maynardi
   iii A. brunneus
   iv A. carolinensis.

b Discuss whether these results support the hypothesis that the various species of Anolis lizards arose from separate colonisations of islands by lizards originating from Cuba.

A2 Suggest how the lizards on the different islands evolved different features.

A3 The dewlaps of male lizards of different species may be different colours. Suggest how this could make one population of lizards reproductively isolated from another population.

A4 How could the researchers determine whether the five types of lizards that they tested really do belong to different species?

A5 Suggest further investigations that could test the hypothesis that the various species of Anolis lizards arose from separate colonisations of islands by lizards originating from Cuba.

9 Survival and response

Ewan McGregor joins Ray Mears on a survival trek.

Ray Mears is no stranger to surviving in extreme circumstances. From an early age growing up in the North Downs, in southeast England, he took an interest in the natural environment around him. His interest in flora and fauna developed well into adulthood when he started to travel the globe to learn more about ancient survival techniques.

It was not long before Ray became an expert in 'bushcraft' and in 1983 he set up his own school to teach others how to survive in the wilderness. Ray believes that in order to survive in nature we need to use our senses in order to learn about the environment we're in. Hearing, smell and sight are all important factors when assessing our immediate surroundings and any danger they may pose to us. Ray believes that once we know about the environment around us, we can apply appropriate survival responses.

All living organisms can detect stimuli. By responding to these stimuli, they increase their chances of survival by avoiding harmful environments.

9.1 Invertebrate responses

Back to basics
First some basic terms:

- **Stimulus** – a change inside or outside an organism that brings about a response in that organism. Examples of stimuli include light, sound, chemicals and pressure.
- **Receptor** – a structure that detects a stimulus and initiates a nerve impulse. Examples of receptors include rod cells in the retina, chemoreceptors on the tongue and in the nose, and pressure receptors in the skin.
- **Effector** – a structure that responds to the arrival of a nerve impulse. Effectors are usually muscles or glands.

Invertebrates, unlike mammals and birds, rely on innate behaviour for many aspects of their lives, such as escaping danger, finding a suitable habitat and locating food. They do this using three types of behaviour pattern:

- a type of orientation movement called a **taxis**
- a type of orientation movement called a **kinesis**
- a **reflex escape response**.

Taxis
Fly maggots use innate behaviour for survival. They avoid bright sunlight, which might harm them or make them visible to predators (Fig. 1).

An animal performs a taxis (plural taxes) when it moves towards or away from a stimulus, such as light, that is coming from a particular direction.

In the case of the maggot the taxis is **negative phototaxis** (taxis away from light). Adult flies have more protective pigments and usually move towards the light (**positive phototaxis**), which warms up their bodies.

Kinesis
Woodlice use innate behaviour to stay in a suitable environment. They live in damp places beneath logs and stones where they are not easily found by predators such as blackbirds and magpies, and are not likely to dry out. At high humidity levels woodlice remain inactive, but any slight drying of the environment is detected and they respond to the harmful stimulus by starting to move about. Once it has started

Fig. 1   Phototaxis

Negative phototaxis

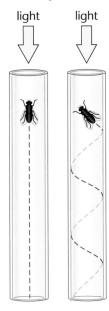

Maggots have light receptors at each side of their heads. As they crawl, they turn their heads alternately to right and left, comparing the light intensity from each side. They always turn towards the darker side, and this takes them away from light.

Positive phototaxis

Adult fruitflies move towards the light by keeping the light intensity the same in each eye. This can be demonstrated with a normally sighted fly and a fly that is blind in one eye. Each fly is placed in a glass tube with the light shining directly down from above. The normally sighted fly moves straight up the inside wall of the tube towards the light whereas the fly that is blind in one eye moves up the tube in a spiral. The attempt to keep the light intensity the same in each eye leads to the fly always having its blind eye turned towards the light.

moving, a woodlouse keeps on moving until it reaches somewhere sufficiently moist, then it slows down or stops completely. In a kinesis, unlike a taxis, the animal does not move in any particular direction with respect to the stimulus.

A slightly different kind of kinesis is seen in flatworms: they respond to chemicals given off by food in the water by increasing their rate of turning, though not in any set direction. For example, if a piece of meat is placed in a pond, you might find several flatworms feeding on it within minutes (Fig. 2).

Fig. 2   Kinesis

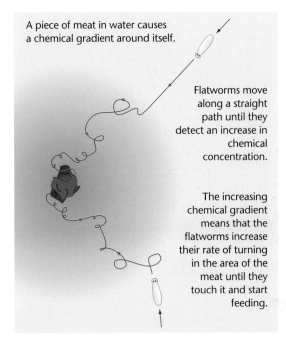

A piece of meat in water causes a chemical gradient around itself.

Flatworms move along a straight path until they detect an increase in chemical concentration.

The increasing chemical gradient means that the flatworms increase their rate of turning in the area of the meat until they touch it and start feeding.

1 Twenty flatworms were placed in a dish of water in a well-lit room. Eventually the flatworms became evenly spread around the dish. A light-proof cover was then placed over half of the dish. One hour later, all the flatworms were in the dark half of the dish.

Explain how the flatworms moved to and remained in the dark half of the dish if they were using:

a a taxis

b a kinesis.

Reflex actions

Earthworms come to the surface of the ground on warm, damp nights to defecate or mate. They respond to the slightest vibration by retreating down their burrows. This is a reflex escape response and it helps them to avoid being taken by a predator such as a shrew or hedgehog. A **reflex action** is a rapid, innate, automatic response to a stimulus.

key facts

- Invertebrates rely on innate behaviour patterns to find food and safety.

- In a taxis, the animal detects the direction of a stimulus, such as light, and moves towards or away from it.

- In a kinesis, the animal moves in a straight line until it meets conditions resembling those it needs. It then responds by either slowing down or increasing its rate of turning until the right conditions are met.

- In a reflex escape response, invertebrates move rapidly away from a stimulus that indicates immediate danger – for example, an approaching predator.

9.2 Tropisms

Plants cannot move from place to place. Usually they are held in position firmly by their roots. They can, however, orientate their stems and roots to obtain favourable conditions.

Phototropism

Two examples of phototropism.

The lentil sprouts in the photo above (left) are growing near a window. They respond by growing towards the direction of light. This response is known as **positive phototropism**.

Roots usually behave in the opposite way to stems: they grow away from the direction of light. They exhibit **negative phototropism**.

> **2** Explain the advantage of positive phototropism to the yellow-flowered plants growing in the shaded conditions (in the photo above, right).

Geotropism

An example of geotropism.

The tree in the photo above originally grew at the top of the hill. Because of erosion, the soil has moved slowly down the hill. Stems usually exhibit **negative geotropism**: they grow away from the direction of the force of gravity. This normally results in stems growing vertically. As the position of the tree changed, so did the direction of growth of the stem, resulting in the curved growth seen in the photo.

Roots usually exhibit **positive geotropism**.

> **3** Explain the advantage to a plant:
>
> **a** of roots exhibiting positive geotropism
>
> **b** of stems exhibiting negative geotropism.

key facts

● Tropic movements maintain the roots and shoots of plants in favourable environments.

● Most plant stems grow towards directional light; they are positively phototropic.

● Most plant roots grow away from directional light; they are negatively phototropic.

● Most plant stems grow away from the direction of the force of gravity; they are negatively geotropic.

● Most plant roots grow towards the direction of the force of gravity; they are positively geotropic.

9.3 The Pacinian corpuscle

Information from our skin allows us to identify several distinct types of sensations, such as tapping, vibration, pressure, pain, heat and cold. What is it that allows us to make these distinctions? Human skin contains different kinds of sensory receptors that respond to various mechanical, thermal or chemical stimuli. These receptors convey information to the brain to areas where we perceive the stimuli. To accomplish this, the nerve endings of the sensory receptors transform mechanical, thermal or chemical energy into electrical impulses. These impulses then travel along parts of **neurones** (nerve cells) called **axons** to the brain. The way we interpret sensations depends not only on the properties of receptors and neurones, but also on previous experiences that are stored in our brains.

Each square centimetre of skin contains about 200 pain receptors, 100 pressure receptors, 12 cold receptors and two warmth receptors.

Table 1 lists the characteristics of the five types of pressure receptors found in human skin. Each type of receptor is sensitive to only one type of stimulus.

The Pacinian corpuscle

The Pacinian corpuscles in the skin are very large receptors, almost 1 mm long, consisting of concentric layers (like onion-scales under the microscope). Each layer is composed of thin, flat cells called lamellae, and fibrous connective tissue separated by gelatinous material. In the centre of the corpuscle is the inner bulb, a fluid-filled cavity with a single axon (Fig. 3).

Fig. 3 Internal structure of a Pacinian corpuscle

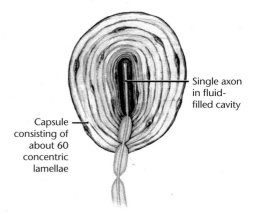

Single axon in fluid-filled cavity

Capsule consisting of about 60 concentric lamellae

Table 1 Skin pressure receptors

Receptor	Stimulus	Sensation
Merkel nerve endings	Steady indentation	Pressure
Meissner's corpuscle	Low-frequency vibration	Gentle fluttering
Ruffini endings	Rapid indentation	Stretch
Pacinian corpuscle	Vibration and pressure	Vibration
Hair receptor	Hair deflection	Brushing

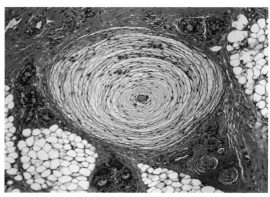

A light micrograph of a Pacinian corpuscle.

Fig. 4   Gated sodium ion channels

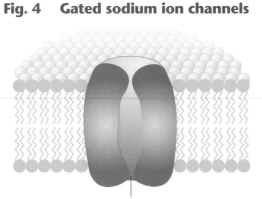

Channel closed. Na⁺ ions unable to pass through

Stretching of membrane causes sodium ion channels to open, allowing Na⁺ ions to pass through

Pacinian corpuscles detect large pressure changes and vibrations. These stimuli cause compression of the capsule, resulting in deformation of the axon. The sodium ion channels in the deformed area open, allowing sodium ions to pass through (Fig. 4).

The influx of sodium ions establishes a generator potential (Fig. 5). Chapter 2 explains how this generator potential leads to an electrical current passing along a sensory neurone.

4   Name the process by which sodium ions pass through the gated channels.

5   Use information from Fig. 5 to describe the relationship between a stimulus and the generator potential.

Fig. 5   Production of a generator potential by a Pacinian corpuscle

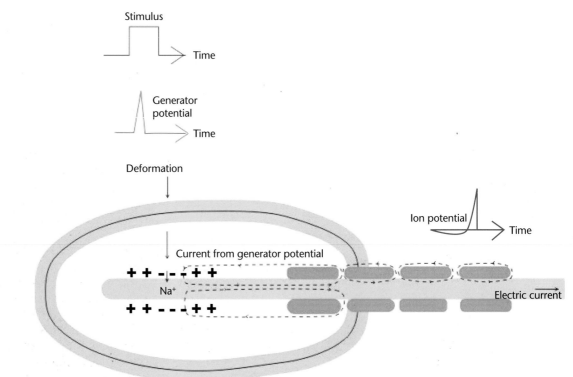

- Receptors respond only to specific stimuli.

- Pacinian corpuscles are sensitive to vibrations and to large changes in pressure.

- Stimulation of the membrane of the axon in the Pacinian corpuscle leads to deformation of stretch-mediated sodium ion channels.

- As sodium ions pass through the sodium ion channels, a generator potential is produced.

9.4 Light receptors

There are two types of light-sensitive cells in the human retina: **rods** and **cones**. The photograph shows how rods and cones appear under the scanning electron microscope. Fig. 6 shows the detailed structure of these two specialised cells.

Fig. 6 Rods and cones

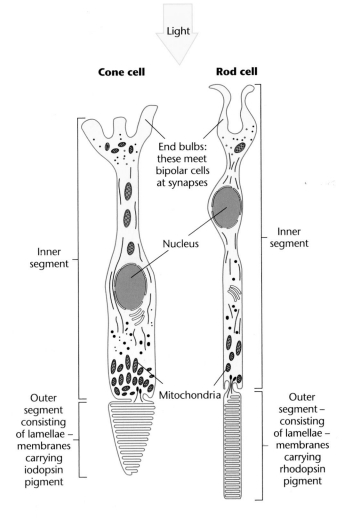

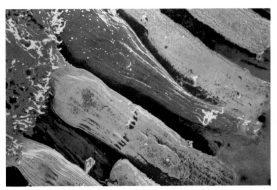

False-colour scanning electron micrograph of rod cells (orange) and cone cells (blue).

Rods and cones have the same basic structure, but they differ in shape and in the pigment they contain. Rod cells contain **rhodopsin**, cone cells have **iodopsin**. Rhodopsin and iodopsin are bleached by light, as are many pigments. (This is why curtains that have been hung in a sunny window fade.) A similar change happens to the pigments in rods and cones, but here the bleaching is rapid and reversible.

Rods

Each rod is packed with about 180 molecules of the light-sensitive pigment rhodopsin arranged in discs called **lamellae**. The rhodopsin molecule breaks down when exposed to light into a pigment called **retinal** and the protein **opsin**. The breakdown of rhodopsin into opsin begins a series of chemical reactions that results in the release of a transmitter substance by the rod cell. The transmitter substance stimulates a chain of events that may eventually result in an impulse being passed to the brain along a nerve fibre in the optic nerve.

When we enter a dark room, the light-bleached pigment is restored to its unbleached form, rhodopsin. This effect is called **dark**

adaptation and explains why we are gradually able to see more as we 'get used to' the lack of light. In the dark, it takes about 30 minutes to resynthesise all the rhodopsin from the retinal and opsin.

In complete darkness, rhodopsin is very stable. If rhodopsin were not so stable, rods might send impulses to the brain when no light was falling on them. In strong light, rhodopsin is broken down quicker than it can be reformed, so the rods are not of much value. In dim light, breakdown of rhodopsin is much slower, so production keeps up with breakdown.

Rod cells are connected to bipolar cells. The rod cell and its **bipolar cell** work in a different way to most other receptors. Other receptors produce a transmitter substance only when they are stimulated. But a rod cell produces a transmitter substance when it is not being stimulated, and stops production when it is stimulated.

Cones

Cones work on the same principle as rods except that their pigment is iodopsin. Cones have a high threshold; the pigment is only broken down by high intensity light, so they operate in daylight. Most of the impulses passing to the brain in bright light come from the cones, where iodopsin is resynthesised much more quickly than rhodopsin in the rods.

6 Why can't we see very well when we first enter a poorly lit room from a brightly lit area?

7 It is often very sunny in winter ski resorts. Skiers wear dark goggles to prevent snow 'blindness'. Explain what causes snow blindness.

Sensitivity and acuity

Fig. 7 shows a section through the retina. The retina is said to be **inverted**, since light has to pass between several layers of cells before it reaches the rods and cones. The rods and cones form synapses with cells called bipolar cells. The

Fig. 7   Light penetration into the retina

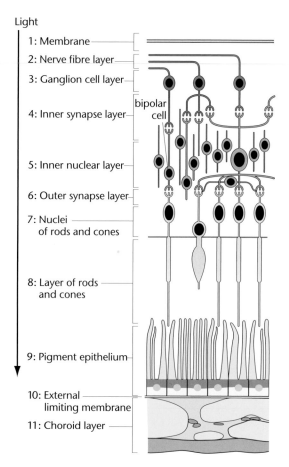

Light

1: Membrane
2: Nerve fibre layer
3: Ganglion cell layer
bipolar cell
4: Inner synapse layer
5: Inner nuclear layer
6: Outer synapse layer
7: Nuclei of rods and cones
8: Layer of rods and cones
9: Pigment epithelium
10: External limiting membrane
11: Choroid layer

bipolar cells in turn synapse with ganglion cells, which synapse with the sensory neurone fibres. These transmit impulses to the brain via the optic nerve.

Cone cells give colour vision and also allow us to see in detail – they give good **acuity**. There is maximum acuity at the **fovea** because here there are many densely packed cones, and each bipolar cell is stimulated by a single cone cell. Elsewhere in the retina there are fewer cones and each bipolar cell is stimulated by several cones (Fig. 8).

Bright light is needed for cones to work well. On the other hand, rods are more effective in dim light, because up to 45 rods synapse with each bipolar cell. Dim light results in the production of only a small amount of transmitter substance by each rod. Individually, this is insufficient to overcome the threshold of the bipolar cell, but the total

Fig. 8 Rods, cones and bipolar cells

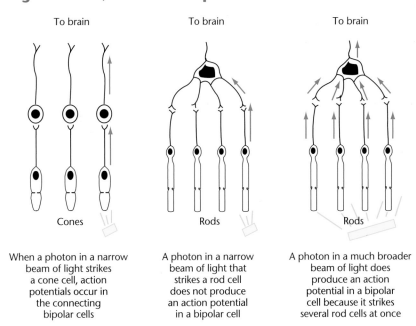

When a photon in a narrow beam of light strikes a cone cell, action potentials occur in the connecting bipolar cells

A photon in a narrow beam of light that strikes a rod cell does not produce an action potential in a bipolar cell

A photon in a much broader beam of light does produce an action potential in a bipolar cell because it strikes several rod cells at once

amount of transmitter substance produced by several rods is sufficient to overcome the threshold and depolarise the rod cell. The result of several rods causing depolarisation of one bipolar cell is to give less acuity, but better **sensitivity**.

When we look directly at something in bright light, light from the object is focused on the fovea, and we see it clearly. But when there is not much light, for example, when we look at a faint star, we can see the object more easily if we look slightly to one side of it. This is because light is then focused to one side of the eye where there are more rods, rather than on the fovea where there are more cones.

A comparison of rods and cones is given in Table 2.

Fig. 9 shows how the relative numbers of rods and cones varies in different parts of the retina.

Table 2 Rods and cones compared

	Rods	**Cones**
Number per eye/millions	120	six
Retinal convergence	15–45 rods to one bipolar cell	in fovea, one cone to one bipolar cell
Sensitivity	good; one photon gives response	poor; several hundred photons needed
Acuity	poor	good

Fig. 9 Relative numbers of rods and cones in the retina

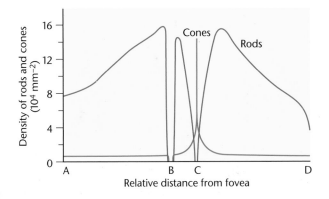

8

a Why can't we see colours in dim light?

b Why do rods give us greater sensitivity in dim light?

c Why do cones give us greater acuity in bright light?

d Light from the centre of our field of vision is focused on the cones at the fovea. What evidence is there, from what we see, that we have cones in other parts of the retina?

9

a Look at Fig. 9. Which region, A to D, is the fovea? Give the reason for your answer.

b Which regions would be most sensitive to dim light?

c What would we see of an image focused at B? Give the reason for your answer.

- Rods and cones are photoreceptors. They contain pigment that is broken down by light. This breakdown produces products that initiate processes leading to impulses in optic nerve fibres.

- Rods are the principal receptors in dim light, cones in bright light.

- 15–45 rods converge with one bipolar cell, giving low visual acuity.

- At the fovea, one cone converges with one bipolar cell, giving high visual acuity.

s&c

stretch and challenge

The trichromatic theory of colour vision

There is only one type of rod and this responds most strongly to bluish-green light. Cones are divided into three types, each of which has a different sensitivity to light. There are red-light receptors, green-light receptors and blue-light receptors, but the ranges of sensitivity overlap and most wavelengths of light stimulate at least two types of cone, as Fig. 10 shows.

The discovery of three types of cone supports the trichromatic theory of colour vision. This theory states that we see all the colours of the visible spectrum by mixing the three primary colours – blue, green and red. A white wall reflects all the colours of the spectrum back to the eye; the white parts of a TV screen are made up of red, green and blue dots. When all three types of cone are stimulated, the brain interprets the impulses from the cones as white light. The brain interprets any combination of messages from the

three types of cone as a particular colour. Look at Fig. 10. Yellow light stimulates the red-light receptors and the green-light receptors and the brain interprets the impulses from the receptors as yellow.

10 Which cones are stimulated by light of wavelength 600 nm?

11 Why do you see an orange colour when light of 600 nm stimulates the cones?

12 Which cones are stimulated by the white paper and black letters of this page?

Fig. 10   Wavelengths of light absorbed by different cones

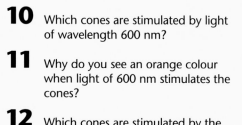

Light absorbed by 'blue' cones

Light absorbed by 'green' cones

Light absorbed by 'red' cones

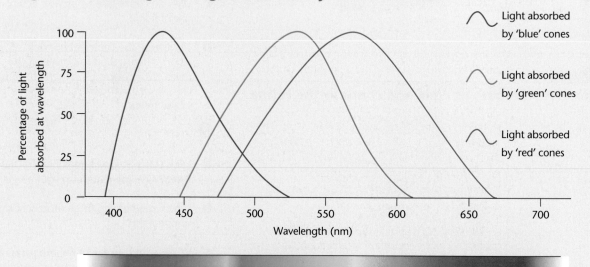

The colour we see depends on the amount of stimulation of each type of cone – so this gives the range of hues that we can see

Colour blindness

About 8% of males and fewer than 1% of females have faulty colour vision. The genes for making the cone receptor pigments (cone opsins) in green-sensitive and red-sensitive cones (but not the blue-sensitive cones) are on the X chromosome.

Many colour-blind males lack either the red-sensitive cones or the green-sensitive cones and so confuse red, blue-green and grey. For example, bright red roses can appear to be the same colour as the leaves, and scarlet clothes may appear to be dark grey.

Complete colour blindness in which all cone pigments are absent is very rare.

The genes for the cone opsins have now been cloned. Their base pair sequences show that the three cone opsins differ only slightly from one another. Since the cone opsins are also similar to rhodopsin, it may be that cone opsins were derived from a modified form of rhodopsin during evolution.

Look at the left part of Fig. 11. If you have normal colour vision you will see a figure seven in reddish brown dots.

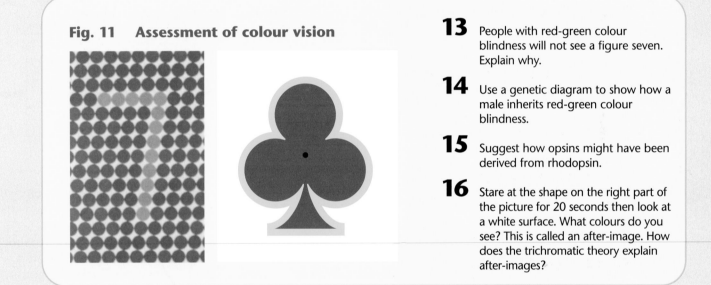

Fig. 11   Assessment of colour vision

13 People with red-green colour blindness will not see a figure seven. Explain why.

14 Use a genetic diagram to show how a male inherits red-green colour blindness.

15 Suggest how opsins might have been derived from rhodopsin.

16 Stare at the shape on the right part of the picture for 20 seconds then look at a white surface. What colours do you see? This is called an after-image. How does the trichromatic theory explain after-images?

9.5 Modifying heartbeat

In your AS course you studied the role of the **sinoatrial node (SAN)** in the control of the heartbeat. This can be summarised as:

- The SAN is the 'natural' pacemaker of the heart.
- The SAN does not require impulses from the nervous system to initiate electric impulses.
- Impulses are conducted from the SAN, first to the atria and then to the ventricles.
- These impulses cause the atria to contract, followed by the ventricles.

Although the heart can beat without nerve impulses, nerves are needed to change the rate of the heartbeat. The heart of a healthy adult beats roughly 70 times per minute. During exercise the rate may rise to over 140 beats per minute. During sleep it may fall as low as 50 beats per minute.

The rate at which the heart beats is modified by nerve signals from two areas of the brain (Fig. 12). One is the **cardioacceleratory centre**, the other is the **cardioinhibitory centre**. Both are located in the cardiovascular centre in the medulla oblongata of the brain (often just called the medulla). Nerve fibres pass from each of these centres to both the SAN and the **atrioventricular node** (AVN).

17 What are the advantages to an athlete of a rise in heart rate during a race?

Fig. 12 Control of heartbeat

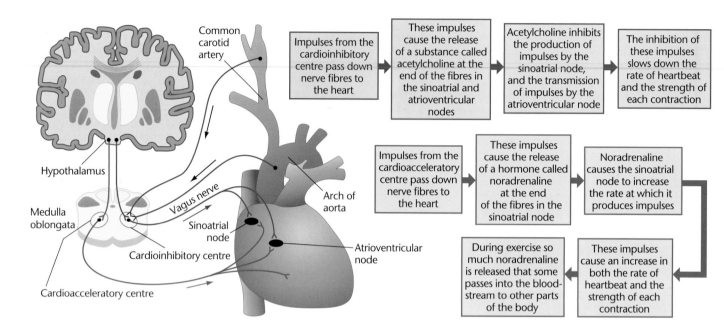

Impulses from the cardioinhibitory centre pass down nerve fibres to the heart

These impulses cause the release of a substance called acetylcholine at the end of the fibres in the sinoatrial and atrioventricular nodes

Acetylcholine inhibits the production of impulses by the sinoatrial node, and the transmission of impulses by the atrioventricular node

The inhibition of these impulses slows down the rate of heartbeat and the strength of each contraction

Impulses from the cardioacceleratory centre pass down nerve fibres to the heart

These impulses cause the release of a hormone called noradrenaline at the end of the fibres in the sinoatrial node

Noradrenaline causes the sinoatrial node to increase the rate at which it produces impulses

During exercise so much noradrenaline is released that some passes into the bloodstream to other parts of the body

These impulses cause an increase in both the rate of heartbeat and the strength of each contraction

Common carotid artery
Hypothalamus
Medulla oblongata
Vagus nerve
Cardioinhibitory centre
Cardioacceleratory centre
Arch of aorta
Sinoatrial node
Atrioventricular node

Chemoreceptors and pressure receptors

We do not know what causes the cardio-acceleratory centre to produce impulses that speed up heart rate. However, we do know that it is usually linked to an increased ventilation rate. If the oxygen concentration of the blood is low, or if the carbon dioxide concentration is high, ventilation rate increases. This somehow also causes the rate of heartbeat to increase.

Chemoreceptors sensitive to oxygen and carbon dioxide concentrations are present in the walls of the aorta and the carotid arteries. However, these chemoreceptors do not control the heart rate directly. The cardioacceleratory and cardioinhibitory centres that affect heart rate are under the direct control of **pressure receptors** in the walls of the aorta and the carotid arteries (Fig. 13).

Fig. 13 Pressure receptors

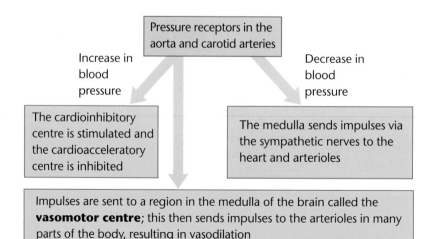

Pressure receptors in the aorta and carotid arteries

Increase in blood pressure

Decrease in blood pressure

The cardioinhibitory centre is stimulated and the cardioacceleratory centre is inhibited

The medulla sends impulses via the sympathetic nerves to the heart and arterioles

Impulses are sent to a region in the medulla of the brain called the **vasomotor centre**; this then sends impulses to the arterioles in many parts of the body, resulting in vasodilation

The heart rate increases when impulses produced by the cardioacceleratory centre pass down **sympathetic nerve fibres** to the SAN and the AVN. Impulses that arrive at the ends of the sympathetic fibres stimulate the release of **noradrenaline**. Noradrenaline causes the SAN to increase heart rate and increases the strength of each contraction.

To slow down heart rate, impulses produced by the cardioinhibitory centre pass down parasympathetic nerves to the SAN and the AVN. When the impulses arrive at the ends of the **parasympathetic fibres**, this releases **acetylcholine**. Acetylcholine causes the SAN to reduce the rate of heartbeat and inhibits the transmission of impulses by the AVN. The actions of the sympathetic and parasympathetic fibres are therefore **antagonistic**.

Although these changes seem to be linked to changes in the ventilation rate, no-one has yet been able to show a direct link between the concentration of respiratory gases in the blood and the production of impulses by the cardioacceleratory and cardioinhibitory centres of the medulla.

18

a   What effect would an increase in blood pressure have on the rate of heartbeat?

b   How would this change in heart rate affect blood pressure?

c   How do sympathetic impulses affect the heart?

d   How do parasympathetic impulses affect the arterioles?

key facts

- The rate at which the sinoatrial node sends out impulses can be modified by the nervous system and by hormones.
- Modification of heartbeat is controlled by the medulla of the brain.
- Impulses via sympathetic nerve fibres speed up the rate at which the sinoatrial node sends out impulses.
- Impulses via parasympathetic nerve fibres slow down the rate at which the sinoatrial node sends out impulses.
- The medulla of the brain is influenced by impulses from pressure receptors and chemoreceptors in the walls of the aorta and carotid sinuses.
- There is no direct relationship between the oxygen and carbon dioxide concentration of the blood and the rate of heartbeat.

1 The highland midge is a tiny blood-sucking insect that is the scourge of people on fishing holidays on the west coast of Scotland. Scientists are investigating claims that an extract from bog myrtle plants acts as a natural midge-repellent. The diagram the shows apparatus used in one of their investigations. In a series of tests more than 80% of the midges flew into branch Q.

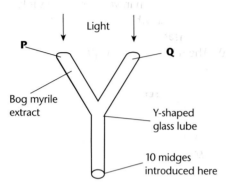

a Suggest why the apparatus was illuminated only from above. (1)

b What type of behaviour did the midges exhibit in response to the bog myrtle extract? (1)

c Outline an investigation to test the effectiveness of bog myrtle extract in preventing human volunteers from being bitten by midges. (3)

Total 5

AQA, B, March 2000, Unit 4, Question 3

2

a The times taken in the various stages of a complete cardiac cycle are shown in the table.

Stage of cardiac cycle	Time taken(s)
Contraction of the atria	0.1
Contraction of the ventricles	0.3
Relaxation of both atria and ventricles	0.4

i Use the information in the table to calculate the heart rate in beats per minute. (1)

ii If the same rate of heartbeat were maintained throughout a 12-hour period, for how many hours would the ventricular muscle be contracting? Show your working. (2)

b Although the heart does have a nerve supply, the role of the nervous system is not to initiate the heartbeat but rather to modify the rate of contraction. The heart determines its own regular contraction.

i Describe how the regular contraction of the atria and ventricles is initiated and coordinated by the heart itself. (5)

ii Describe the role of the nervous system in modifying the heart rate in response to an increase in blood pressure. (5)

c An interventricular septal defect is an opening in the wall (septum) that separates the left and right ventricles. Suggest and explain the effect of this defect on blood flow through the heart. (2)

Total 15

AQA, B, June 2002, Unit 3, Question 7

3 **Figure 1** shows a section through a human eye.
Figure 2 shows the distribution of rods and cones in the retina of the human eye.

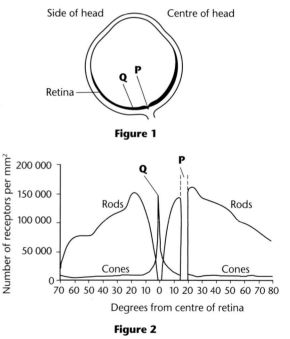

Figure 1

Figure 2

a Use **Figures 1** and **2** to explain why:

i no image is perceived when rays of light strike the retina at the point marked **P** (1)

ii most detail is perceived when rays of light strike the part of the retina labelled **Q**. (1)

b Rod cells allow us to see objects in dim light. Explain how the connections of rod cells to neurones in the retina make this possible. (3)

Total 5

AQA, January 2003, Unit 6, Question 1

4

a **i** The core temperature of a human foetus is approximately 0.5 °C higher than that of its mother. Explain what causes this higher body temperature. (3)

ii Under the same environmental conditions, a newborn human infant is more susceptible to heat loss than an older child. Suggest **two** reasons for this. (2)

b Protein is essential for growth. Different proteins, however, have different nutritive values. One method of measuring protein quality is to calculate its biological value from the equation:

$$\text{Biological value} = \frac{\text{Amount of protein used in maintenance and growth}}{\text{Amount of protein digested and absorbed}}$$

In order to do this, the following quantities are measured:

I = nitrogen intake

F = nitrogen in faeces

U = nitrogen in urine

Use the letters **I**, **F** and/or **U** to show how each of the following is calculated:

i the amount of protein digested and absorbed (1)

ii the amount of protein used in maintenance and growth. (1)

c When food passes through the gut, only a certain amount of the protein it contains is digested. The rest passes out of the body in the faeces. The digestibility coefficient is one way of measuring this.

$$\frac{\text{Digestibility}}{\text{coefficient}} = \frac{\text{Nitrogen intake} - \text{Nitrogen in faeces}}{\text{Nitrogen intake}}$$

i Explain why nitrogen intake is a useful measure of protein intake. (2)

ii Other than protein in the food which has not been digested, give **two** sources of protein in faeces. (2)

iii A lot of the protein in a vegetarian diet comes from cereals and vegetables. Use your knowledge of plant cells to explain why the digestibility coefficient for protein in vegetarian diets is lower than that for diets in which most of the protein comes from meat. (3)

d The table shows how protein requirement varies with age.

Age / years	Sex	Mean body mass (kg)	Estimated protein requirement (g day^{-1})
11–14	Male	43.0	33.8
15–18		64.5	46.1
50+		71.0	42.6
11–14	Female	43.8	33.1
15–18		55.5	37.1
50+		62.0	37.2

i Explain the difference in the estimated protein requirements for males and females between 15 and 18 years of age. (2)

ii In people of 50 years and over, protein is used entirely for maintenance. Explain how dietary protein is essential for maintenance of the oxygen-carrying capacity of the blood. (2)

iii It has been suggested that maintenance requirements for protein are approximately the same for males and for females. Do you agree with this statement? Use suitable calculations from figures in the table to support your answer. (2)

Total 20

AQA, June 2003, Unit 9, Question 1

5

a Give **two** fishing regulations and explain how each helps to maintain fish stocks. (2)

b Fishing effort is a measure of the size of a fishing fleet and the number of days the fleet spends fishing. The bar charts overleaf show the ages of one species of fish caught in areas where there have been different fishing efforts.

High fishing effort

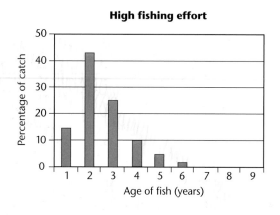

Low fishing effort

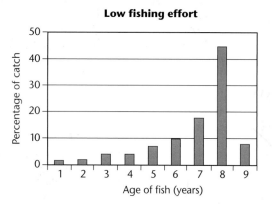

Describe and explain the effect of reducing fishing effort on stocks of this species of fish. (2)

Total 4

AQA, June 2006, BYB6/A, Question 1

6 Chitons are small animals that live on the seashore. When the tide is out they are found on the lower surfaces of stones. When the stones are turned over, the chitons move to the new lower surface as shown in the diagram.

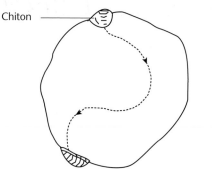

a Suggest **two** advantages to the chitons of this response. (2)

b Give **two** factors, other than light, to which the chitons might be responding. (2)

c A student investigated the response of chitons to light. Three covered dishes, **X**, **Y** and **Z**, were arranged as shown in the diagram. One half of the top of each dish was painted black, the other half was transparent. The dishes were placed on a table outside at noon on a bright day with light clouds in the sky. Ten chitons were placed in the light half of each dish. The number of chitons in each half of the dishes was recorded every five minutes for the next hour.

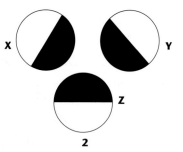

i Suggest why the dishes were arranged as shown in the diagram. (1)

ii The results of the investigation are shown in the table below.

Time / minutes	Number of chitons					
	Dish X		Dish Y		Dish Z	
	Light	Dark	Light	Dark	Light	Dark
0	10	0	10	0	10	0
5	9	1	7	3	8	2
10	6	4	7	3	7	3
15	5	5	7	3	6	4
20	3	7	5	5	5	5
25	3	7	4	6	5	5
30	3	7	3	7	5	5
35	3	7	3	7	5	5
40	2	8	3	7	4	6
45	1	9	3	7	4	6
50	1	9	2	8	4	6
55	1	9	3	7	2	8
60	1	9	1	9	1	9

What conclusions may be drawn from these results? (2)

iii Explain how kinesis could account for the results shown in the table. (1)

Total 8

AQA, June 2000, Unit 4, Question 1

Using Skylab to investigate tropisms

Skylab is an orbiting space laboratory. As part of its science education programme, NASA allows students to suggest experiments to be performed in Skylab. A student, Donald W. Schlack, suggested an experiment for Skylab wherein the effects of light on a seed developing in zero gravity would be studied.

With the help of engineers he constructed the apparatus shown in the photograph below. This consisted of eight compartments arranged in two parallel rows of four. The growth container was similar to cardboard potting cartridges found at plant nurseries. Each compartment had two windowed surfaces, which allowed periodic photography of the developing seedlings from both a front and side view.

The container in which the rice seeds were planted.

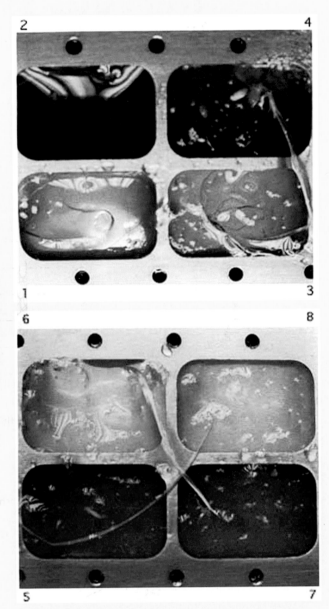

The seedlings in the compartments of the container after 30 days.

The study of light intensities on plants was accomplished by using light filters. For this purpose:

- Five windows were covered with special filters with different degrees of light transmittance.
- Two windows were blocked to prevent any light from reaching the seeds.
- The remaining window had no filter, allowing 100% transmission of light.

Three rice seeds were inserted into each compartment through covered holes. Photographs of the apparatus were taken at regular intervals for 30 days.

The photographs opposite shows the appearance of the seedlings after 30 days.

Of the 24 seeds planted, only 10 developed. This is close to the germination ratio of 12 out of 24 observed in the control group planted on Earth.

The three largest plants grew in the container in compartments 1, 4 and 6 with light transmission of 100%, 3%, and 2%, respectively.

Plant growth was extremely irregular and inconsistent. Some plant stems made 180-degree turns away from the light and many plant tips demonstrated curled patterns.

The longest stems to develop in testing on Earth were approximately 2 inches (5 cm) long. The stem on the plant grown on board Skylab grew to 4.2 inches (10.7 cm).

A1 Evaluate the design of the experiment.

A2 Did the experiment produce valid results? Explain the reason for your answer.

A3 Use the results to explain the plants' response to light in zero gravity.

10 | Coordination by the nervous system

Christopher Reeve as Superman (right) and at a neuroscience conference at MIT in 2003 (left)

As Superman, Christopher Reeve performed stunts that tested his muscles and nerves to the limit. However, in 1995, his lifestyle was shattered by a riding injury that left him as a quadriplegic – unable to move his arms or legs by himself.

Nerve fibres in Christopher Reeve's spinal cord had been damaged. The fibres, called axons, are long thin 'strings' of cytoplasm inside a plasma membrane; axons carry electrical impulses along the spinal cord. The nerve fibres are assembled in bundles. Some axons carry sensory information about touch, temperature and pain to the brain; if these are damaged the person can no longer feel any of these sensations. Other axons run downwards from the brain to control the voluntary movements of the body; if these are damaged the muscles they serve will be paralysed.

Part of Christopher's treatment involved the use of a 'shocker-cycle' – a contraption that looks like a cross between an exercise bicycle and a type of interrogation device. Electrodes pasted to his thighs and calves delivered a series of 50 volt shocks, making the muscles contract in sequenced spasms that brought his legs

back to life and sent the pedals flying. It preserved his muscle tone, moved the blood and stopped the flesh from withering on his arms and legs. Christopher tolerated this treatment because he believed that science would find a way to reconnect the shattered nerves in his spine and he wanted his body to be fit for that time.

Without realising his dream of walking again, Reeve died in 2004 of heart failure after having treatment for an infected pressure wound. Just before he died, he had regained some movements in his fingers, elbows, knees and toes and was able to feel hot and cold.

Christopher established the Christopher Reeve Paralysis Foundation, a non-profit research organisation, and used his fame to raise millions of dollars for spinal cord injury research.

Regeneration of nerves in the spinal cord was thought to be impossible, but recent scientific discoveries hold great hope that human nerve cells can be encouraged to regenerate and make new connections in shattered human spinal cords.

10.1 The nervous system

The nervous system is a complex network of specialised cells that allow us to sense our surroundings and to react to it. In this chapter we take a brief look at the structure of the **central nervous system**, and then we look in detail at an individual nerve cell, known as a **neurone**. The text introduces the main features of the cell, and later in the chapter we see how nerve impulses are transmitted along a neurone, and between neurones.

Back to basics

First, some more basic terms:

- **Sensory neurone** – a nerve cell that carries impulses from a receptor to the central nervous system.
- **Motor neurone** – a nerve cell that carries impulses from the central nervous system to an effector.
- **Reflex action** – a rapid automatic response to a stimulus, for example, the pain withdrawal reflex that makes you pull your hand away quickly when you touch something hot.

Overview of the central nervous system

The nervous system consists of two regions: the **central nervous system** and the **peripheral nervous system**. The central nervous system consists of the brain and the spinal cord. The peripheral nervous system consists of nerves made up of neurones that carry information to and from the central nervous system. There are two regions in the central nervous system, called **white matter** and **grey matter**. Grey matter consists of nerve cells – you have countless billions of these in the grey matter of your brain and they control most of your body's activities. Grey matter is also where nerve cells meet and pass information to one another. White matter consists of the fibres of long nerve cells that carry information within the brain and up and down the spinal cord, as shown in Fig. 1.

The motor neurone

The nervous system contains about 10 billion (10^{12}) neurones. Neurones all have the same basic structure, but the different types of nerve cells are specialised for their particular functions. Motor neurones show typical features of neurones. Fig. 2 shows the structure of a motor neurone that carries a nerve impulse from the spinal cord to a muscle. If impulses do not reach this neurone, the muscle it serves will be paralysed.

A neurone consists of a cell body, containing the nucleus of the cell, with one or more long thin structures called processes. These processes are extensions of the cytoplasm, surrounded by a plasma membrane. The processes that conduct nerve impulses towards the cell body are called **dendrons**.

Fig. 2   The motor neurone

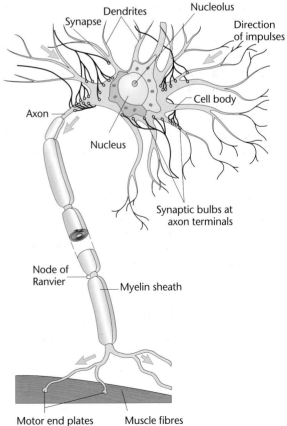

Fig. 1   The central nervous system

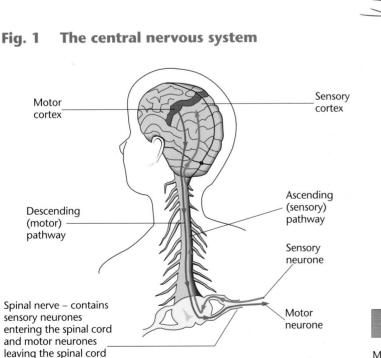

Dendrons have very fine processes called **dendrites**. The dendrites on the cell body of this motor neurone receive impulses from other neurones in the spinal cord.

The processes that carry nerve impulses away from the cell body of a neurone are called axons. The motor neurone in Fig. 2 has one long axon, called a nerve fibre, which carries nerve impulses to a muscle. At the junction between the neurone and the muscle fibres is the **motor end plate**.

You will notice that this motor neurone has an insulating cover over the axon, called a **myelin sheath**. This sheath is formed by **Schwann cells**, which twist around the axon several times as they grow, as Fig. 3 shows. Most of the myelinated sheath is composed of the cell

membranes of Schwann cells. The junctions between adjacent Schwann cells are called **nodes of Ranvier**.

> **1**
>
> The myelinated sheath is made up mainly of the membranes of Schwann cells. These membranes contain phospholipid molecules that contain fatty acids. These fatty acids prevent the movement of charged water-soluble ions. There are several layers of membranes in the sheath. Suggest the function of the sheath.

Fig. 3   Schwann cell growing round an axon

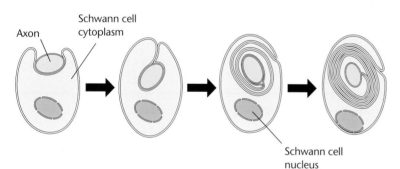

Axon

Schwann cell cytoplasm

Schwann cell nucleus

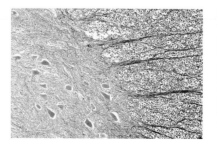

False-colour transmission electron micrograph of part of the myelin sheath (orange and green layers) of the human auditory nerve.

Light micrograph of normal human spinal cord. This cross-section shows the junction between the grey matter (bottom, orange) and the white matter (top). Grey matter contains the cell bodies of the neurones; the white matter consists of axons that form nerve fibres.

Where neurone meets neurone

A junction between two neurones is called a **synapse**. At this junction is a narrow gap, only about 20 nm wide. Transmission across this gap is by movement of chemical substances, as discussed in this chapter on page 182.

Nerves

Nerves are bundles of nerve fibres, as shown in Fig. 4.

Types of nerve

A spinal nerve usually contains both sensory and motor nerve fibres. Myelinated sheaths insulate the plasma membrane of the fibres from tissue fluid. This enables myelinated fibres to transmit impulses much faster than non-myelinated fibres. The outside of a nerve is made up of connective tissue – a tissue adapted for binding structures together.

> **2**
>
> **a** What is the difference between a nerve and a neurone?
>
> **b** Which type of neurone in a spinal nerve is damaged when:
>
>   **i** a person cannot feel a pin prick on the leg?
>
>   **ii** a person cannot move a leg?

Fig. 4   Cross-section through a nerve

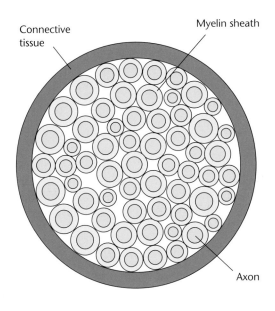

Connective tissue

Myelin sheath

Axon

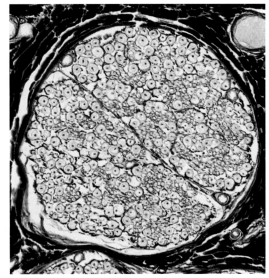

In a nerve, the axons are separated by myelin sheaths and the bundle is held together by connective tissue.

Light micrograph of a cross-section through the human sciatic nerve showing myelinated nerve fibres of different sizes (yellow circles). The nerve is surrounded by loose connective tissue (dark brown).

hsw

how science works

Myelinated and non-myelinated nerves

Fig. 5   Myelinated and non-myelinated nerves

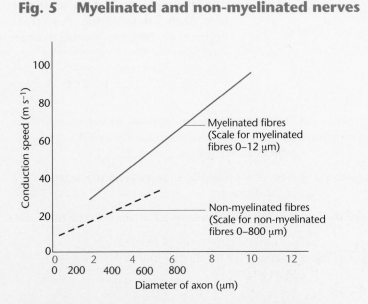

Myelinated fibres
(Scale for myelinated
fibres 0–12 μm)

Non-myelinated fibres
(Scale for non-myelinated
fibres 0–800 μm)

The graph shows the relationship between nerve impulse conduction speed and axon diameter in myelinated and non-myelinated nerve fibres.

3 What is the maximum conduction speed of a non-myelinated nerve fibre?

4 Calculate the percentage increase in conduction speed when the diameter of a non-myelinated fibre increases from 200 μm to 400 μm.

5 How much faster is the myelinated fibre conducting at maximum speed than the non-myelinated fibre transmitting at maximum speed?

6 Vertebrate animals possess myelinated fibres in many parts of the nervous system, but invertebrates do not have myelinated nerve fibres. Give two advantages to vertebrates having myelinated nerve fibres, in addition to conduction speed.

The reflex arc

The simplest responses to stimuli are reflex actions. A reflex action is a rapid automatic response to a stimulus. Reflex actions are rapid because there is a direct link between the receptor and the effector; the nerve impulses do not need to be processed by the higher centres of the brain to bring about a response. The shortest route between a receptor and an effector is called the **reflex arc**.

The nerve pathway of a reflex arc can go through the brain, for example, widening and narrowing the pupil, or sneezing. Or it can go through the spinal cord, for example, the pain withdrawal reflex shown in Fig. 6, or the knee jerk reflex.

Usually three neurones make up a reflex arc, but in the knee jerk reflex there are just two, the sensory neurone and the motor neurone. A spinal nerve contains both sensory neurones and motor neurones.

As Fig. 6 shows, a spinal nerve has two roots where it joins the spinal cord. The fibres of sensory neurones enter the spinal cord via the **dorsal root** of the spinal nerve. The cell bodies of the sensory neurones are found in this dorsal root. The motor neurones leave via the **ventral root** of the spinal nerve.

Fig. 6 The pain withdrawal reflex

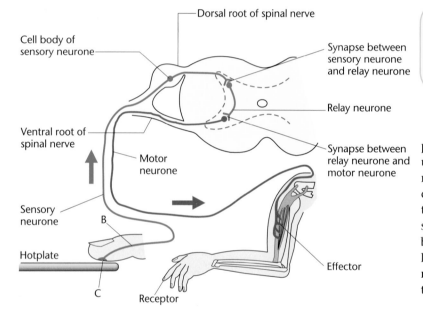

- Dorsal root of spinal nerve
- Cell body of sensory neurone
- Synapse between sensory neurone and relay neurone
- Relay neurone
- Ventral root of spinal nerve
- Motor neurone
- Synapse between relay neurone and motor neurone
- Sensory neurone
- B
- Effector
- Hotplate
- C
- Receptor

7 Look at Fig. 6. Which part of the sensory neurone is the axon? Give the reason for your answer.

Reflexes do not have to be learned: the nerve pathways are built into the nervous system under genetic control. They control simple responses needed for survival or avoiding danger. Neither do reflex actions need conscious thought. When we stand on a pin or touch something hot or cold, the response occurs before we are aware of the pain. The fact that we later become aware of pain shows that the neurones in reflex arcs also have connections to the brain.

key facts

- There are three types of neurones in most reflex arcs: sensory neurones, relay neurones and motor neurones.

- These neurones have the same basic structure:
  - a cell body containing the nucleus
  - dendrons that carry impulses towards the cell body
  - an axon that carries impulses away from the cell body.

- A myelin sheath, made up of Schwann cells, acts as an insulator of nerve fibres.

- The white matter of the spinal cord consists of myelinated nerve fibres; the grey matter consists mainly of cell bodies.

- The junctions between neurones are called synapses; transmission across synapses is by movement of chemical substances.

- A nerve is a collection of nerve fibres (the axons of the neurones).

- A reflex action is a rapid automatic response to a stimulus.

- A reflex arc is the shortest route from a receptor to an effector.

Reversing paralysis?

Roger Fenn fell and broke his neck. The accident damaged his spinal cord above the point where spinal nerves branch off to the arms. Nerve tissue, unlike bone, cannot repair itself easily, so Roger's arms and legs are permanently paralysed. Fortunately, he can still move the muscles in his shoulders normally and this has given him the chance to be much less dependent on the help of others.

Electronics and biology came together and, using a thorough understanding of the structure of the nervous system, designers came up with the electronic grip system, shown in Fig. 7. The electronics mimic the natural signals that pass along nerve cells to operate the muscles. Roger had an operation to implant the system in his right arm. After 14 years of living without the use of his hands, Roger now uses his shoulder muscles to operate his lower arms and hands and he can feed himself, clean his teeth, comb his hair, and let himself in and out of the house with an ordinary door key.

8 In what form is the electric current carried along the wires in the electronic grip system?

9 What structure in the nervous system does the joystick mimic?

10 What type of neurone has the same function as the wires that take electronic signals to Roger's muscles?

Fig. 7   Reversing paralysis

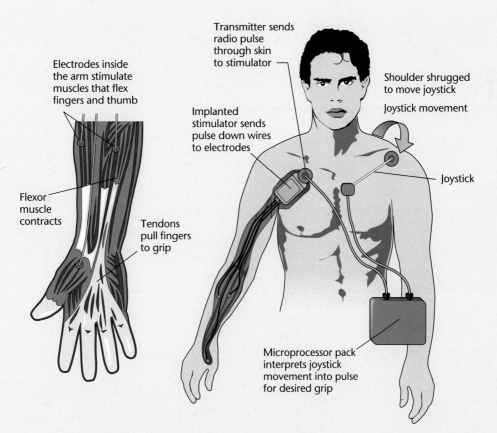

Electrodes inside the arm stimulate muscles that flex fingers and thumb

Flexor muscle contracts

Tendons pull fingers to grip

Transmitter sends radio pulse through skin to stimulator

Implanted stimulator sends pulse down wires to electrodes

Shoulder shrugged to move joystick

Joystick movement

Joystick

Microprocessor pack interprets joystick movement into pulse for desired grip

10.2   Nerve impulses

In this section we see how nerve impulses are transmitted down the axon of a neurone. Before we see how this happens, it is important to find out what is happening inside a nerve cell at rest, when no impulse is being transmitted.

The resting potential

To describe a neurone as 'resting' is a bit misleading as it gives the impression that the cell is inactive. In fact, it is using energy to maintain a difference in electrical charge between the inside and the outside of the axon. Fluid on the inside of the axon is negatively charged, compared with the fluid that bathes the outside. This bathing fluid has the same concentration of ions as tissue fluid. The difference in charge is called the resting potential of the neurone. It gives the *resting* neurone the *potential* to transmit a nerve impulse.

Measuring potential differences

But how do we know this about nerve cells? Scientists first observed electrical activity in nerves over 200 years ago, but it is only in the last 50 years that the underlying mechanisms have been understood. During the 1950s, scientists developed micropipettes with points as narrow as 0.5 μm. (You would have to split a human hair at least a hundred times before it would fit into this pipette.) These can be used to insert microelectrodes into individual axons. Two electrodes are needed to measure the potential difference (the voltage) between the outside and inside of an axon; one in the neurone and one immersed in the fluid outside the axon.

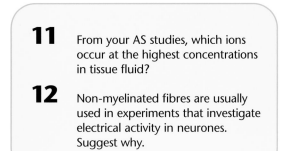

11   From your AS studies, which ions occur at the highest concentrations in tissue fluid?

12   Non-myelinated fibres are usually used in experiments that investigate electrical activity in neurones. Suggest why.

Fig. 8 shows how electrodes can be used to investigate electrical activity in a giant squid axon. When both electrodes are positioned in the bathing fluid there is no potential difference between them. When the micropipette is inserted into the axon there is a potential difference of –70 mV between the inside of the axon and the outside. This difference in potential is called the resting potential.

Positively charged ions such as sodium (Na^+) are called cations (because they move towards a negative electrode called a cathode); negatively charged ions such as chloride (Cl^-) are called anions (because they move towards an anode). The resting potential is caused by the movement of cations through the cell membrane of the axon. Since the inside of the axon is negatively charged with respect to the outside, the net movement of cations required to produce a resting potential is *outwards* through the cell membrane.

How is the resting potential produced?

Fig. 9 shows the mechanisms that produce the resting potential across the membrane of the axon. Two mechanisms are involved:

- active transport by Na^+/K^+ pumps – specialised carrier protein molecules
- facilitated diffusion through channel protein molecules.

Fig. 8   Investigating electrical activity in an axon

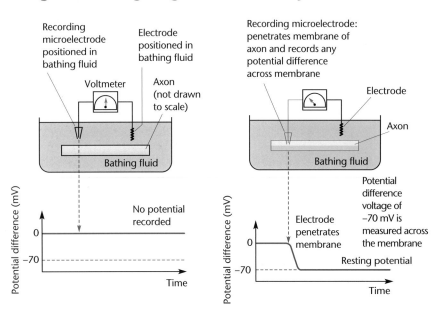

Fig. 9   Producing the resting potential

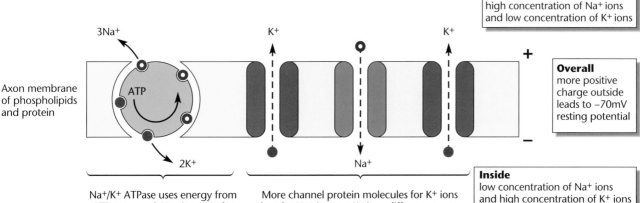

Outside
high concentration of Na⁺ ions and low concentration of K⁺ ions

Overall
more positive charge outside leads to −70mV resting potential

Inside
low concentration of Na⁺ ions and high concentration of K⁺ ions

Na⁺/K⁺ ATPase uses energy from ATP to move 3 Na⁺ ions out for every 2 K⁺ ions moved in

More channel protein molecules for K⁺ ions than for Na⁺ ions, so K⁺ ions diffuse out more rapidly than Na⁺ ions diffuse in

The Na⁺/K⁺ pump consists of the enzyme **ATPase**, which is also a carrier protein. For this reason, the pump is known as the Na⁺/K⁺ ATPase pump. Every square micrometre of the axon membrane contains up to 200 of these pumps. Each one moves about 200 Na⁺ ions out of the axon and about 130 K⁺ ions inwards every second – a ratio of approximately 3 Na⁺ ions outwards for every 2 K⁺ ions inwards. The two types of ions also move through the axon membrane by facilitated diffusion through channel proteins. However, the channel proteins move ions in the opposite direction to the Na⁺/K⁺ ATPase pump. Na⁺ ions diffuse in through the membrane and K⁺ ions diffuse out.

So what is the net result of all this ion transport in different directions? There are two important facts to note:

- The channel proteins that carry K⁺ ions outnumber those that carry Na⁺ ions, so more K⁺ ions diffuse out than Na⁺ ions diffuse in.
- The Na⁺/K⁺ ATPase pump moves more ions than are moved by facilitated diffusion.

The net result is therefore that there are more Na⁺ ions outside the membrane than K⁺ ions inside. This gives a net positive charge outside the membrane – and therefore a net negative charge inside. This gives the potential difference that is the resting potential.

13 Why do the Na⁺ ions and K⁺ ions move in opposite directions through the channel proteins?

14 Dinitrophenol (DNP) is a metabolic poison that inhibits respiration. If DNP is added to the solution bathing an axon, the axon does not develop a resting potential. Explain why.

The action potential

When a neurone is stimulated, information passes down the length of its axon. This information takes the form of a wave of **depolarisation** – instead of having a negative resting potential, one small stretch of axon after another develops a positive charge with respect to the fluid outside it. Depolarisation happens because of channels in the membrane of the axon that can change shape. In the 'open' position, they allow ions to pass through; in the 'closed' position, they block ion movement (See Fig. 4 of Chapter 9).

Depolarisation

Fig. 10 shows the movements of ions during the depolarisation and repolarisation that occurs during an action potential. The gated channels for Na⁺ ions are normally closed in the resting neurone. If the neurone is stimulated, these gated channels open. Since there is a higher concentration of Na⁺ ions outside the axon membrane than inside, they move along their concentration gradient and rush in through the membrane. This movement results in a higher concentration of cations inside the membrane than outside, giving the inside of the axon a positive charge. This part of the axon membrane is depolarised. This change of potential from negative to positive, which travels down the axon membrane, is known as the **action potential**.

A stimulus can set up an action potential only if it is greater than a minimum level known as the **threshold level**. If a stimulus exceeds the threshold level, the gated sodium channels in the axon membrane open. If the stimulus is below the threshold level, they remain closed and no depolarisation occurs.

Depolarisation of the membrane takes only a few milliseconds but it can be studied by attaching microelectrodes to an axon, then connecting them, to a cathode ray oscilloscope (CRO) (Fig. 11). The CRO can record, then 'freeze' events that occur in each tiny fraction of a second.

Repolarisation

Depolarisation occurs in a short stretch of membrane, and then this recovers its resting potential, and the action potential moves on down the axon. The membrane recovers its negative resting potential because of gated potassium channels. Shortly after, the gated sodium channels in the axon open to let Na⁺ ions in, thus depolarising the membrane; gated potassium channels also open (Fig. 12).

Because there is a greater concentration of K⁺ ions inside the membrane than outside, they move along their concentration gradient to the outside. This movement of K⁺ ions results in the net positive charge outside the membrane being restored, once again forming a resting potential. The restoration of the resting potential is called **repolarisation**. For a short period after repolarisation both the gated channels for sodium ions and the gated channels for potassium ions remain closed. During this period the membrane cannot be depolarised and therefore no impulse can pass. This period is known as the **refractory period**.

Fig. 10 An action potential

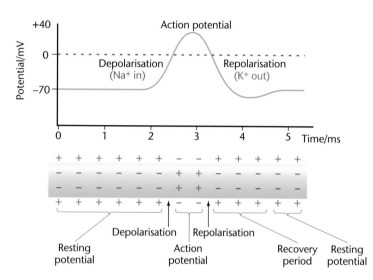

Fig. 11 Observing an action potential

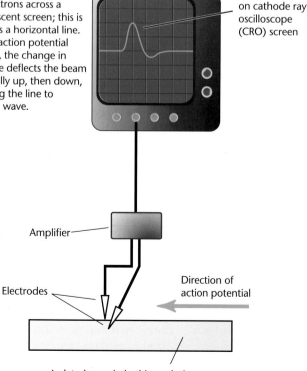

The CRO sweeps a beam of electrons across a fluorescent screen; this is seen as a horizontal line. As an action potential passes, the change in voltage deflects the beam vertically up, then down, causing the line to form a wave.

Action potential on cathode ray oscilloscope (CRO) screen

Amplifier

Electrodes

Direction of action potential

Isolated axon in bathing solution

Fig. 12 Permeability to sodium and potassium ions

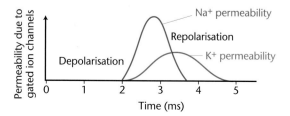

The refractory period is important for the following reasons:

- Each action potential is kept discrete – there is no overlapping of potentials.
- It ensures the action potentials pass in only one direction.

The action potential generated in a neurone is always the same size. The size of the action potential does NOT depend on the size of the stimulus. This is known as the **all-or-nothing principle**.

hsw

how science works

Fig. 13 The action potential graph

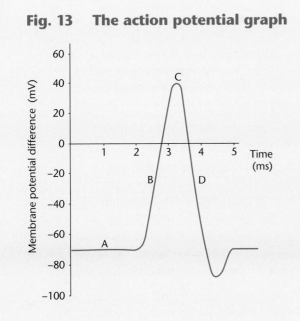

Fig. 13 shows changes in potential across an axon membrane during and after an action potential.

15 Which of the letters A to D labels:

    i repolarisation?
    ii action potential?
    iii resting potential?
    iv depolarisation?

16 By how much did the potential difference change when the action potential was produced?

17 How long is the refractory period?

Fig. 14 Conduction along myelinated and non-myelinated fibres

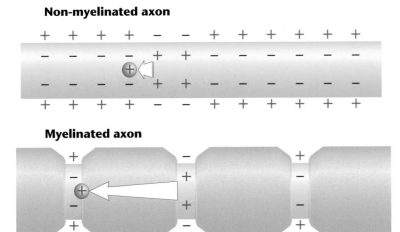

Non-myelinated axon

Myelinated axon

Transmission of impulses along nerve fibres

In non-myelinated fibres, the impulse is transmitted along the fibre when each tiny segment of the membrane causes the next segment to depolarise. Transmission is relatively slow since the whole membrane must be depolarised in successive sections.

Depolarisation in myelinated fibres occurs only at the nodes of Ranvier, since the sheath cells are pressed tightly against the axon (Fig. 14). In these myelinated nerve fibres, the impulse 'jumps' from one node of Ranvier to the next, resulting in very rapid conduction. This is known as **saltatory conduction**.

The myelin sheath is an adaptation for rapid transmission of nerve impulses. Non-myelinated fibres are not adapted in this way and therefore transmit impulses much more slowly.

Transmission speed is also influenced by the diameter of the axon. The wider the axon, the larger its surface area and the greater the speed of conduction. Squid do not have myelinated fibres, but have evolved fibres with giant axons, up to 1 mm in diameter, which conduct impulses very rapidly.

In invertebrates, transmission speed is affected by temperature – the higher the temperature the greater the transmission speed. Mammalian nerve fibres are at body temperature, so these fibres do not normally experience changes in temperature.

> **18**   Suggest why increasing temperature increases the rate of conduction in invertebrate nerve fibres.

key facts

- A resting potential gives a neurone the potential to transmit a nerve impulse.

- The resting potential is produced mainly by the active transport of sodium ions to the outside of the nerve fibre. This makes the outside of the membrane positively charged. The membrane in this state is said to be polarised.

- The pump that does this consists of the enzyme ATPase, which is also a carrier protein in the cell membrane.

- As sodium ions are pumped out of the cell, potassium ions are pumped in, but at a slower rate.

- Depolarisation occurs when closed sodium channels in the membrane suddenly open. The concentration of sodium ions is higher outside than inside, therefore sodium ions move into the cell by facilitated diffusion.

- This movement of sodium ions into the cell gives the inside of the membrane a higher positive charge than the outside – the membrane is depolarised.

- This change in potential of the membrane is called the action potential.

- After depolarisation, the membrane is repolarised by the outflow of potassium ions. The resting potential is then restored by the sodium/potassium (Na^+/K^+) pump.

- While the membrane is being repolarised, the membrane cannot be depolarised. The period when this happens is called the refractory period.

- Conduction along myelinated fibres is very rapid since the impulse 'jumps' from one node of Ranvier to the next. This is known as saltatory conduction.

- Conduction speeds are increased by increased diameter of axons and by increased temperature.

10.3   Synapses

The junction where two neurones meet is called a **synapse**. In all but the most powerful electron micrographs, neurones appear to be actually touching, but there is in fact a very narrow gap between them called the **synaptic cleft**. Information is carried across the synaptic cleft by chemicals called **neurotransmitters**. Some drugs interfere with this chemical transmission; it is by studying the effects of drugs on synapses that we have obtained much of our understanding of how synapses work.

The scanning electron micrograph (SEM) shows several nerve fibres forming synapses with a cell body. A neurone may have synapses that connect with up to 10 000 other neurones!

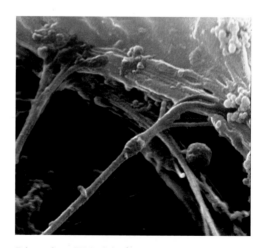

False-colour SEM of the junction sites (synapses) between nerve fibres (purple) and a neurone cell body (yellow).

Fig. 15 Transmission of an impulse across a synapse

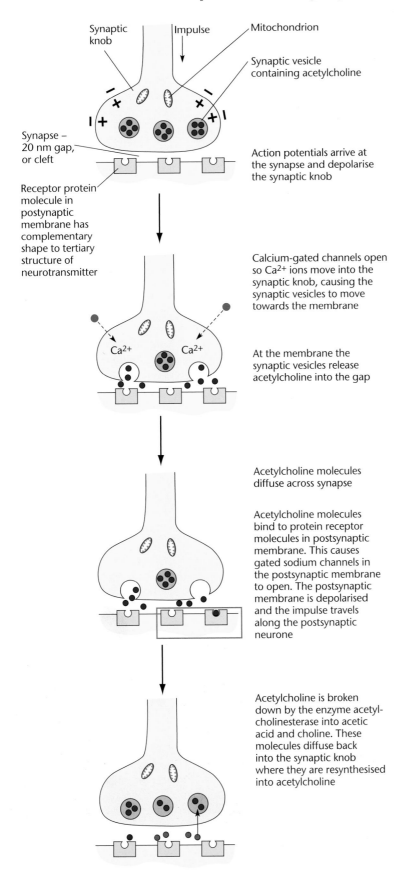

Synaptic knob

Impulse

Mitochondrion

Synaptic vesicle containing acetylcholine

Synapse – 20 nm gap, or cleft

Action potentials arrive at the synapse and depolarise the synaptic knob

Receptor protein molecule in postynaptic membrane has complementary shape to tertiary structure of neurotransmitter

Calcium-gated channels open so Ca²⁺ ions move into the synaptic knob, causing the synaptic vesicles to move towards the membrane

At the membrane the synaptic vesicles release acetylcholine into the gap

Acetylcholine molecules diffuse across synapse

Acetylcholine molecules bind to protein receptor molecules in postsynaptic membrane. This causes gated sodium channels in the postsynaptic membrane to open. The postsynaptic membrane is depolarised and the impulse travels along the postsynaptic neurone

Acetylcholine is broken down by the enzyme acetyl-cholinesterase into acetic acid and choline. These molecules diffuse back into the synaptic knob where they are resynthesised into acetylcholine

Transmission of an impulse across a synapse

Neurotransmitters are stored at nerve endings in tiny bags of membranes called **synaptic vesicles**. Fig. 15 shows the main stages in the transmission of impulses across the junction between two neurones, as described below.

- An action potential arrives at the end of the presynaptic neurone.
- This action potential causes calcium-gated channels in the membrane to open, resulting in an influx of calcium ions. The calcium ions cause the synaptic vesicles to move to the membrane and release the neurotransmitter substance into the synaptic cleft. At cholinergic synapses the neurotransmitter is acetylcholine.
- The neurotransmitter diffuses across the synaptic cleft to the postsynaptic membrane.
- The neurotransmitter molecules bind to protein receptor molecules in the postsynaptic membrane.
- This causes depolarisation of the postsynaptic membrane, starting a nerve impulse in the postsynaptic neurone.
- There is a delay of about half a millisecond at a synapse. This is the time needed for the neurotransmitter to diffuse across the gap, bind to a receptor protein and so activate or inhibit the next cell.

The different ways in which synapses connect neurones to other neurones or to muscle cells are used to produce a range of responses. The response can also be affected by the type of neurotransmitter, which can be either excitatory (they increase the activity of the next cell) or inhibitory (they decrease the activity of the next cell), enabling synapses either to pass signals onward or to block signals.

Synapses can become 'fatigued'. This happens when so many action potentials arrive in so short a time that the cell runs out of neurotransmitter (or its components). The impulse can no longer cross the synapse. This loss of response at a synapse is known as **adaptation** and it means that animals ignore stimuli that go on for a long time. For example, once we are used to them we do not notice stimuli such as the tick of a clock or the feel of our clothes brushing against our skin. Such stimuli are irrelevant to our survival. Sudden changes in our environment, like the unexpected approach of a car in a quiet lane, are more important.

Synapses between neurones might direct an impulse to just one other neurone, or spread it out to several. Synapses also occur at the junctions between nerves and muscle fibres, where they are called neuromuscular junctions. Besides their role in controlling deliberate movements such as writing or lifting a cup, synapses are just as important in regulating involuntary movement such as heartbeat.

Cholinergic synapses

Over 40 different neurotransmitters have now been identified. One of the most common in voluntary nerves is **acetylcholine**. Synapses that have acetylcholine as their neurotransmitter are called **cholinergic synapses**.

Once the nerve impulse has been passed on to the next neurone, the neurotransmitter is quickly removed from the synaptic cleft, otherwise the neurone would keep on firing uncontrollably. A very fast-acting enzyme called **acetylcholinesterase** breaks down acetylcholine into acetic acid and choline. These substances are reabsorbed through the presynaptic membrane. ATP from mitochondria is used to provide the energy to resynthesise acetylcholine, which is then returned to the vesicles.

Summation

Sometimes, the amount of neurotransmitter resulting from one impulse in the presynaptic neurone is not sufficient to cause depolarisation of the postsynaptic membrane. This may be because the neurotransmitter at a synapse is removed almost as quickly as it is released. To produce depolarisation, several action potentials might be needed to produce enough neurotransmitter to overcome the threshold of

19 Use information from Fig. 15 to explain why transmission across a synapse is unidirectional.

20 Eserine is a drug used after eye surgery. It helps to prevent swelling of the eye caused by fluid accumulation. Muscle contractions help to remove excess fluid from the eye. Eserine works by inhibiting acetylcholinesterase. What is the effect of eserine at a cholinergic synapse?

the postsynaptic membrane. This can be done in two ways: temporal summation and spatial summation. They are shown in Fig. 16.

- In temporal summation, the action potentials from one presynaptic neurone arrive in rapid succession.
- In spatial summation, several action potentials from different neurones arrive at the postsynaptic membrane at the same time. Spatial summation occurs in the retina of the eye. In dim light an individual receptor cell might not produce sufficient transmitter substance to depolarise a sensory neurone, but several receptors will produce enough, thus enabling us to see.

Summation is important for the brain to work properly. For example, it might be better to ignore a weak stimulus, such as mild pressure on the skin, but a stronger stimulus might cause injury and so require a response. A strong stimulus sends a high frequency of nerve impulses to the brain.

hsw

how science works

Synapses that use noradrenaline

Synapses that use noradrenaline affect heart rate, breathing rate and brain activity. This is similar to the effect of the hormone adrenaline, which prepares the body for emergencies.

In Chapter 1 you learned that noradrenaline speeds up the rate of heartbeat when it is released by nerve fibres. In some people, the release of too much noradrenaline causes the heart to race. One way to treat this is to use drugs known as beta-blockers. These drugs have molecular shapes similar to noradrenaline.

21 Which part of the heart is directly affected by the release of noradrenaline?

22 Suggest how beta-blockers work.

Fig. 16 Temporal and spatial summation

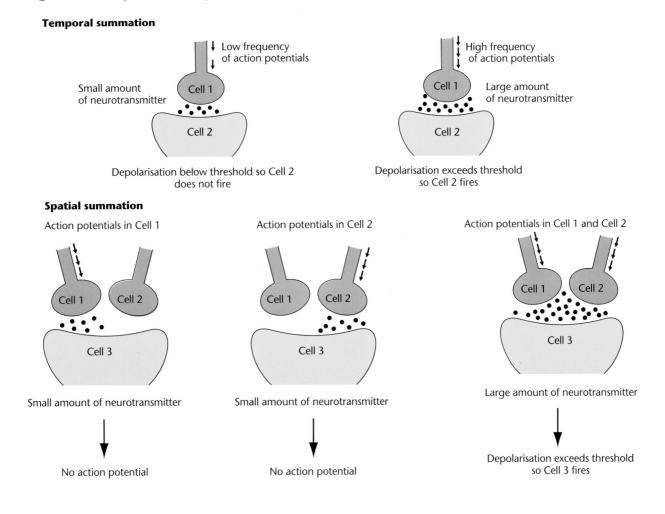

Temporal summation

↓ Low frequency of action potentials

Cell 1

Small amount of neurotransmitter

Cell 2

Depolarisation below threshold so Cell 2 does not fire

↓ High frequency of action potentials

Cell 1

Large amount of neurotransmitter

Cell 2

Depolarisation exceeds threshold so Cell 2 fires

Spatial summation

Action potentials in Cell 1

Cell 1   Cell 2

Cell 3

Small amount of neurotransmitter

↓

No action potential

Action potentials in Cell 2

Cell 1   Cell 2

Cell 3

Small amount of neurotransmitter

↓

No action potential

Action potentials in Cell 1 and Cell 2

Cell 1   Cell 2

Cell 3

Large amount of neurotransmitter

↓

Depolarisation exceeds threshold so Cell 3 fires

Fig. 17 A neuromuscular junction

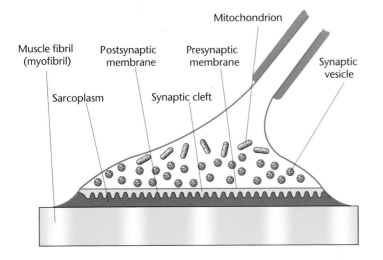

Muscle fibril (myofibril)

Postsynaptic membrane

Mitochondrion

Presynaptic membrane

Synaptic vesicle

Sarcoplasm

Synaptic cleft

The neuromuscular junction

Fig. 17 shows the junction between a motor neurone and a muscle fibre.

Exactly the same sequence of events happens at this synapse as when two neurones connect, but in this case the postsynaptic membrane is the muscle fibre membrane, the sarcolemma. The sarcolemma is depolarised when acetylcholine released from the presynaptic membrane of the neurone diffuses across the synapse and binds to protein receptors. Depolarisation sets in motion a sequence of events that lead to contraction of the muscle fibre. These events are described in Chapter 11.

key facts

- The sequence of events at cholinergic synapses and neuromuscular junctions is:
  ○ Action potential arrives at the presynaptic membrane.
  ○ Ca^{2+} channels open and Ca^{2+} ions enter the presynaptic membrane.
  ○ Vesicles containing acetylcholine move towards the presynaptic membrane.
  ○ Acetylcholine is released from synaptic vesicles into the synaptic cleft.
  ○ Acetylcholine diffuses across the synaptic cleft to the postsynaptic membrane.
  ○ Acetylcholine binds with specific receptor proteins of the postsynaptic membrane.

- Depolarisation of the postsynaptic membrane results in:
  ○ an action potential in the postsynaptic neurone;
  or
  ○ depolarisation of the sarcolemma, leading to contraction of the muscle fibre.

10.4   Drugs and synapses

Transmitters are released in tiny amounts: only 500–1 000 molecules from each synaptic knob are required to transmit an impulse. So, drugs that affect transmitters or their binding sites can have powerful effects when given in fairly small doses. Some chemicals, many of them from plants, have a dramatic effect on the nervous system.

Fig. 18   Drugs and transmitter binding sites

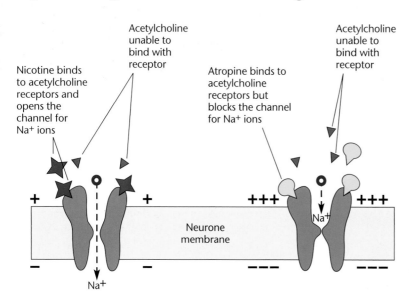

Nicotine binds to acetylcholine receptors and opens the channel for Na+ ions

Acetylcholine unable to bind with receptor

Atropine binds to acetylcholine receptors but blocks the channel for Na+ ions

Acetylcholine unable to bind with receptor

Neurone membrane

The effect of nicotine

A nicotine molecule is a similar shape to acetylcholine, so it can bind with acetylcholine receptors and open sodium channels, as shown in Fig. 18.

Nicotine receptors are found throughout the nervous system and there are at least five different types. Some of these are more sensitive to nicotine than others. Some activate quickly, and then turn off. Others stay active as long as nicotine is present. Nicotine is a major component of tobacco and is the main reason for smokers becoming addicted to their habit. Once the nervous system gets used to nicotine, doing without it creates unpleasant symptoms such as depression and anxiety. The average smoker inhales smoke from about 20 cigarettes per day, which averages at about 200 puffs per day or 80 000 puffs per year. This delivers a constant supply of nicotine to the nervous system – a smoker has to puff on a cigarette every five minutes during waking hours to maintain nicotine levels that will keep the nicotine receptors active.

There is a powerful correlation rate between smoking and depression. Do people take up smoking to relieve an underlying depression or anxiety?

23

a   A caffeine molecule has a similar shape to a molecule of an excitatory brain transmitter called adenosine. How does caffeine increase alertness?

b   How could someone become addicted to caffeine?

The effect of atropine

Atropine also binds to acetylcholine receptors, but does not open the sodium channels. It blocks the receptors, and prevents acetylcholine from binding to them, as shown in Fig. 18. When this happens in motor neurones, it causes muscle paralysis.

Atropine is extracted from the deadly nightshade plant, *Atropa belladonna*. 'Atropa' comes from the Greek Fate, Atropa, who severs the strand of life; 'belladonna' means 'beautiful woman' in Italian. The latter alludes to the Renaissance when a 'doe-eyed' expression was considered to be the utmost of beauty. Ladies would drop juice from deadly nightshade leaves into their eyes, dilating their pupils and giving them the desired look.

The deadly nightshade plant (*Atropa belladonna*) is very poisonous. It contains the alkaloids atropine, solanine and hyoscyamine.

key facts

- Drugs that have shapes similar to neurotransmitter substances can affect protein receptors in postsynaptic membranes.

- Some drugs bind to postsynaptic receptors, causing gated ion channels to remain open.

- Some drugs bind to postsynaptic receptors, causing gated ion channels to remain closed.

- Some drugs inhibit the action of acetylcholinesterase.

- Some drugs inhibit the reabsorption of the neurotransmitter substance from the synapse.

More than you think

In the introduction to his book *The Astonishing Hypothesis*, Francis Crick writes:

"The Astonishing Hypothesis is that 'You', your joys and your sorrows, your memories and your ambitions, your sense of personal identity and free will, are in fact no more than the behaviour of a vast assembly of nerve cells. One of the most difficult aspects of the nervous system to explain is: how can ion movements in neurones give us consciousness and self-awareness? Yet we accept that certain other complex structures have qualities that we would not predict by looking at their component parts. For example, the components of a motorcar do very little until they are fitted together – but then the whole assembly can be driven down the motorway at 70 mph."

Conditioning

The pets sharing our homes are quick to learn where food can be found. The sound of a tin being opened brings a cat or dog into the kitchen very quickly, because the animal has learned to associate the sound of the tin opener with food. This type of learning is called conditioning. There are two types of conditioning: classical conditioning and operant conditioning.

Classical conditioning

Conditioning interested the physiologist Pavlov, who worked in Russia at the end of the 19th century. Pavlov wanted to test the effects of different types of food on the salivation reflex. He collected and measured the amount of saliva produced by the dogs when food was placed in their mouths. However, he also noticed that his dogs started to salivate as soon as they heard his approaching footsteps, before he had given them any food. Pavlov then found that by giving the dogs a reward of food, he could train them to salivate in response to a stimulus such as a flashing light or a ringing bell. Pavlov called this type of learning, in which the usual stimulus (food) is replaced by a new stimulus (light or a bell), a conditioned reflex. This was the first type of conditioning to be described. As a result, it is now known as classical conditioning.

Operant conditioning

Pavlov's work encouraged others to study the way animals learn. Skinner developed a special apparatus (the Skinner box) for training an animal, usually a rat. By pressing a lever in the box, the rat gains a reward of a food pellet. At first, the rat presses the lever just by accident, but it soon learns to associate lever pressing with a reward. This is called conditioning, because the animal is *rewarded* for an operation (movement) that it does naturally from time to time. The food reward makes it more likely that the rat will press the lever again; in other words the food is a positive reinforcement of the behaviour. All kinds of movement can be reinforced in operant conditioning. For example, if a pigeon is rewarded with food every time it preens its feathers, the rate of preening quickly increases. Reinforcers need not be food: monkeys learn for the reward of seeing another monkey, chickens for the reward of straw to nest in, and dogs for the reward of attention from their owner.

Animals can also be conditioned using punishment. A stern telling off works as a punishment and prevents a dog from jumping up to greet people, so long as the training is done consistently for a few months.

Conditioning is part of the natural lives of animals. For example, caterpillars of the mullein moth have a nasty taste and bright orange spots. The bad taste works as a punishment for any bird that attempts to eat the caterpillar, and the bird learns to associate the orange spots with a bad taste. This is an example of warning coloration.

Learned human behaviour

Learning plays a major part in our daily lives. For example, conditioning is used by advertisers. Some advertisements try to familiarise us with a particular sign or logo, so that we are more likely to select it from a range of similar logos. More commonly, advertisements try to make us associate their product with being attractive, successful or intelligent. We are encouraged to think that buying the product will give us these qualities. Parents often use money to reward children for good behaviour. They also use various forms of punishment. This is rather like operant conditioning in rats, but humans can often see the motives behind other people's behaviour and resist being 'conditioned'. Sociologists and psychologists study the effects of reward and punishment on human behaviour, for example, the effects of prison sentences as a deterrent against crime. While most people agree that too much punishment can make a child withdrawn and unresponsive, some think that too many rewards can encourage selfishness. Skilled parents and teachers are good at giving the right amount of control and encouragement to each child in a home or classroom.

'99% of our behaviour is automatic.'

24 Discuss this statement, using information from the extracts above and your own ideas.

1
a The graphs in **Figure 1** show the relationship between the membrane potential of an axon membrane and the numbers of Na^+ (sodium ion) channels and K^+ (potassium ion) channels that are open.

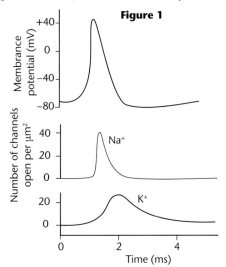

Figure 1

Using the information in the graphs, explain how:
  i the action potential is generated;   (2)
  ii the axon membrane is repolarised.   (2)
b The secretion of gastric juice by the stomach is stimulated by nerves and by hormones.
Figure 2 shows the volume of gastric juice secreted following nervous stimulation and hormonal stimulation.

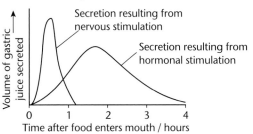

Figure 2

Give **two** differences between the nervous and hormonal control of the secretion of gastric juice. Use information in the graph to illustrate your answer.   (2)
c Gastric juice contains pepsin.
  i Pepsin is produced as inactive pepsinogen. What is the advantage of this?   (2)
  ii Pepsin is an endopeptidase. What is an *endopeptidase*?   (1)
  iii What is the advantage in endopeptidases acting on proteins before exopeptidases do?   (2)
d When pepsin leaves the stomach it enters the small intestine. Explain how pepsin is inactivated by the high pH in the small intestine.   (4)
Total 15
AQA, January 2003, Unit 6, Question 8

2 Acetylcholine is a neurotransmitter which binds to postsynaptic membranes and stimulates the production of nerve impulses. GABA is another neurotransmitter. It is produced by certain neurones in the brain and spinal cord. GABA binds to postsynaptic membranes and inhibits the production of nerve impulses. The diagram shows a synapse involving three neurones.

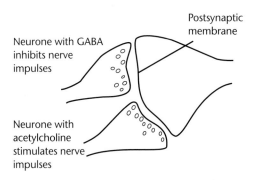

a Describe the sequence of events leading to the release of acetylcholine and its binding to the postsynaptic membrane.   (4)
b The binding of GABA to receptors on postsynaptic membranes causes negatively charged chloride ions to enter postsynaptic neurones. Explain how this will inhibit transmission of nerve impulses by postsynaptic neurones.   (3)
c Epilepsy may result when there is increased neuronal activity in the brain.
  i One form of epilepsy is due to insufficient GABA. GABA is broken down on the postsynaptic membrane by the enzyme GABA transaminase. Vigabatrin is a new drug being used to treat this form of epilepsy. The drug has a similar molecular structure to GABA. Suggest how Vigabatrin may be effective in treating this form of epilepsy.   (2)
  ii A different form of epilepsy has been linked to an abnormality in GABA receptors. Suggest and explain how an abnormality in GABA receptors may result in epilepsy.   (3)
d During an epileptic seizure muscular contractions may occur. In which part of the brain would neuronal activity produce muscular contractions of the right leg?   (2)
Total 14
AQA, June 2006, Unit 4, Question 9

3
a The amount of light entering the eye is controlled by a reflex involving the iris.
  i Where are the receptors for this reflex?   (1)
  ii Explain the role of the autonomic nervous system in the control of this reflex.   (2)

b   The human retina contains rods and cones. Describe the similarities and differences between the pigments in the rods and cones and the ways in which these pigments respond to light. (5)

c   In dogs, 10–20% of the photoreceptors in the central region of the retina are cones. Two types of cone are present. The graph shows their absorption spectra.

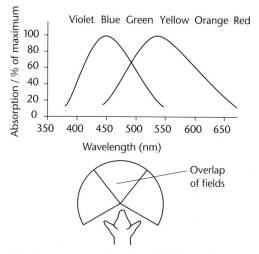

A dog's eyes are on the sides of its head. This gives the dog a visual field of about 240 degrees, with an overlap of the fields from each eye of about 60 degrees.

i   Apart from the number of types of cone, give **one** way in which the retina of a dog differs from that of a human. (1)

ii   Using the information given, suggest and explain how a dog's vision differs from human vision. (6)

Total 15

AQA, January 2005, Unit 4, Question 9

4

a   Explain how a resting potential is maintained in a neurone. (4)

b   In an investigation, an impulse was generated in a neurone using electrodes. During transmission along the neurone, an action potential was recorded at one point on the neurone. When the impulse reached the neuromuscular junction, it stimulated a muscle cell to contract. The force generated by the contraction was measured. The results are shown in the graph.

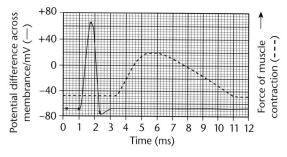

The distance between the point on the neurone where the action potential was measured and the neuromuscular junction was exactly 18 mm.

i   Use the graph to estimate the time between the maximum depolarisation and the start of contraction by the muscle cell. (1)

ii   Use your answer to part **i** to calculate the speed of transmission along this neurone to the muscle cell. Give your answer in mm per second. Show your working. (2)

iii   Give **one** reason why the value calculated in part **ii** would be an underestimate of the speed of transmission of an impulse along a neurone. (1)

Acetylcholine is the neurotransmitter at neuromuscular junctions.

c   Describe how the release of acetylcholine into a neuromuscular junction causes the cell membrane of a muscle fibre to depolarise. (3)

d   Use your knowledge of the processes occurring at a neuromuscular junction to explain each of the following.

i   The cobra is a very poisonous snake. The molecular structure of cobra toxin is similar to the molecular structure of acetylcholine. The toxin permanently prevents muscle contraction. (2)

ii   The insecticide DFP combines with the active site of the enzyme acetylcholinesterase. The muscles stay contracted until the insecticide is lost from the neuromuscular junction. (2)

Total 15

AQA, January 2004, Unit 4, Question 7

5

a   **Figure 3** shows the changes in membrane potential at one point on an axon when an action potential is generated.

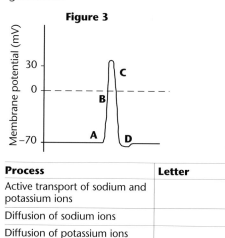

Process	Letter
Active transport of sodium and potassium ions	
Diffusion of sodium ions	
Diffusion of potassium ions	

The changes shown in **Figure 3** are due to the movement of ions across the axon membrane. Complete the table by giving the letter (**A** to **D**) that shows where each process is occurring most rapidly. (2)

b **Figure 4** shows the relationship between axon diameter, myelination and the rate of conduction of the nerve impulse in a cat (a mammal) and a lizard (a reptile).

Figure 4

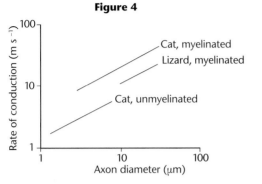

i Explain the effect of myelination on the rate of nerve impulse conduction. (2)

ii For the same diameter of axon, the graph shows that the rate of conduction of the nerve impulse in myelinated neurones in the cat is faster than that in the lizard. Suggest an explanation for this. (2)

Figure 5 shows how a stimulating electrode was used to change the potential difference across an axon membrane. Two other electrodes, **P** and **Q**, were used to record any potential difference

Figure 5

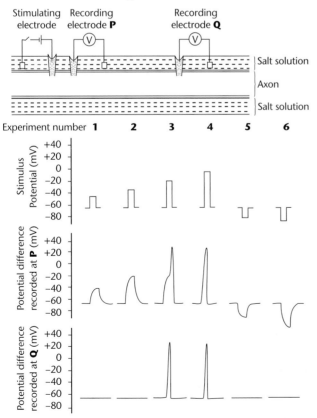

produced after stimulation. The experiment was repeated six times, using a different stimulus potential each time.

In experiments **1** to **4**, the stimulating voltage made the inside of the axon less negative.

In experiments **5** and **6**, it made the inside of the axon more negative.

Figure 6

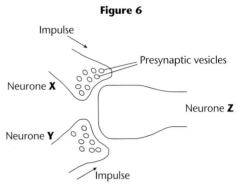

c Explain the results of experiments **1** to **4**. (5)

d **Figure 6** shows two neurones, **X** and **Y**, which each have a synapse with neurone **Z**.

Neurone **X** releases acetylcholine from its presynaptic vesicles. Neurone **Y** releases a different neurotransmitter substance which allows chloride ions (Cl^-) to enter neurone **Z**. Use this information, and information from **Figure 5**, to explain how neurones **X** and **Y** have an antagonistic effect on neurone **Z**. (4)

Total 15

AQA, June 2006, Unit 6, Question 8

6

a The table shows the membrane potential of an axon at rest and during the different phases of an action potential. Complete the table by writing in each box whether the sodium ion (Na^+) channels and potassium ion (K^+) channels are open or closed.

	Resting	**Starting to depolarise**	**Repolarising**
Membrane potential/mV	–70	–50	–20
Na^+ channels in axon membrane			
K^+ channels in axon membrane			

b Describe how the resting potential is established in an axon by the movement of ions across the membrane. (2)

c Sodium and potassium ions can only cross the axon membrane through proteins.

Explain why. (2)

Total 4

AQA, June 2005, Unit 6, Question 4

how science works assignment

Illegal drug use

Cocaine is a class A drug. Possession of class A drugs can carry sentences of up to seven years in prison or an unlimited fine. Judges can sentence class A dealers to life in prison or give them an unlimited fine.

The 'pleasure' people get from taking cocaine is caused by the accumulation of a substance called dopamine in synapses in a 'reward' centre in the brain.

Fig. 19 shows the normal way in which dopamine acts at a synapse in a reward centre in the brain.

A1 Use information from Fig. 19 to describe the cycling of dopamine at a synapse.

A2 Cocaine blocks the dopamine re-uptake transporter. Suggest how this will affect the working of the synapse.

Fig. 19 A dopamine-activated synapse in the brain

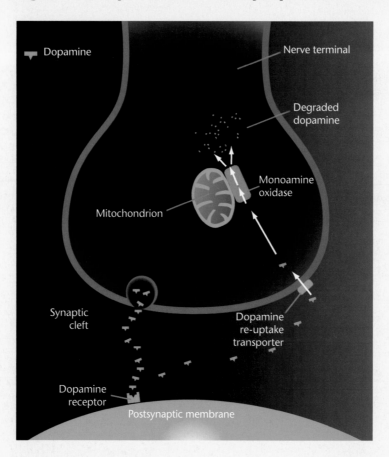

The British Crime Survey (BCS) is a large national survey of adults who live in a representative cross-section of private households in England and Wales. In addition to asking respondents about their experiences of crime, the BCS also asks about a number of other crime-related topics. Since 1996, the BCS has included a comparable module of questions on illegal drug use. This examines the prevalence and trends of illegal drug use among 16–59 year-olds.

Table 1 below shows trends in the use of drugs between 1998 and 2006.

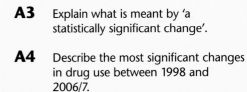

A3 Explain what is meant by 'a statistically significant change'.

A4 Describe the most significant changes in drug use between 1998 and 2006/7.

A5 What conclusions can be drawn from the data about the relationship between illegal drug use and age?

Table 1   Trends in the use of drugs between 1998 and 2006

Drug	Proportion (%) of people in England and Wales reporting to having used drugs in the last year (British Crime Survey)			
	16–24 year olds		16–59 year olds	
	1998	2006/07	1998	2006/07
Cocaine	3.1	6.0	1.2	2.6↑
Crack	0.3	0.4	0.1	0.2
Ecstasy	5.1	4.8	1.5	1.8
LSD	3.2	0.7	0.8	0.2↓
Magic mushrooms	3.9	1.7	0.9	0.6
Heroin	0.3	0.2	0.1	0.1
Methadone	0.6	0.1	0.1	0.1
Amphetamines	9.9	3.5	3	1.3
Tranquillisers	1.5	0.6	0.7	0.4
Anabolic steroids	0.5	0.2	0.3	0.1
Cannabis	28.2	20.9 ↓	10.3	8.2
Ketamine*	n/a	0.8	n/a	0.3
Amyl nitrate**	5.1	4.2	1.5	–1.4
Glues**	1.3	0.6	0.2	0.2
Class A	**8.6**	**8**	**2.7**	**3.4**
Any drug	**31.8**	**24.1↓**	**12.1**	**10.0↓**

↑↓ Statistically significant change 1998 to 2006/07
*   Only included in BCS since 2006/07
**   Not an illegal drug

Many criminologists and social workers complain that the BCS data is very inaccurate. Evidence for this is provided by a Home Office report. Instead of asking people what drugs they have used, an estimate of the number of opiate, crack, cocaine and injecting users was extrapolated from drug treatment, probation, police and prison data. The figures are over six times higher than those from the BCS covering the same period.

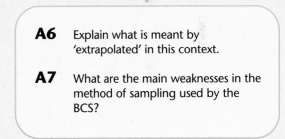

A6 Explain what is meant by 'extrapolated' in this context.

A7 What are the main weaknesses in the method of sampling used by the BCS?

11 Muscle power

The 2007 Tour de France was ruined by the abuse of performance drugs by the competitors.

Athletes are always looking for new and quicker ways to build muscle. While anabolic steroids may have been the thing in the 1980s, the 'hormonal manipulator' in the 21st century is human growth hormone (hGH). Some athletes are using this to build bigger muscles without extra training, but at the price of a variety of side effects.

HGH is a polypeptide produced by the pituitary gland. HGH affects many body tissues. In adolescents it stimulates growth and the maturation of the bones. There is increased protein synthesis in muscle cells. In addition, hGH stimulates the intracellular breakdown of body fat so that more fat is used for energy. The synthesis of collagen (the sticky substance that is the glue of the body) is stimulated, which is necessary for strengthening of cartilage, bones, tendons and ligaments.

Some individuals naturally produce too much hGH, which can have two effects – gigantism and acromegaly. Gigantism results in an abnormally large physical stature; acromegaly leads to enlarged hands, toes, nose and face. Athletes using hGH risk considerable side effects!

The hormone can now be produced synthetically by genetic manufacturing techniques, resulting in its widespread use among athletes who yearn for an extra competitive edge.

In this chapter we consider the structure of muscle fibres, the mechanism of muscle contraction and how muscle contraction moves parts of the body.

Competing in the Tour de France involves the extreme use of muscles. David Millar in the UK stage of the 2007 Tour de France.

11.1 Skeletal muscle

There are three types of muscle tissue in the body: **involuntary muscle** – also called **unstriped muscle**, **cardiac muscle**, and **voluntary muscle** – also called **striped muscle** or **skeletal muscle**. Unstriped muscle brings about peristalsis in the gut and is also found in the iris of the eye. The muscles attached to the skeleton are voluntary muscles – you can decide whether or not to lift your arm, but you have no control over your iris.

The structure of skeletal muscle

Skeletal muscles are made up of overlapping striped muscle fibres. These are held together by **connective tissue**, with a **tendon** at each end. The tendons attach the muscle to bones. When the fibres contract, the muscle shortens, pulls the tendons and moves the bones.

The stripes on skeletal muscle fibres can be seen through a light microscope, as Fig. 1 shows.

Striped muscles are very different from the rest of the cells in the body. Like other cells, a striped muscle fibre has cytoplasm, called **sarcoplasm**, and is surrounded by a plasma membrane, called a **sarcolemma**. However, there is an important difference: striped muscle fibres have many nuclei in the cytoplasm, which lie near the surface of the fibre.

The central part of a striped muscle fibre consists mainly of much thinner fibre-like structures called **myofibrils**, which are composed mainly of protein molecules. It is these myofibrils that transfer chemical energy from food into movement.

Fig. 1   Structure of skeletal muscle

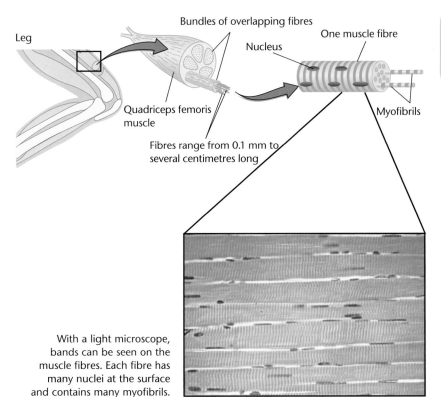

With a light microscope, bands can be seen on the muscle fibres. Each fibre has many nuclei at the surface and contains many myofibrils.

1 List the main differences between a striped muscle fibre and a typical animal cell.

Ultrastructure of skeletal muscle

Although it is possible to see the stripes in skeletal muscle using a light microscope, it is only when we look at muscle fibres using an electron microscope that it becomes clear what the stripes actually are. Each myofibril is made up of protein filaments of two types – thick filaments composed of the protein **myosin** and thin filaments composed of the protein **actin**.

Fig. 2 shows the structure of a myofibril as determined by early electron microscopes. The light bands consist of actin filaments only. The dark bands consist of overlapping actin and myosin filaments. Discs hold the groups of actin filaments together.

Fig. 2   The myofibril

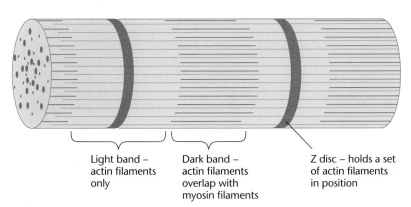

Light band – actin filaments only

Dark band – actin filaments overlap with myosin filaments

Z disc – holds a set of actin filaments in position

This false-colour electron micrograph shows that the bands on the myofibrils are lined up to give a banded appearance to the whole fibre.

As the resolving power of electron microscopes increases, more detail can be seen in the myofibril:

- Where the two sets of filaments overlap the myofibril appears dark.
- The region where there are only myosin filaments is slightly lighter.
- The region where there are only actin filaments appears lighter still.
- In the centre of each light band is the Z disc, which holds actin filaments in position. Similarly, M discs hold myosin filaments in position.

On this false-colour electron micrograph, the straight green lines are the Z discs, the myosin filaments are orange-pink and the actin filaments are blue. The green bubble-like structures are sarcoplasmic reticula.

Fig. 3 Structure of striped muscle as revealed by electron microscopy

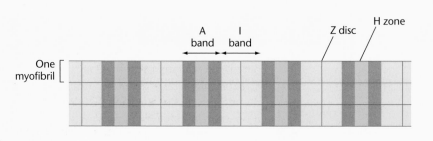

2 Scientists use the letters A, H, I and Z to identify the different parts of a myofibril. Look at Fig. 3.

a Why do the A bands look dark?

b What is found:

   i in the H zone?

   ii in the I band?

s&c

stretch and challenge

Delayed-onset muscle soreness (DOMS)

Having trouble getting out of bed? Having to walk backwards down the stairs because your legs feel as though they are on fire? You are probably suffering from DOMS.

Every person, regardless of his or her fitness level, has experienced sore and stiff muscles after unaccustomed moderate or strenuous exercise. These symptoms usually occur after 24–48 hours. During these activities your muscles become more susceptible to structural damage, resulting in muscle soreness. This is DOMS.

DOMS is not a new phenomenon. Research investigating the cause of DOMS dates back to the early 1900s, and several theories have been put forward to explain the underlying cause. Probably one of the most popular explanations is the accumulation of lactic acid in the muscles. Blood and muscle lactate levels typically return to normal values after 30–60 minutes of recovery.

Eccentric exercise, such as running downhill, produces the most severe muscle soreness but requires relatively low energy expenditure. We know from bitter experience that the pain associated with DOMS peaks after 24–72 hours.

Subsequent research has suggested that damage to the muscle ultrastructure and connective tissue may be responsible for DOMS. It is suggested that a sequence of events starting with exercise causes muscle damage and then breakdown of muscle protein, resulting in cell inflammation and increased local muscle temperature. As a result, pain receptors are activated, causing the sensation of DOMS. Further research suggests that muscle damage alone may not be the best explanation for the cause of DOMS. Inflammation and swelling should also be considered as they also activate and sensitise pain sensors around the muscle fibres.

3 Most scientists now discount the accumulation of lactic acid as the cause of DOMS. Use evidence from the passage to explain why.

4 Relate the symptoms of DOMS to the ultrastructure of muscle.

11.2 Muscle contraction

Now that you know the structure of skeletal muscle it is possible to go on to look at how muscle fibres are able to contract to bring about movement. When a myofibril contracts:

- The sarcomeres (lengths of muscle fibres between Z discs) become shorter.
- The light bands become shorter.
- The dark bands stay the same length.

In the 1950s, Jean Hanson and Hugh Huxley at London University put forward the **sliding filament theory of muscle contraction,** which suggested that muscular contraction comes about by the actin filaments sliding between the myosin filaments, using energy from ATP. This theory is still accepted. In the region of overlap, six actin filaments are arranged neatly around each myosin filament, as Fig. 4 shows.

Fig. 4   The sliding filament theory of muscle contraction

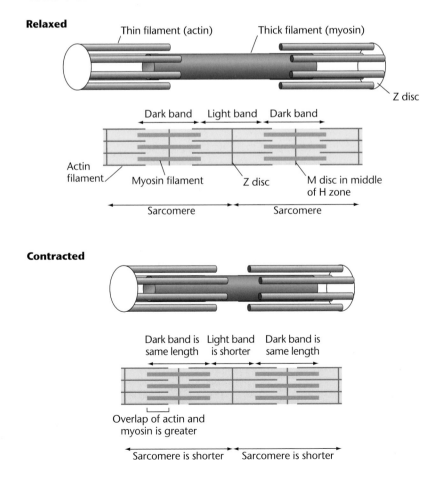

Relaxed

Thin filament (actin)   Thick filament (myosin)

Z disc

Dark band   Light band   Dark band

Actin filament   Myosin filament   Z disc   M disc in middle of H zone

Sarcomere   Sarcomere

Contracted

Dark band is same length   Light band is shorter   Dark band is same length

Overlap of actin and myosin is greater

Sarcomere is shorter   Sarcomere is shorter

5

a Do the actin filaments or the myosin filaments change their length when a sarcomere shortens?

b Why does the A band widen and the H zone shorten when a sarcomere contracts?

Electron micrographs of the dark bands show cross-bridges between the myosin and actin filaments. These bridges are part of the myosin molecules, and push on the actin filaments to make the myofibril shorten. This is called the **ratchet mechanism** because the actin molecules are moved along one step at a time by the myosin heads. Fig. 5 overleaf shows how this mechanism operates.

Myosin cross-bridges point in six different directions and are arranged in the form of a spiral. Each myosin cross-bridge has a wider 'head'. This head engages in a 'binding site' on the actin filament. The myosin cross-bridge then performs its 'power stroke' – it moves rather like an oar, remaining stiff and therefore moving the actin filament. The myosin head then disengages and moves through a 'recovery stroke' to engage with the next binding site, rather like returning an oar to the starting position. The cycle is then repeated. Because of their spiral arrangement, cross-bridges will be at different positions in the power stroke/recovery stroke cycle, so muscle contraction is smooth rather than jerky.

The role of tropomyosin, calcium ions and ATP in muscle contraction

Muscle contraction is turned on and off by a calcium switch. When a nerve impulse arrives at the muscle fibre, calcium channels open and calcium ions diffuse in. This allows actin and myosin to bind. Each myosin cross-bridge can bind to, push against and release an actin filament many times a second. Each swing of a cross-bridge uses the energy from one

Fig. 5 The ratchet mechanism

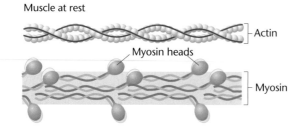

Muscle at rest

Actin

Myosin heads

Myosin

A nerve impulse reaches the muscle fibre, releasing calcium ions. This causes myosin binding sites in the actin myofilament to be exposed. The myosin heads use the energy gained from ATP hydrolysis to move towards the binding sites

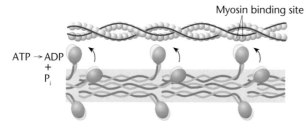

Myosin binding site

ATP → ADP
+
P$_i$

As the myosin heads bind to the actin myofilament they tilt, pulling the actin myofilament past them

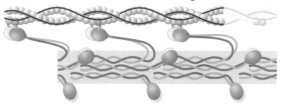

← Sliding movement of actin filament

As the actin filament moves, the myosin heads detach and re-attach to the next binding site

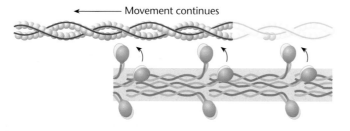

← Movement continues

molecule of ATP. Many thousands of cross-bridges working together create the power of the muscle.

Why don't cross-bridges form in resting muscle? Although the actin filaments have myosin binding sites, these are blocked by protein molecules called tropomyosin, as shown in Fig. 6.

When calcium ions bind to tropomyosin molecules, this alters the shape of the protein and it can no longer block the binding site, so the fibre is ready to start the process of muscle contraction. The calcium ions also activate the myosin molecules to break down ATP which releases the energy needed to bring about contraction.

Fig. 6 Tropomyosin, calcium ions and ATP

Transverse section

Myosin cross-bridges point in six directions, in a spiral

Actin

Myosin cross-bridge

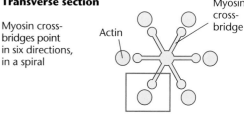

Calcium switch

Ca^{2+} ions bind to the tropomyosin switch protein and move it away from the binding site so the myosin can bind to the actin and make a power stroke

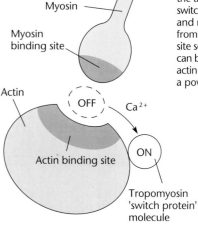

Myosin

Myosin binding site

Actin

OFF

Ca^{2+}

ON

Actin binding site

Tropomyosin 'switch protein' molecule

The neuromuscular junction

In muscles, calcium ions are stored in the endoplasmic reticulum of the cytoplasm that surrounds each myofibril. When the muscle is resting, calcium ions are pumped into the endoplasmic reticulum by active transport. The arrival of an action potential at a neuromuscular junction sets off the chain of events leading to the contraction of myofibrils, shown in Fig. 7.

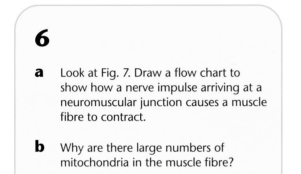

6

a Look at Fig. 7. Draw a flow chart to show how a nerve impulse arriving at a neuromuscular junction causes a muscle fibre to contract.

b Why are there large numbers of mitochondria in the muscle fibre?

Fig. 7   The neuromuscular junction

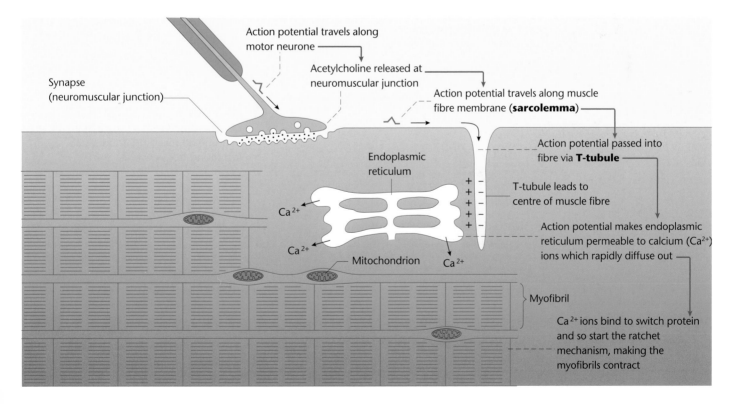

Controlling muscle contraction

Fig. 8   Controlling muscle contraction

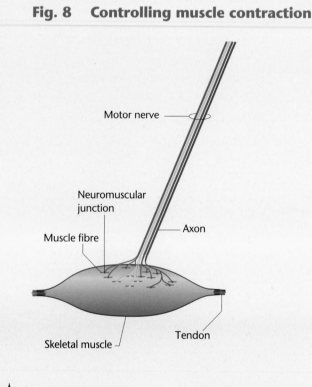

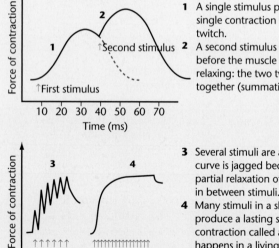

1   A single stimulus produces a single contraction called a twitch.
2   A second stimulus is applied before the muscle has finished relaxing: the two twitches add together (summation).

3   Several stimuli are applied. The curve is jagged because of partial relaxation of the muscle in between stimuli.
4   Many stimuli in a short time produce a lasting smooth contraction called a tetanus. This happens in a living muscle when nerve impulses arrive rapidly at the neuromuscular junction.

Source: Tortora and Grabowski, *Principles of Anatomy and Physiology*, Pearson, 1993

In order to control movement, the brain can control how strongly a muscle contracts. It does this by changing:

- how much each fibre contracts
- how many fibres contract.

When one action potential arrives at a neuromuscular junction, it causes one brief contraction or twitch of the fibre. If a second action potential arrives before the fibre has fully relaxed, the second twitch adds to the effect of the first, and a greater contraction occurs. The 'adding' of the effects of action potentials is called summation.

7 In Chapter 10, you learned that two types of summation can occur at neurone–neurone synapses: spatial summation and temporal summation. Which of these types of summation is occurring in a muscle fibre to cause a stonger contraction?

A rapid sequence of action potentials causes a continuous strong contraction of the muscle fibre called a tetanus. The more nerve impulses there are per second in a motor nerve, the stronger the contraction of each muscle fibre will be.

8 What are the main differences in behaviour between a muscle fibre and a neurone on receiving impulses at increasing frequencies?

Each motor neurone serves about 150 muscle fibres. One motor neurone and its associated muscle fibres are together called a motor unit. Action potentials in a motor neurone cause all the muscle fibres in its motor unit to contract together. If a stronger contraction is needed, the brain recruits more motor units by sending action potentials along more motor neurones within the motor nerve.

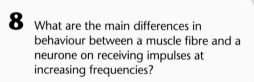

9 What type of summation is this?

- There are three types of muscle tissue in the body: voluntary (skeletal) muscle, which we use mainly to move our limbs; cardiac muscle, which makes up most of the heart; and involuntary (unstriped) muscle, which is found in organs such as the intestines, ureters and blood vessels.

- Skeletal muscle is composed of bundles of striped fibres, which have features that distinguish them from typical body cells.

- Each fibre contains thick myosin filaments and thinner actin filaments.

- The sliding filament theory of muscle contraction states that the contraction of myofibrils is brought about by a ratchet mechanism, in which processes from the myosin filaments bind to the actin filaments and move the actin filaments.

- The sequence of events causing contraction of myofibrils is:

  ○ The nerve impulse arrives at the neuromuscular junction.

  ○ Acetylcholine is released into the synaptic cleft.

  ○ An action potential is produced in the plasma membrane.

  ○ The action potential is transmitted down T-tubules.

  ○ Calcium ions are released from the endoplasmic reticulum.

  ○ Calcium ions bind to tropomyosin switch proteins.

  ○ Binding sites on the actin filaments are exposed.

  ○ Myosin processes bind to exposed sites on the actin filaments.

  ○ Energy from ATP is used to move actin filaments.

11.3 Muscles as effectors

ATP and phosphocreatine

Although ATP is the energy currency for muscle cells, only about 85 grams of ATP is stored in the body at any one time. This provides only enough energy for performing intense exercise for several seconds. ATP must therefore be constantly resynthesised to supply energy for muscle contraction. Some energy for ATP resynthesis is supplied directly and rapidly by the splitting of a phosphate molecule from another energy-rich compound called **phosphocreatine**, or creatine phosphate (CP). CP is similar to ATP because a large amount of energy is released when the bond is split between its creatine and phosphate molecules. Resynthesis of ATP occurs if sufficient energy is available to rejoin an ADP molecule with one phosphate (P) molecule. The breakdown of CP can supply this energy.

Cells store CP in much larger quantities than ATP. The mobilisation of CP for energy is almost instantaneous, so CP is known as a 'reservoir' of high-energy phosphate bonds. A low concentration of ADP in a cell stimulates the activity of creatine kinase, the enzyme that breaks down CP.

$$\text{Creatine kinase}$$
$$\downarrow$$
$$CP \rightarrow C + P + energy$$
$$\downarrow$$
$$ADP + P \rightarrow ATP$$

The energy released from the breakdown of ATP and CP can sustain all-out exercise, such as running, for five to eight seconds. In the 100-metre race, the body cannot maintain maximum speed for longer than this duration. During the last few seconds of a race the runners are actually slowing down. The winner is usually the one who slows down least!

10 Suggest four sports in which short intense bursts of energy are essential.

Fast-twitch and slow-twitch muscle fibres

Have you ever sat down for Christmas dinner and found yourself wondering why turkeys have some dark meat and some white meat? Well, you were not the first! A scientist named Ranvier reported differences in muscle colour in 1873. The explanation for the colour differences has a basis in physiology. The dark meat of the turkey is 'red' or **slow-twitch** muscle. The white meat is 'white' or **fast-twitch** muscle. Most animals have some combination of these two fibre types, though the distinctions may be less obvious.

Muscle physiologists use biopsies to study the make-up of muscles from athletes who compete in different types of events. A biopsy involves using a special type of needle to remove about 50 mg of living muscle tissue from a muscle. Studying samples has shown that all muscles contain two types of fibres, fast-twitch and slow-twitch.

Fast-twitch fibres can produce ATP very quickly via glycolysis. Their contraction speed is very rapid. Fast-twitch fibres are used mainly for sprinting and jumping or escaping from danger.

Slow-twitch fibres contract at about half the speed of fast-twitch fibres. These slow-twitch fibres contain more mitochondria than fast-twitch fibres, so they produce ATP mainly via aerobic respiration. They are used mainly during walking or jogging. Middle-distance running, and sports like football and hockey use both types.

Table 1 below shows some of the properties of fast-twitch and slow-twitch fibres. Most muscles contain a mixture of the two sorts of fibre, but people differ in their ratios of the two. The ratio is controlled by genes. Exercise increases the number of both types of fibre in muscles, but in the same ratio. So, you are born either a sprinter or a long-distance runner.

11 You will need to revise the information about respiration in Chapter 3 before answering these questions.

a Apart from muscles, where is most of the body's glycogen stored?

b Where are triglycerides stored in the body?

c **i** What is the advantage to a sprinter of having phosphocreatine as the main energy source in muscles?

ii What is the main disadvantage?

d What is the end product of anaerobic respiration?

e Explain how lactic acid contributes to muscle fatigue.

f What is the advantage to slow-twitch fibres in having abundant mitochondria?

Table 1   Features of slow-twitch and fast-twitch muscles

Type	Speed of contraction	ATP source	Respiration	Fatigue	Motor unit
Slow-twitch	Slow	Glycogen and triglyceride	Aerobic	Slow	Small
Fast-twitch	Fast	Phosphocreatine	Anaerobic	Rapid	Large

key facts

- Phosphocreatine is an accessory energy store in muscle cells.
- Energy from the breakdown of phosphocreatine is used by muscle cells to synthesise ATP from ADP and phosphate.
- Slow-twitch muscle fibres derive most of their energy from aerobic respiration, are slow to fatigue and are most useful in long-lasting exercise.
- Fast-twitch muscle fibres derive most of their energy from phosphocreatine, are quick to fatigue and are most useful during short bursts of intense activity.

1 The diagram shows part of a myofibril from a relaxed muscle fibre.

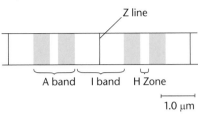

a When the muscle fibre contracts, which of the A band, I band and H zone:
  i remain unchanged in length (1)
  ii decrease in length? (1)
b Explain what caused the decrease in length in part **a ii**. (2)
c The whole muscle fibre is 30 mm long when relaxed. Each sarcomere is 2.25 μm long when contracted. Use the scale given on the diagram to calculate the length of the contracted muscle fibre in millimetres. (2)
d The table gives some properties of the two different types of muscle fibre found in skeletal muscle.
  i Complete the table by writing the words 'high' or 'low' for the remaining three properties of each type of muscle fibre.

	Type of muscle fibre	
	Type 1	Type 2
Speed of contraction	high	low
Force generated	high	low
Activity of the enzymes of glycolysis	high	low
Number of mitochondria		
Activity of Krebs cycle enzymes		
Rate of fatigue		

(3)

  ii The myosin-ATPase of **Type 1** muscle fibres has a faster rate of reaction than that in **Type 2** fibres. Use your knowledge of the mechanism of muscle contraction to explain how this will help **Type 1** muscle fibres to contract faster than **Type 2**. (4)
  iii The blood leaving an active muscle with a high percentage of **Type 1** muscle fibres contained a higher concentration of lactate than that leaving a muscle with a high percentage of **Type 2** muscle fibres. Explain why. (2)

Total 15

AQA, A, June 2006, Unit 7, Question 7

2 **Figure 1** shows part of a single myofibril from a skeletal muscle fibre as it appears under an optical microscope.

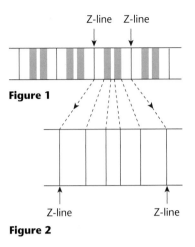

Figure 1

Figure 2

a i Complete **Figure 2** to show the arrangement of actin and myosin filaments in this part of the myofibril as they would appear under an electron microscope. Label the actin and myosin filaments. (2)
  ii Why are the details you have drawn in **Figure 2** visible under the electron microscope but not under the optical microscope? (1)
b The myofibril in **Figure 1** is magnified x8000. A muscle fibre is 40 μm in diameter. Calculate the number of myofibrils which would fit side by side across the diameter of the muscle fibre. Show your working. (2)

Total 5

AQA, A, January 2005, Unit 7, Question 2

3 **Figure 1** shows part of a sarcomere.

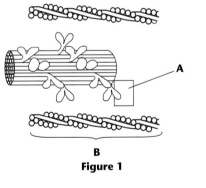

Figure 1

a i Name the main protein in structure **B**. (1)
  ii Name the structure in box **A**. (1)
b i Describe how calcium ions cause the myofibril to start contracting. (2)
  ii Describe the events that occur within a myofibril which enable it to contract. (3)

Slow-twitch and fast-twitch skeletal muscle fibres differ in a number of ways. Slow twitch muscle fibres get their ATP from aerobic respiration while anaerobic respiration provides fast-twitch muscle fibres with their ATP. **Figure 2** shows a bundle of fast-twitch and slow-twitch muscle fibres as seen through an optical microscope. The muscle fibres have been stained with a stain that binds to the enzymes which operate in the electron transport chain.

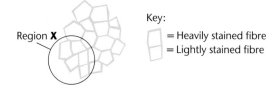

Figure 2

c i Describe how you could calculate the percentage of fast-twitch muscle fibres in this bundle. (1)

ii The figure calculated by the method in part **c i** may not be true for the muscle as a whole.
Explain why. (1)

d The muscle fibres in **Figure 3** correspond to those in region **X** of **Figure 2**. They were stained with a substance that binds to enzymes involved in glycolysis. Shade **Figure 3** to show the appearance of the muscle fibres. Use the shading shown in the key. (2)

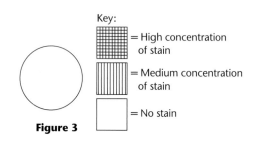

Figure 3

e Recent research has hown that the difference in muscle fibre types is due in part to the presence of different forms of the protein myosin with different molecular shapes. Explain how a new form of myosin with different properties could have been produced as a result of mutation. (4)

Total 15

AQA, A, January 2004, Unit 7, Question 7

4 The diagram shows the arrangement of some of the proteins in a myofibril from a skeletal muscle. The myofibril is shown in the relaxed state.

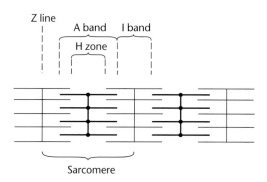

Sarcomere

a Name the protein found in the H zone. (1)

b When the muscle contracts, what happens to the width of:
i the A band
ii the I band? (2)

c The distance between two Z lines in a myofibril is 1.6 μm. Calculate the magnification of the diagram. Show your working. (2)

Total 5

AQA, A, January 2003, Unit 7, Question 3

5 Skeletal muscle is made of bundles of fibres.

a Describe the roles of calcium ions, ATP and phosphocreatine in producing contraction of a muscle fibre. (4)

b The table shows some properties of slow-twitch and fast-twitch muscle fibres.

Property of muscle fibre	Type I (slow-twitch fibres)	Type II (fast-twitch fibres)
Number of mitochondria per fibre	Many	Few
Concentration of enzymes regulating glycolysis	Moderate	High
Resistance to fatigue	High	Low

Endurance athletes, such as marathon runners, nearly always have a high proportion of slow-twitch fibres in their muscles. Explain the benefit of this. (6)

c During exercise, much heat is generated. Describe the homeostatic mechanisms that restore normal body temperature following vigorous exercise. (5)

Total 15

AQA, A, June 2002, Unit 7, Question 9

Can muscle fibre types be changed?

If you want to win an Olympic medal in the 100 metres, you had better be born with about 80% fast-twitch fibres! Want to win the Olympic marathon? Put in an order for 80% slow-twitch fibres in your calf muscles!

Table 2 shows the percentage of slow-twitch fibres in three muscle groups in male athletes.

Table 2   Percentage of slow-twitch fibres in three muscle groups in male athletes

Event	Percentage of slow-twitch fibres in muscle group		
	Shoulder muscles	Calf muscles	Thigh muscles
Long-distance runner		79	
Canoeist	71		
Triathlete	60	59	63
Swimmer	67		
Sprinter		24	
Cyclist			57
Weight lifter	53	44	
Shot putter		38	

A1 Do the data in the table above support the hypothesis that there is a relationship between event and the ratio of slow-twitch fibres to fast-twitch fibres in the relevant muscle group? Use data from the table to support your answer.

Consider the two pieces of evidence, **A** and **B** below.

A Six men participated in a five-month programme of aerobic bicycle training. Muscle biopsy specimens taken from the calf muscles before and after training indicated no change in fibre composition, although all the men improved considerably in work capacity and aerobic power.

B In another study of four men who took part in 18 weeks of aerobic and 11 weeks of anaerobic training, the anaerobic training caused an increase in the percentage of fast-twitch fibres and a decrease in the percentage of slow-twitch fibres.

A2 Evaluate the contribution of the evidence from A and B in supporting the hypothesis that training can affect the proportions of fast-twitch and slow-twitch fibres in muscle.

Table 3 at the bottom of the page shows the results of a survey into the effects of types of training on skeletal muscle.

A3 From the information in the table below, evaluate the status of the evidence supporting the hypothesis that training influences muscle fibre properties.

Table 3   Results of effects of types of training on skeletal muscle

Muscle property	Slow-twitch muscle fibres		Fast-twitch muscle fibres	
	Anaerobic training	Aerobic training	Anaerobic training	Aerobic training
Percentage composition	0 or ?	0 or ?	0 or ?	0 or ?
Size	+	0 or ?	++	0
Oxidative capacity	0	++	0	+
Anaerobic capacity	? or +	0	? or +	0
Glycogen content	0	++	0	++

  0 = no change
  ? = unknown
  + = moderate increase
++ = large increase

12 Hormonal control

In ancient Egypt, diagnosing pregnancy involved the woman drinking a mixture of pulped watermelon and breast milk from a mother who had borne a son. Unfortunately, a positive result was indicated by a period of violent sickness. Another, less harrowing, idea of the ancient Egyptians was to pour a woman's urine over corn seeds to see if they would germinate.

The first reliable biological tests were introduced just 60 years ago. It was discovered that when urine from pregnant women was injected into certain animals, it stimulated them to produce eggs. At first mice were used, but it took five days to produce a result. Rabbits were quicker and gave the result in 24–48 hours, but the Hogben test, which used a female South African clawed toad, worked within just a few hours. The Hogben test was carried out in some specialist laboratories until the 1960s. During the 1980s, reliable home pregnancy tests were developed, in which an immunological reaction was detected by a colour change. These tests enabled the detection of pregnancy from the first day of the missed period. The biological tests all depend on the presence of a hormone in the urine.

Source: adapted from Lesley Foster, Unipath Ltd. Pregnancy Testing. NCBE Newsletter Winter 1990

This woman is delighted by the results of her pregnancy test.

It is difficult to imagine a world without hormones. There would be no oral contraceptive pills and no babies born by IVF (*in vitro* fertilisation). People with type 1 diabetes would still die because there would be no insulin therapy to treat them.

It was Ernest Starling who first coined the term 'hormone' in 1905. He was working at Cambridge University and needed a word to describe an agent released into the bloodstream that caused activity in a different part of the body. It is thought that a colleague who was an authority on Greek poetry suggested the Greek verb for 'excite' or 'arouse', and the deed was done.

Scientists had been aware of such chemicals earlier than this. A French doctor called Brown-Sequard believed extract of testicles had a rejuvenating effect in men and tested it on himself. George Oliver, a physician working in Harrogate in 1893, believed extracts of the adrenal glands might raise low blood pressure and used his son as a guinea pig.

Since then, more than 30 different hormones have been discovered and have changed the course of medicine. The most dramatic discovery was insulin. Years ago, if you developed type 1 diabetes you would die within a few weeks.

But there is a darker side to some of the discoveries. In the 1950s and 60s, thousands of children born with a condition that meant they would never reach adult height were given injections of extracts of human pituitary glands from dead bodies. We now know that some of those bodies had the human form of mad cow disease, CJD, so a small but significant percentage of those children died. Since sex hormones like testosterone were found, athletes have abused these steroids to out-perform opponents. The animals we eat are also fed hormones to make them plumper and meatier. Female sex hormones are entering our water supplies via the urine of the millions of women using the contraceptive pill around the world. It is thought that this may be changing the sex of male fish and potentially affecting human fertility, although other research suggests that hormone-like compounds from plastics may be responsible.

In this chapter we will consider the principles of hormonal control in both humans and plants.

12.1 Chemical controllers

Many cells produce chemicals that affect the activity of other cells. There are two main groups of these:

- **chemical mediators**
- **hormones**.

Chemical mediators

Have you ever been 'stung' by nettles? Then you will have seen a rash similar to that in the photograph below.

The rash is partly caused by a chemical called **histamine** present in the nettle leaves.

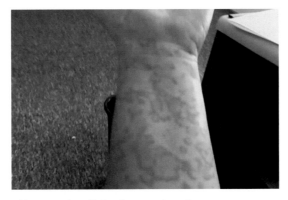

This person is suffering from nettle rash.

Histamine is a biologically active substance found in a great variety of living organisms. It is distributed widely throughout the animal kingdom and is present in many plants and bacteria and in insect venom.

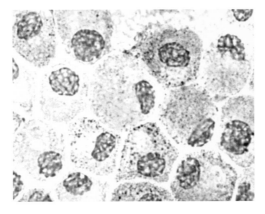

A mast cell containing histamine granules.

In humans, histamine is found in nearly all tissues of the body, where it is stored primarily in the granules of tissue **mast cells**.

Once released from its granules, histamine produces many effects within the body, including:

- the contraction of smooth muscle tissues of the lungs, uterus and stomach
- the dilation of blood vessels
- the acceleration of heart rate
- acting as a neurotransmitter, carrying chemical information between nerve cells.

Injured tissue mast cells release histamine, causing the surrounding blood vessels to dilate and increase in permeability. This allows fluid and cells of the immune system, such as leucocytes and blood plasma proteins, to leak from the bloodstream through the vessel walls into the surrounding tissue, where they begin to destroy pathogens and nourish and heal the injured tissues.

The effects of histamine are usually localised to a small area of the body. For this reason histamine is known as a **chemical mediator**.

Many substances cause allergic responses in the body. One of the most common allergic responses is hay fever, mostly caused by pollen grains from grass.

Histamine works by binding to histamine receptors on the surface of cells. The activity of histamine can be blocked by various drugs called antihistamines, which prevent the binding of histamine to these receptors.

1 What does this suggest about the molecular shape of antihistamines?

Another group of chemicals that act as chemical mediators are the **prostaglandins**. First discovered in semen during the 1930s by Swedish scientist Ulf von Euler, prostaglandins received their name from the prostate gland, where they were first thought to originate. Early studies of prostaglandins revealed that these chemicals were capable both of lowering blood pressure and instigating the contraction of uterine tissue. We now know that prostaglandins are found in most animal tissues. Prostaglandins are only active for a short period of time before they are broken down and excreted from the body. Thus, tissues cannot store prostaglandins, but must instead synthesise them when needed. Synthesis takes place in the membranes of various cells.

Recent research has shown that prostaglandins play an important role in blood clotting, essentially controlling whether or not coagulation occurs. In addition, many prostaglandins induce inflammation.

Hormones

Hormones are chemical messengers that affect almost all aspects of the functioning of the body. They regulate growth, metabolism and reproduction. Hormones are produced by **endocrine glands** – glands that do not have a duct; they pass their secretions into the blood stream for circulation. As a result, the actions of hormones are often widespread, but occur more slowly than reflex actions. However, the effects of hormones are usually more long-lasting than a nervous response such as blinking.

Hormones are usually steroids, polypeptides or lipids.

The major function of hormones is to alter the rates of specific cellular reactions of specific '**target cells**'. They accomplish this by:

- altering the rate of intracellular protein synthesis
- changing the rate of enzyme activity
- modifying cell membrane transport
- inducing secretory activity.

A target cell's ability to respond to a hormone depends mainly on the presence of specific receptors on its outer membrane. Some hormones affect most cells in the body whereas others only affect cells in one specific organ.

> **2** Suggest an explanation for the difference in the range of cells affected by different hormones.

Endocrine glands can be stimulated to produce hormones in three ways:

- by hormones produced by other endocrine glands
- by changes in the concentrations of ions or nutrients in the blood and other body fluids
- by nerve fibres of the autonomic nervous system.

key facts

- Chemical mediators, including histamine and prostaglandins, affect only cells in the immediate vicinity.
- Hormones are produced by endocrine glands and are distributed by the bloodstream to their target cells.

- This results in hormone responses being slow, long lasting and widespread.
- In comparison, nervous responses are rapid, short-lived and localised.

12.2 Feedback systems

Negative feedback

Many of the processes in the body are controlled by feedback systems.

The classic example of a feedback system is a thermostat connected to a boiler. The thermostat compares the temperature of the air in the room to a pre-determined setting. If the temperature in the room is below that setting, the thermostat signals the boiler to send warm water to the radiators. After the boiler has been

running for a while and the air in the room becomes warmer than the desired temperature, the thermostat stops signalling to the boiler. The boiler switches off. If it is cold outside the room, however, eventually the air in the room will grow cool again. When it gets cool enough to be detected by the thermostat as being different from the ideal setting, the boiler switches on again. The cycle repeats indefinitely, as information about the system's status is constantly *fed back* into the system. Because of their cyclical nature, **feedback systems** are often referred to as feedback *loops*.

Fig. 1 shows how most processes in the body are controlled. The response reverses the effect of the original factor increase. This effect is called **negative feedback**.

In the next chapter you will study homeostatic mechanisms in the body such as temperature regulation. These mechanisms are controlled by negative feedback.

Fig. 1   Negative feedback

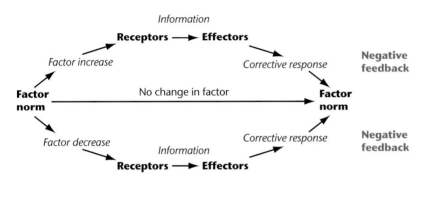

3 From your GCSE studies, suggest one other biological system that is controlled by negative feedback.

Positive feedback

Microphone feedback is that high-pitched, unpleasant sound that can pierce through a musician's speakers when he or she least expects or wants it. It is caused by a signal that travels in a continuous loop. In a typical feedback situation (Fig. 2), a microphone feeds a signal into a sound system, which then amplifies and outputs the signal from a speaker, which is then picked up again by the microphone. Thus the system responds by further changes in the same direction. This is known as **positive feedback**.

Fig. 2   Positive feedback in a sound system

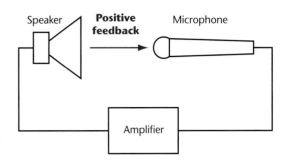

One example of a biological positive feedback loop is the onset of contractions in childbirth. When a contraction occurs, the hormone oxytocin is released into the body, which stimulates further contractions. This results in contractions increasing in amplitude and frequency.

Fig. 3   Positive feedback in biological systems

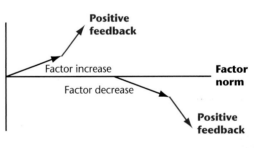

Positive feedback can also be harmful, for example, when a fever causes a positive feedback that pushes the temperature continually higher. During a severe fever, body temperature can reach extremes of 45 °C, a temperature at which some cellular proteins denature.

In this chapter you will see the effects of both positive and negative feedback in the control of the human oestrus cycle.

---

key facts

● Negative feedback restores systems to their original level.

● In positive feedback situations, the system responds to deflection from the norm by causing further change in the same direction.

Fertilisation

During intercourse (or coitus), the penis ejaculates 3–5 cm^3 of semen at the base of the cervix. This semen contains 20–100 million spermatozoa (sperms) per cm^3, which are produced by meiosis in the testes, plus seminal fluid made by the prostate gland, seminal vesicle and other accessory glands. Seminal fluid has several important roles:

- It is slightly alkaline and helps to neutralise the acidity of the vagina.
- It contains fructose, a sugar that provides energy for the sperm.
- It provides a medium in which sperm can swim, and causes them to become motile.
- It stimulates small uterine contractions which help to move sperm towards the oviduct.
- It coats the sperm heads with a protein that protects against the acidity and immune response of the vagina.

Capacitation

Sperm head membranes have a protective coating of proteins and glycoproteins provided by secretions in the epididymis and by the seminal fluid. The coating protects sperm from the hostile environment of the reproductive tract, and enables them to survive longer as they swim towards the oviduct. However, the coating must be removed by enzymes in the uterus before a sperm has the capacity to fertilise an egg. This is called capacitation and it takes six to seven hours.

The acrosome reaction

Fertilisation usually occurs in the upper part of the oviduct. There is no clear evidence that human sperm follow a chemical signal to the oocyte (taxis, Chapter 9) as occurs in some other species. The vast number of sperm and the relatively large size of the oocyte ensure that some sperm reach their target by chance.

Contact between a sperm and the granulosa cells around the oocyte triggers the acrosome reaction – the release of enzymes stored in the acrosome in the sperm head (Fig. 4).

One of the enzymes released is hyaluronidase, which digests the matrix of hyaluronic acid surrounding the granulosa cells. After passing between the granulosa cells, the sperm meets the zona pellucida. The zona pellucida is composed of glycoproteins, one of which, ZP3, is a sperm receptor. The inner acrosomal membrane, which is exposed during the acrosome reaction, contains a receptor protein that binds to ZP3. This forms a neat slit through the zona pellucida, allowing entry of the sperm, which lashes its tail to push through into the perivitelline space between the oocyte and the zona pellucida. The posterior part of the sperm head fuses with the oocyte membrane, which is covered with thousands of microvilli that close around the sperm head. The tail of the sperm continues to move, causing the oocyte to rotate inside the zona pellucida until all the tail is pulled into the perivitelline space (Fig. 5).

The cortical reaction

It is essential that only one sperm fertilises each oocyte. If two or more sperm enter (polyspermy), normal development does not occur and the embryo dies after three or four cell divisions. Polyspermy is prevented by the cortical reaction, in which the zona pellucida forms a barrier to further sperm entry.

- Fusion of sperm and oocyte causes an increase in calcium ion concentration in the oocyte cytoplasm.
- Increased calcium ion concentration causes exocytosis of oocyte lysosomes, called cortical granules, into the perivitelline space.
- Enzymes from the cortical granules cause the zona pellucida to thicken (forming the fertilisation membrane) and also destroy ZP3, so no more sperm can bind to the zona pellucida.

Fig. 4   The acrosome reaction

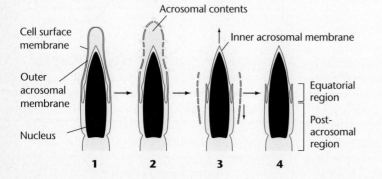

1 Before the acrosome reaction. 2 Changes in the permeability of the membranes to calcium ions cause the cell surface membrane to 'point fuse' with the outer acrosomal membrane. This allows the acrosomal contents to diffuse out, releasing enzymes to help sperm penetrate the outer coats of the oocyte. 3 The fused membranes are sloughed off, leaving the inner acrosomal membrane exposed. 4 Physiological changes occur in the equatorial region of the sperm head, rendering it capable of fusion with the oocyte surface membrane.

Source: adapted from Baggott, *Human Reproduction*, Cambridge University Press, 1997

Fig. 5   Fertilisation

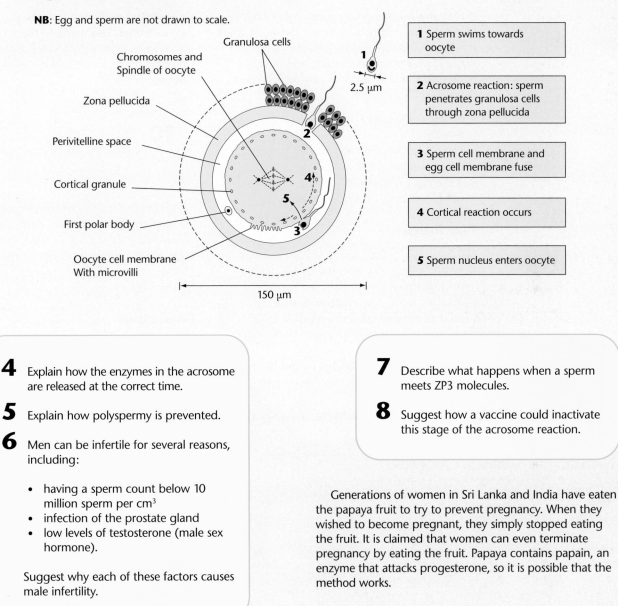

NB: Egg and sperm are not drawn to scale.

Granulosa cells

Chromosomes and
Spindle of oocyte

Zona pellucida

Perivitelline space

Cortical granule

First polar body

Oocyte cell membrane
With microvilli

2.5 µm

150 µm

1 Sperm swims towards oocyte

2 Acrosome reaction: sperm penetrates granulosa cells through zona pellucida

3 Sperm cell membrane and egg cell membrane fuse

4 Cortical reaction occurs

5 Sperm nucleus enters oocyte

4 Explain how the enzymes in the acrosome are released at the correct time.

5 Explain how polyspermy is prevented.

6 Men can be infertile for several reasons, including:

- having a sperm count below 10 million sperm per cm^3
- infection of the prostate gland
- low levels of testosterone (male sex hormone).

Suggest why each of these factors causes male infertility.

7 Describe what happens when a sperm meets ZP3 molecules.

8 Suggest how a vaccine could inactivate this stage of the acrosome reaction.

Generations of women in Sri Lanka and India have eaten the papaya fruit to try to prevent pregnancy. When they wished to become pregnant, they simply stopped eating the fruit. It is claimed that women can even terminate pregnancy by eating the fruit. Papaya contains papain, an enzyme that attacks progesterone, so it is possible that the method works.

A male contraceptive?

In about 5% of men, antigens on the surface of the sperm trigger an immune response in the man's body; these men are infertile. This suggests the possibility of using a vaccine as a means of contraception. The vaccine would be targeted at the sperm head's receptor proteins or at the receptor molecules (ZP3) on the zona pellucida, which are involved in the acrosome reaction.

9

a Explain how papaya could act as a contraceptive.

b Suggest a reason why ingestion of a hormone-digesting enzyme might not work as a means of contraception.

12.3 The human oestrus cycle

A woman's oestrus cycles begin in her early teens and finishes with the menopause, at about the age of 50 years. The length of each cycle varies from woman to woman, and may be between 25 and 35 days, averaging 28 days. During each cycle:

- An egg is released from one of the woman's ovaries.
- The lining of her uterus (womb) is made ready to accept and support an embryo.
- If the egg is not fertilised, **menstruation** occurs, and the cycle begins again.

The uterus
The uterus consists of two layers of tissue:

- an outer layer, the **myometrium**, with smooth muscle, important during birth
- an inner lining, the **endometrium**, which is shed and regenerated during each cycle.

The ovaries
The ovaries contain **follicles** in which ova develop (Fig. 6). At birth there are thousands of **primordial follicles** in each of a female baby's ovaries. The ovum in these primordial follicles is not yet fully formed, but is halted at prophase 1 of meiosis and is called a **primary oocyte**.

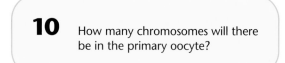

10 How many chromosomes will there be in the primary oocyte?

After puberty a few of the primordial follicles develop and mature during each cycle. The cells surrounding the oocyte divide to form cells that form a follicle, which surrounds the oocyte. Meiosis is not completed unless a sperm fertilises the ovum.

Ovulation
The mature follicle is 10–25 mm in diameter and is known as a **Graafian follicle**. It moves towards the surface of the ovary where it bursts and releases the **secondary oocyte**; this is

Fig. 6 Development of a Graafian follicle during the oestrus cycle

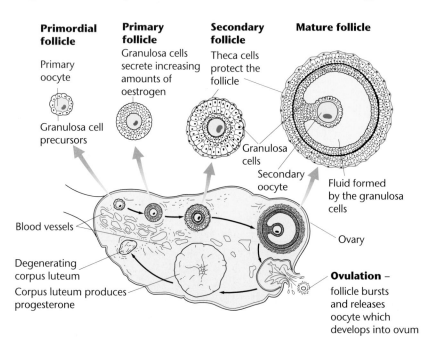

Primordial follicle

Primary oocyte

Granulosa cell precursors

Blood vessels

Degenerating corpus luteum

Corpus luteum produces progesterone

Primary follicle

Granulosa cells secrete increasing amounts of oestrogen

Secondary follicle

Theca cells protect the follicle

Granulosa cells

Secondary oocyte

Mature follicle

Fluid formed by the granulosa cells

Ovary

Ovulation – follicle bursts and releases oocyte which develops into ovum

Source: adapted from Berne and Levy, *Principles of Physiology*, Wolfe, 1990

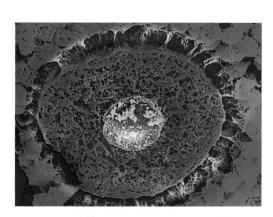

In this false-colour scanning electron micrograph of a maturing primary oocyte, the oocyte (green) contains a nucleus (yellow) and is surrounded by the zona pellucida (blue). Round the outside are granulosa cells (red).

called **ovulation**. Only one follicle usually reaches this stage during each cycle, the rest degenerate.

The oocyte is surrounded by the **zona pellucida** (jelly coat). Many granulosa cells from the follicle remain surrounding the zona pellucida after ovulation. These extra cells may help the cilia to move the oocyte into the funnel and down the oviduct.

Fig. 7   Hormonal control of the oestrus cycle

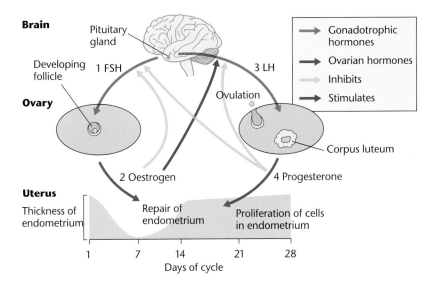

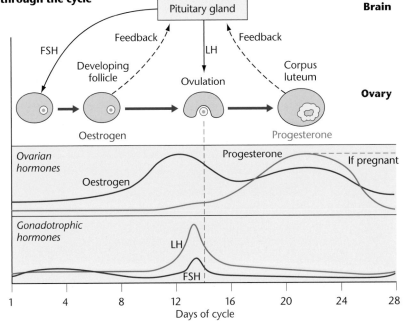

11

a   A woman reached puberty at age 13. She later had three children and reached the menopause at age 45. After each child, her menstrual cycle stopped for three months while the baby breast-fed. Calculate how many secondary oocytes were ovulated by this woman in her lifetime. State any assumptions you make.

b   State the longest time, in years, between the formation of a primary oocyte and its development into a secondary oocyte in this woman.

Hormonal control of the menstrual cycle

The release of an oocyte by the ovary is linked to the repair of the endometrium. This means that, if fertilisation occurs, the uterus is ready to receive the embryo. These changes in the ovary and uterus are coordinated by the interaction of four hormones:

* two protein hormones, **follicle stimulating hormone (FSH)** and **luteinising hormone (LH),** made by the anterior lobe of the pituitary gland
* two steroid hormones, **oestrogen** and **progesterone**, made by the ovaries.

The pituitary gland is attached by a stalk to the **hypothalamus**. The hypothalamus contains specialised neuro-secretory cells, which secrete hormones rather than neurotransmitters. The hormones travel through small capillaries from the hypothalamus to the nearby pituitary gland to control the pituitary's secretion of FSH, LH and other hormones. The **gonadotrophic** hormones, FSH and LH, travel in the bloodstream to the ovaries (female gonads) where they stimulate production of oestrogen and progesterone. These two ovarian hormones in turn have **feedback effects** on the hypothalamus and pituitary gland.

The start of the menstrual cycle is signalled by the beginning of menstruation on day 1 (Fig. 7).

The lining of the endometrium passes out via the cervix and vagina over four to five days. The inner layer of the endometrium remains attached to the myometrium, allowing a new endometrium to grow by mitosis later in the cycle.

Follicle development in the ovaries is triggered by the release of FSH from the pituitary gland in the first week of the cycle. As the follicle cells begin to divide, they develop a better blood supply into which they release oestrogen. The oestrogen has two effects:

- It binds to specific **receptor proteins** in the surface cells of the uterus, causing the repair and thickening of the endometrium.
- It has a feedback effect on the secretion of the pituitary hormones FSH and LH.

During days 5–10, oestrogen levels rise *slowly*. Oestrogen inhibits the release of FSH and LH, an example of **negative feedback**. FSH levels decline and LH levels remain low. During days 10–14 there is a *rapid* rise in oestrogen levels. This has a **positive feedback** effect on FSH and LH. There is a surge in FSH and LH levels around days 12–14. These surges trigger ovulation at day 14. LH also causes the transformation of the empty follicle cells into the **corpus luteum** (yellow body).

The corpus luteum is a mass of cells whose role is to produce progesterone and oestrogen in the second half of the cycle. From about day 15, progesterone levels start to rise, reaching a peak at about day 21. Progesterone has two effects:

- Like oestrogen in the first half of the cycle, progesterone binds to specific receptor proteins in cells on the surface of the endometrium. The endometrium cells respond by becoming glandular and secreting glycoproteins, which make the endometrium highly receptive to an embryo on days 20–21.
- Progesterone has a feedback effect on FSH and LH:
  - At *low* concentrations, progesterone has a **negative feedback** effect on FSH, so preventing further follicles from developing during the second half of the cycle.
  - At *high* concentrations, progesterone has a **negative feedback** effect on LH, which effectively turns off its own production (since LH stimulates the corpus luteum to make progesterone) – the resultant fall in progesterone from day 21 results in constriction of blood vessels in the uterus, and the start of menstruation at the end of the 28-day cycle.

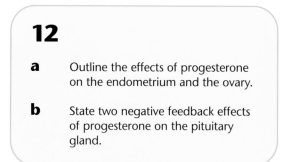

12

a   Outline the effects of progesterone on the endometrium and the ovary.

b   State two negative feedback effects of progesterone on the pituitary gland.

key facts

- The oestrus cycle involves the regular building up and shedding of the endometrium (uterus lining).

- The ovaries release an oocyte every 28 days.

- The oocyte develops inside a group of cells called a follicle. Several follicles develop in each ovary during every cycle, but usually only one matures into a Graafian follicle and ovulates.

- The oestrus cycle coordinates the release of an oocyte with the preparation of the uterus wall to receive an embryo.

- The menstrual cycle is controlled by the interaction of two pituitary hormones, FSH and LH, and two ovarian hormones, progesterone and oestrogen. These four hormones interact through positive and negative feedback effects. The effects of these four hormones depend on both their relative concentrations and their rate of change of concentration in the bloodstream.

how science works

What brings on puberty in girls?

The oestrus cycle in women is controlled by hormones produced by the pituitary gland and by the ovaries. During childhood the pituitary gland secretes small amounts of two hormones, FSH and LH, and the ovaries secrete small amounts of the hormone oestrogen. Puberty begins at about the age of nine when the pituitary gland increases the secretion of FSH. FSH causes oestrogen secretion by the ovaries to increase. The first outward sign of increased oestrogen secretion is the appearance of breast buds; later effects include the growth of pubic hair and of the uterus.

Increased oestrogen secretion between the ages of 10 to 12 has a positive feedback effect, increasing the production of both FSH and LH. Similarly, the more FSH, the more oestrogen that is produced, so the concentrations of all three hormones in the blood rise sharply between the ages of 10 and 12. However, the positive feedback effect of oestrogen on FSH production is only temporary. As the pituitary gland matures, its sensitivity changes and, by the age of 13 to 14, oestrogen starts to have a negative feedback effect on FSH production (but not on LH production). The effect of this is to kick-start the menstrual cycle.

13 Which of the hormones stimulates the development of secondary sexual characteristics such as pubic hair?

14 Fertility drugs contain FSH. Birth-control pills contain a cocktail of hormones that mimic the effects of oestrogen on FSH production. Suggest how fertility drugs and birth-control pills work.

15 Although examples of positive feedback are relatively rare, they are involved in several other aspects of the reproductive process, as well as those mentioned in this text – the process of labour, for example. Why could these processes not be controlled by negative feedback?

16 What broadly different roles do negative and positive feedback play in the human body?

12.4 Plant hormones

Phototropism

In Chapter 9 you learned that plant roots and stems respond to gravity and light; and that their responses to these stimuli are called tropisms. In this section you will learn how these responses are controlled by plant hormones.

Most of the experiments that have led to an understanding of the mechanism of tropisms have been done using oat seeds. The oat seed, like other grass seeds, germinates to produce a cylindrical **coleoptile**. The coleoptile encloses the primary (first) leaf of the young oat plant.

Charles Darwin and his son Francis performed some of the first experiments on phototropism. In one experiment (**A** of Fig. 8) they placed tinfoil caps on coleoptiles. The coleoptiles were then placed in unidirectional light.

Fig. 8   The tip of a coleoptile is the light-sensitive region

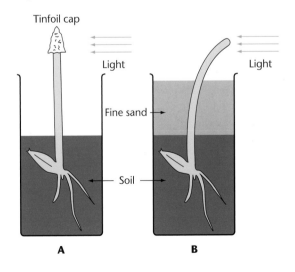

The coleoptiles continued to grow vertically, whereas a control group of coleoptiles grew towards the direction of the light stimulus.

In a second experiment (**B** of Fig. 8) they placed sand around the coleoptile, leaving only the tip exposed to unidirectional light. This coleoptile showed a normal phototropic response.

18   Do the results of this experiment support Boysen-Jensen's hypothesis? Explain the reasons for your answer.

17   What can you conclude from these two experiments about:

a   the position of the photoreceptor in an oat coleoptile?

b   the effector region of the coleoptile?

The response of the coleoptiles to unidirectional light is caused by the cells on the shaded side of the coleoptile growing faster than those on the illuminated side.

In the early 20th century a scientist called Boysen-Jensen hypothesised that a chemical messenger carried information from the tip of a coleoptile to the growing cells. Fig. 9 shows one of his experiments. Mica is a material that is totally impermeable. He inserted wafers of mica just below the tips of coleoptiles, as shown in the diagram.

Boysen-Jensen's explanation of the results was that a substance from the tip caused the cells on the shaded side of the coleoptile to grow faster than those on the illuminated side.

The next step was to isolate this chemical substance. A scientist called Went cut the tips of coleoptiles and placed them on agar blocks. After a few hours he placed these agar blocks on decapitated coleoptiles, as shown in Fig. 10.

Fig. 10   Auxin diffuses into agar blocks

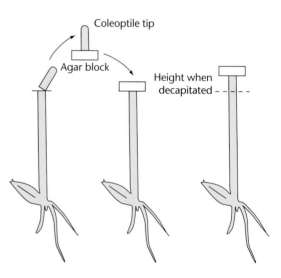

A control group of decapitated coleoptiles showed no growth in the period of the experiment. His conclusion was that a growth-promoting substance had diffused from the coleoptile tips into the agar blocks, and then from the agar blocks into the decapitated coleoptiles.

Went then repeated the experiment, but this time he placed the agar blocks eccentrically on decapitated coleoptiles, as shown in Fig. 11.

Fig. 9   Mica prevents the translocation of auxins

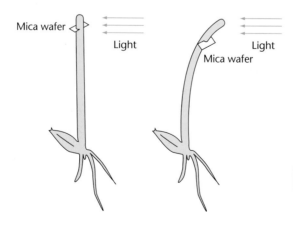

19   Explain the results of this experiment.

Fig. 11   An eccentrically placed agar-auxin block causes curvature

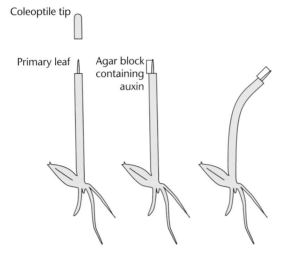

The amount of the growth-promoting substance was too small for it to be analysed at that time, so Went coined the term **auxin** for this substance.

> **20** This experiment could be used to compare the amount of auxin that diffuses from coleoptile tips into agar blocks. Suggest how this could be done.

In a further experiment Went illuminated coleoptiles from one side for a few hours, then decapitated them and placed the tips on agar blocks, as shown in Fig. 12.

Fig. 12   Differential auxin distribution in an unequally illuminated coleoptile tip

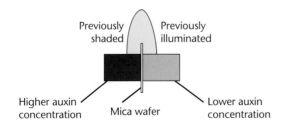

> **21** What was the function of the mica wafer in this experiment?

On average, twice as much auxin diffused into the agar block under the non-illuminated side than the block under the illuminated side.

> **22** Suggest two possible explanations for this unequal distribution of auxin.

Later in the 20th century a substance called **indolylacetic acid (IAA)** was identified as the principal auxin involved in tropisms.

There is still much debate among scientists about the mechanism of phototropism, but the most widely accepted hypothesis is as follows:

- The tip of the shoot detects the direction of the light stimulus.
- The photoreceptor is a protein called **phototropin**.
- IAA is synthesised by the tip and moves down the stem.
- **Auxin transporter proteins** are moved into the plasma membrane at the sides of cells of the shoot.
- Auxin moves through these transporter proteins and accumulates in the cells on the shaded side.
- This stimulates elongation of the cells on the shaded side, causing the shoot to grow towards the light.

Geotropism

Similar techniques have been used to investigate geotropism. Fig. 13 overleaf shows an experiment to investigate the effect of gravity on auxin distribution.

Similar results are obtained with both root tips and shoot tips, but the increased concentration of auxin on the lower side has different effects on roots and shoots. The cells at the lower side of shoots are stimulated to grow by the increased auxin concentration, but the growth of root cells is inhibited.

Fig. 13   Movement of auxin in geotropism

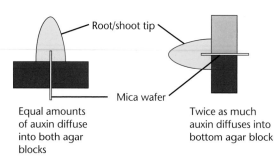

Equal amounts of auxin diffuse into both agar blocks

Twice as much auxin diffuses into bottom agar block

Fig. 14 shows the reason for this. Root cells and shoot cells have very different sensitivities to auxins.

Fig. 14   Differential effect of auxin on growth of roots and stems

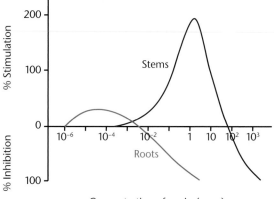

23 Explain in terms of auxins and cell growth why stems are negatively geotropic.

In phototropism, there is a photodetector that causes auxin transporter proteins to move into the sidewalls of the stem cells. The mechanism for geotropism cannot involve a photoreceptor, but must involve a structure that is sensitive to the force of gravity. Again, there is still much debate among scientists about the mechanism of geotropism, but the most widely accepted hypothesis is as follows:

When a root is placed on its side:

- **Amyloplasts** (organelles containing starch grains) settle by gravity to the bottom of cells in the root tip.
- The amyloplasts cause auxin transporter proteins to be inserted in the plasma membrane of the root cells.
- The auxin transporter proteins move auxins downwards out of the cells.
- Auxins accumulate in the cells on the underside of the root.
- This inhibits root cell elongation.
- The cells at the top surface of the root elongate, causing the root to grow down.

key facts

- Auxins affect the growth of plant cells.
- Auxins are produced at the tips of the stems and the roots.
- Auxins are translocated away from the tips to the rest of the root/shoot.
- Shoot and root cells have different sensitivities to auxins.

- Unidirectional light results in auxin moving towards the shaded side of a vertical shoot, causing the cells there to elongate.
- The force of gravity results in auxins moving towards the underside of a horizontal root, inhibiting the growth of root cells there.

1

a The graph shows the concentrations of two hormones during one sexual cycle of a human female. The diagram shows structures that produce these hormones.

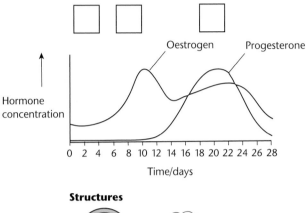

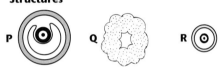

Structures

i Write the appropriate letters in the boxes on the graph to show the order in which the structures labelled **P** to **R** appear during the cycle. (1)
ii Name the hormone that causes structure **Q** to develop. (1)
b Describe **two** effects of progesterone on the uterus. (2)
c Explain how oestrogen in contraceptive pills prevents fertilisation from taking place. (2)
d The sexual cycles of some female farm animals can be synchronised by giving them low doses of progesterone. When this treatment is stopped the animals come into oestrus a few days later. Explain how the withdrawal of progesterone causes them to come into oestrus. (2)

Total 8

AQA, A, June 2006, Unit 2, Question 7

2 **Figure 1** shows different stages in the development of a follicle and corpus luteum in a human ovary.

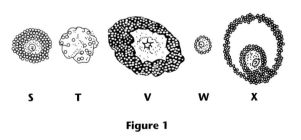

Figure 1

a **i** Give the letter which shows a mature ovarian follicle. (1)
ii Give the sequence of letters which shows the correct order of these stages. (1)
b **Figure 2** shows a sperm about to fertilise an oocyte.

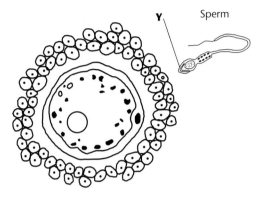

Figure 2

i Describe the role of structure **Y**. (1)
ii The diameter of the oocyte nucleus is 24 μm. Calculate the magnification of **Figure 2**. Show your working. (2)

Total 5

AQA, A, June 2003, Unit 7, Question 1

3 The diagram shows the events that occur in the human ovary during one menstrual cycle.

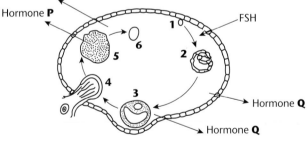

a **i** Describe the role of follicle stimulating hormone (FSH) in the menstrual cycle. (2)
ii FSH is a globular protein and specifically targets cells in the ovary. Explain how the structure of FSH accounts for this specific targeting. (3)
b **i** Name the hormones **P** and **Q**. (2)
ii After the menopause, when the menstrual cycle ceases, hormone secretion by the ovary is much reduced. Explain why the blood of post-menopausal women contains high levels of FSH. (2)

Total 9

AQA, B, January 2005, Unit 8, Question 2

4 Contraceptives containing progesterone may be used to control fertility. Progesterone acts as a contraceptive by making the mucus at the entrance to the uterus more dense. At high concentrations it inhibits LH production.

a Explain why the probability of conception is reduced when

   i the mucus at the entrance to the uterus is more dense; (1)

   ii the production of LH is inhibited. (2)

b The contraceptive may be supplied in three different ways. These are shown in the table.

Method of supplying progesterone	Length of time that contraceptive is effective
Daily pill	27–30 hours
Injection	12 weeks
Implant	5 years

The graph shows the concentration of progesterone in the blood in users of the three types of contraceptive over a period of several weeks.

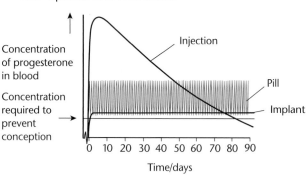

   i Fewer pregnancies occur in women using the injection than in women using daily pills. Use the information given to suggest an explanation for this. (1)

   ii Suggest and explain the advantages and disadvantages of using implants rather than injections as a method of contraception. (3)

Total 7

AQA, B, June 2004, Unit 8, Question 3

5

a Give **one** function of LH in females. (1)

In males, FSH stimulates sperm production and LH causes the release of testosterone. A hormone stimulates the release of FSH and LH by attaching to receptor molecules in the surface membrane of cells in the pituitary gland. The diagram shows one receptor molecule for this hormone.

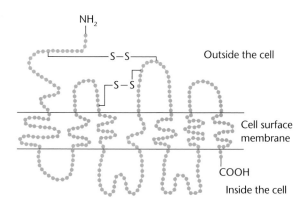

b **i** Give **two** pieces of evidence from the diagram which suggest that the receptor molecule is a protein. (2)

   ii Explain how the tertiary structure of this protein is important for its function as a receptor molecule. (2)

c Research has identified a substance which could be used as a male contraceptive pill.

This substance binds to the receptor molecules in the pituitary gland and stops the release of FSH, but allows the release of LH to continue.

   i Explain **one** advantage of the substance not inhibiting LH release. (2)

   ii This substance is not a protein. Explain why a protein could **not** be used as an oral contraceptive. (2)

Total 9

AQA, B, January 2004, Unit 8, Question 6

Fertility treatment

One of the most common fertility treatments is *in vitro* fertilisation (IVF). *In vitro* means 'in glass', and indicates that fertilisation occurs outside the body, usually in a glass dish. IVF is used to treat both male and female infertility. More than 1% of children born in the UK are 'test-tube' babies.

A couple that seek IVF treatment are interviewed and assessed before medical tests begin. IVF can use the oocytes a woman produces in her natural cycle, but a higher success rate is achieved if she is given hormone-based drugs to induce super-ovulation – the production of several ripe follicles in one cycle. The combination of drugs used depends on the individual, and may include clomiphene. The normal oestrus cycle is suppressed, and then ovulation is induced at a time controlled by the doctor.

> **A1**   Clomiphine blocks the effect of oestrogen on the pituitary gland. Explain how this will affect the oestrus cycle.

An ultrasound probe is used to guide a needle towards the mature follicles in the ovary. The oocytes are removed from the follicles by suction. The man now provides a semen sample. The semen is treated to select the most vigorous sperm, and a sample containing about 100 000 sperm is added to each oocyte. The sperm and eggs are cultured together in a Petri dish in a suitable fluid medium overnight. The eggs are then inspected by a microscopist who looks for 2–4-celled embryos. The healthiest embryos are selected and placed into the upper uterus using a catheter. With luck, one or more embryos will develop.

Embryos that are not used may be stored for up to five years, and semen for 10 years. About 75% of frozen embryos survive the thawing process. Embryo selection can now involve pre-implantation genetic diagnosis (PGD) whereby one or two cells are removed from an 8-cell embryo for genetic analysis. The remaining part of the embryo will grow normally and the removed cells can be screened for factors such as:

Table 1 below shows pregnancy rates achieved at a particular clinic by IVF in relation to the cause of infertility.

> **A2**   What is the most common reason for fertility treatment at this clinic?
>
> **A3**   Suggest why:
>
> **a**   the number of egg collections was lower than the number of cycles treated
>
> **b**   the number of embryo transfers was lower than the number of egg collections
>
> **c**   the number of pregnancies was lower than the number of embryo transfers.
>
> **A4**   Which stage of the process is the least successful? Use figures to support your answer.
>
> **A5**   The government is considering banning the transfer of more than one embryo into a woman. Suggest the pros and cons of transferring more than one embryo.

- chromosome abnormalities
- sex
- genetic markers, such as for mutations in the breast cancer gene
- a compatible tissue donor.

> **A6**   Discuss the ethical issues raised by PGD.

Table 1   Pregnancy rates achieved by IVF in relation to cause of infertility

Diagnosis	Cycles	Egg collections	Embryo transfers	No. of pregnancies	Pregnancy rate per cycle (%)	Pregnancy rate per egg collection (%)	Pregnancy rate per transfer (%)
Unexplained	155	144	119	30	19.4	20.8	25.2
Male factor	248	230	169	42	16.9	18.3	24.8
Miscellaneous	149	138	111	26	17.4	18.8	23.4
Male and female	58	54	46	13	22.4	24.1	28.3
Tubal damage	199	185	164	40	20.1	21.6	24.4
Total	809	752	609	151	18.7	20.2	24.85

Source: Bourne Clinic Website (*From Behaviour and populations p. 48*)

13 Homeostasis

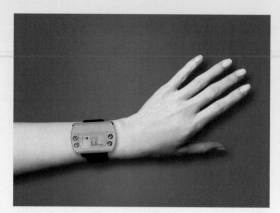

For some people with diabetes, this watch might be better than a Rolex!

People with diabetes have trouble controlling their blood glucose level. If the problem is severe, someone with diabetes needs to check his or her blood glucose levels every few hours, sometimes up to 10 times a day. This involves the finger-prick test – taking a drop of blood from the finger and placing it on a test-strip that indicates the concentration of glucose in the blood. It doesn't sound so bad, but imagine having to do that 10 times a day. The good news is that a new type of watch (see photographs above), the Glucowatch® Biographer, can monitor blood glucose levels painlessly, without the need to draw blood. It also has the advantage that it can take measurements much more frequently.

The Glucowatch® Biographer checks blood glucose levels every 20 minutes by sending tiny electric currents through the skin. This attracts glucose to the surface of the skin. The sensor in the Glucowatch® Biographer has tiny gel discs that sense the level of glucose in the blood. The monitor gives a reading of the blood glucose level, and sounds an alarm if the level becomes dangerously high or low. In the future, this system could be linked to an automatic insulin pump, permanently connected to the person's blood supply. This would allow the pump to respond directly to the blood glucose level, mimicking the body's normal method of maintaining homeostasis.

13.1 What is homeostasis?

All cells in the body need to be in the best physical and chemical conditions possible if they are to function properly. Keeping every cell at the optimum temperature, with enough water and nutrients is a complex task that involves many systems in the body; biologists call this overall control mechanism **homeostasis**.

The word *homeostasis* means *steady state* and refers to the fact that conditions inside the body need to be maintained within certain limits. Some factors – such as temperature and blood pH – are kept within very narrow limits. Others, such as blood glucose, can vary within a wider range without adversely affecting the individual.

For all homeostatic mechanisms you should focus on three basic aspects:

- What causes the change? For example, a rise in external temperature causes the body's temperature to start to rise.
- What detects the change? If body temperature starts to rise, this is detected by the hypothalamus in the brain.

- How is the change corrected? To reduce its temperature, the body sweats, blood vessels in the skin dilate, etc.

You have already come across some important examples of homeostasis. In the AS book you learned how important it is to keep the water potential of the blood plasma constant. Then, your work on exchange surfaces in the AS book gave you some idea of how various animals have become adapted to keep their body temperature within narrow limits. In this chapter you will learn about the homeostatic mechanisms in the human body that keep blood glucose concentration and core body temperature within limits.

Stimulus and response

Homeostasis involves mechanisms that correct changes occurring in the body. These changes are usually brought about by a stimulus. A stimulus can be *external* – it can come from outside the body, or it can be *internal*. For example, when we eat a carbohydrate-rich meal, the absorption of glucose into the blood raises blood glucose concentration. This is an **internal stimulus** that changes conditions inside the body. In a healthy person, the body then responds by producing a hormone that reduces the amount of glucose in the blood, readjusting the balance back towards normal.

The way that the Glucowatch® Biographer monitors changes in blood glucose concentrations also involves a stimulus and a response, although these are obviously mechanical rather than physiological. The stimulus is an increase in blood glucose concentration – if this is great enough, the watch responds by sounding an alarm.

s&c

stretch and challenge

Treating liver failure

The liver has important functions in homeostasis. Some of these include getting rid of metabolic waste products.

Liver failure is one of the most common causes of death in people who have taken an overdose of either medicinal or 'social' drugs. Some drugs may kill large numbers of liver cells. As the normal functions of the liver grind to a halt, metabolic by-products begin to accumulate in the blood. One of these, ammonia, is highly toxic. Other by-products and drugs can be particularly dangerous if not removed – they poison the brain and they inhibit regeneration of liver cells. The tranquilliser Valium is a benzodiazepine, so an overdose of Valium is particularly bad news for the liver.

Fortunately, the liver is a tough organ that fights back. If someone who has taken a drug overdose doesn't die of the initial effects of the poison, his or her liver cells may grow again naturally. The liver has tremendous powers of regeneration – more than any other part of the body. A surgeon can remove 90% of the liver in a cancer patient, for example, and it will grow back to within 1% of its original size.

Keeping people alive until this happens can be tricky. An early method involved passing the patient's blood through an isolated pig's liver. This worked well for a short time but, after a few days, antibodies produced by the patient's white blood cells started to destroy the pig's liver cells. Today, a more likely treatment is to transplant part of a donor's liver into the overdose patient. The transplanted liver removes metabolic by-products and drugs from the blood and carries out the liver's other normal functions. When the patient's own liver cells have regenerated, the transplanted liver is removed. The problem with this method is obtaining livers for transplant. To get around this, scientists have been trying for some time now to develop 'liver machines' that contain cultures of human liver cells. The idea is to use this artificial but living structure to carry out most, if not all, normal liver function until the patient's own liver recovers. Bioartificial devices are proving increasingly successful in treating liver failure. One such device is shown in Fig. 1.

The plasmapheresis device separates the patient's plasma from the blood cells. The patient's plasma flows through a bioreactor, a series of hollow-fibre membranes surrounded by hepatocytes (liver cells). The cells in this bioreactor are from pigs' livers, but other similar devices use genetically engineered human liver cells. As the plasma flows through the bioreactor, the liver cells remove metabolic waste products that build up as a result of liver failure. These devices have been used successfully to keep patients alive until their own livers regenerate.

Fig. 1   A bioartificial device for the treatment of liver failure

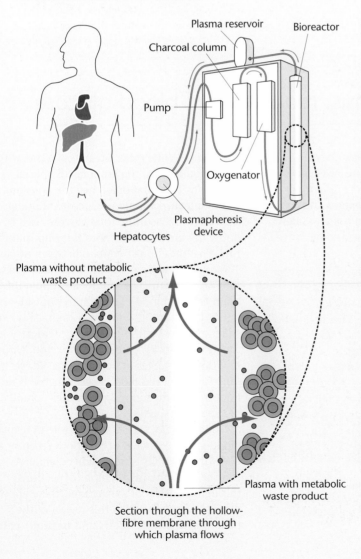

Plasma reservoir
Bioreactor
Charcoal column
Pump
Oxygenator
Plasmapheresis device
Hepatocytes
Plasma without metabolic waste product
Plasma with metabolic waste product
Section through the hollow-fibre membrane through which plasma flows

s&c

stretch and challenge

1 The plasmapheresis device shown in Fig. 1 separates blood cells from blood plasma.

a What are the main constituents of plasma?

b Suggest one reason why blood cells are not allowed to flow through the bioreactor.

2 Suggest one function of the charcoal column.

3 The plasma is oxygenated by the device. How is oxygen transported in plasma?

4 Suggest how metabolic waste products enter the hepatocytes.

5 Hepatocytes extracted from the body can be kept alive and functioning in vitro, but they do not divide. A clone of hepatocytes that does divide has been produced by genetic engineering. This is called an immortalised cell line.

a What is the disadvantage of using hepatocytes that do not divide?

b Name the type of cell division in dividing hepatocytes.

c Suggest how the immortalised cell line was genetically engineered.

d Suggest one possible danger of inserting genes for cell division into cells.

13.2 An example of homeostasis: regulation of blood glucose

The control of blood glucose is a classic example of homeostasis. Two negative feedback systems work together to detect and correct a rise or fall in blood glucose, maintaining it within limits that the body can tolerate safely. But why does blood glucose need to be regulated? While other organs can cope with greater variations, the brain needs special attention. This vital organ needs a constant and well-controlled supply of glucose. Although it uses about 25% of the total glucose used by the body, the brain cannot store this important food fuel. The blood must contain between 60 and 120 mg of glucose per cubic decimetre of blood (60–120 mg dm^{-3}) so that it can supply the brain cells. If blood glucose strays outside these limits, even for a few minutes, brain cells can suffer and the central nervous system may not function correctly. In the AS book you learned how small changes in volume affected the functioning of red blood cells. Glucose is an important contributor to the water potential of blood and tissue fluid.

The pancreas plays a central role in regulating blood glucose since it acts as both receptor and effector; the brain is not involved at all. As you can see from the micrographs of sections through the islets of Langerhans, there are two distinct types of cells in this region of the pancreas, **alpha (α) cells** and **beta (β) cells**.

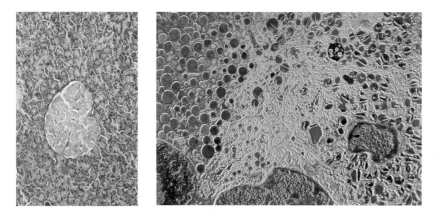

The light micrograph (left) shows a section through the islets of Langerhans in the pancreas. The electron micrograph on the right shows insulin-secreting beta cells (the green, yellow and brown cells on the right) and glucagon-secreting alpha cells (the red cells on the left).

Fig. 2 An overview of the control of blood glucose concentration

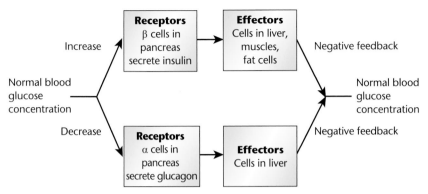

The two types of cells secrete different hormones, **glucagon** and **insulin**, which have opposite effects on blood sugar concentration. Fig. 2 shows an overall summary of how these two hormones control blood glucose.

What happens when blood glucose falls?

The alpha cells in the islets of Langerhans in the pancreas are receptor cells. They detect a *decrease* in blood glucose concentration. When the concentration of glucose in blood flowing through the pancreas falls, the alpha cells respond by secreting glucagon. Glucagon has its effect mainly on the liver. It activates enzymes inside cells that convert glycogen into glucose, a process called **glycogenolysis**. The glucose then diffuses into the blood, increasing the blood glucose concentration towards its normal level. This process is therefore an example of negative feedback.

If glycogen supplies are depleted, glucose can be formed from pyruvate by **gluconeogenesis**. This process is essentially the reverse of glycolysis, but the process involves different enzymes. During starvation, the body's proteins are broken down into amino acids which are then used to produce pyruvate for use in gluconeogenesis. Gluconegenenesis ensures that blood glucose does not fall below the critical concentration.

> **6** Will gluconeogenesis require energy or release energy? Give the reason for your answer.

What happens when blood glucose rises?

The beta cells in the islets of Langerhans in the pancreas are also receptor cells but they are sensitive to an *increase* in blood glucose concentration. When the concentration of glucose in blood flowing through the pancreas rises, the beta cells respond by secreting the hormone **insulin**. This travels to all parts of the body in the blood, but it mainly affects cells in the muscles, liver and adipose (fat storage) tissue. Insulin increases the rate at which liver cells, muscle cells and fat cells absorb glucose.

Insulin stimulates glucose carrier proteins in the cytoplasm to move to the cell surface membrane. The more carrier molecules that reach the surface, the greater the quantity of glucose that can be absorbed. Insulin also activates enzymes in the liver to convert glucose into glycogen, a process called **glycogenesis**. This decreases the glucose concentration in the blood passing through the liver. Insulin secretion therefore causes a fall in blood glucose concentration to bring it down to normal level – another example of negative feedback.

Fig. 3 The effect of insulin on glucose uptake by cells

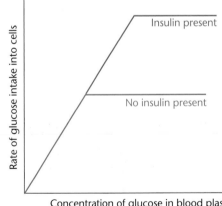

Fig. 3 shows how the rate of glucose intake into the liver is affected by the presence of insulin.

As insulin and glucagon have opposite effects on blood sugar level they are said to be **antagonistic**. Insulin acts to decrease blood glucose concentration while glycogen acts to increase it. Many other hormones that are involved in homeostasis also operate as **antagonistic pairs**.

7

a   Only 2% of cells in the pancreas are involved in blood glucose regulation. What is the function of the other 98%?

b   What type of carbohydrate is glycogen?

c   How is glycogen adapted to its storage function by its shape?

d   What type of reaction causes the breakdown of glycogen into glucose?

e   What type of reaction converts glucose to glycogen?

f   Glucose enters liver cells by facilitated diffusion.
What are the main factors that affect the rate of facilitated diffusion of a substance into a cell?

g   How does an increase in the number of carrier proteins in the cell surface membrane increase the rate of glucose uptake by cells?

Another hormone, **adrenaline**, affects blood glucose concentration. Adrenaline is secreted by the **adrenal glands**, two small glands situated just above the kidneys. This hormone has a similar effect on liver cells as glucagon – it stimulates glycogenolysis.

Adrenaline and glucagon do not actually enter their target cells. Instead they form weak bonds with receptor sites on the cell membrane. This means that they act via a **'second-messenger' model** of hormonal control where the hormone itself acts as the first messenger going from the endocrine gland to the target cell and binding to a receptor on the target cell membrane.

This binding, in turn, stimulates a second chemical messenger **cAMP** (cyclic adenosine monophosphate) *inside* the target cell. The cAMP then activates specific enzymes within the cell and thus initiates glycogenolysis.

8   Glucagon only affects liver cells. Suggest an explanation for this.

Diabetes: when blood glucose regulation fails

One form of diabetes is called **type I diabetes**, or early-onset diabetes. Symptoms usually appear in young people, under the age of 20. The affected person has a problem with the beta cells in the pancreas – they may have been damaged or destroyed by an autoimmune response. As a result, the pancreas cannot release insulin when blood glucose rises.

Sugary drinks and foods seem like fun but eating too much of them when you are a child and a young adult is thought to increase your risk of developing diabetes in later life.

Until insulin treatment was developed in the 1930s, this form of diabetes was usually fatal. Since insulin treatment for diabetes was developed 75 years ago, millions of diabetics have been able to live relatively normal lives – but they have had to inject themselves daily with insulin. The hormone cannot be taken as a pill because it is digested by the enzymes in the gut.

A modern insulin pen.

In order to stay healthy, people with diabetes need to balance what they eat and the amount of insulin they inject very carefully. If they inject too much insulin, skip a meal or take too much strenuous exercise, their blood glucose can fall too low. Low blood glucose is known as **hypoglycaemia**, and this is often described as 'having a hypo'. The symptoms include sweating, trembling, hunger, blurring of vision and difficulty in concentration. Eating starchy foods soon raises the blood sugar concentration and the symptoms disappear. An untreated 'hypo' can lead to unconsciousness.

The opposite can also happen. If the person is not keeping to their diet and eats too much carbohydrate, or if they are taking too little insulin, blood glucose can rise to dangerously high levels. This is called **hyperglycaemia**. The symptoms include feeling sick, drowsiness and stomach pain. It can also lead to unconsciousness if left untreated.

Another form of diabetes is the more common **type 2 diabetes**. This accounts for about 90% of all cases of diabetes in the industrialised world. Unlike type I diabetes, it tends to appear later in life, typically around the age of 40 in people who are overweight. It is not caused by damage to the islet cells; it occurs because cells in the body become unable to respond to insulin. When the blood glucose level rises, insulin is released normally, but something goes wrong with the way it is detected by cells. They no longer increase the number of glucose carrier proteins on their cell membranes, and blood glucose can remain higher than normal.

This link between diabetes and obesity intrigued scientists for years but they could not prove it. In January 2001, scientists in the USA discovered a new hormone in mice – called resistin. This hormone is produced by fat cells and makes cells in the body *resistant* to the effects of insulin. Humans also produce resistin, but further research will be needed to find out if it really is the important link between being overweight and developing diabetes.

9

a Explain why people with diabetes are encouraged to eat starchy foods rather than sugary foods.

b Suggest an explanation for the symptoms of hypoglycaemia.

---

key facts

- The pancreas both monitors and controls blood glucose concentration; the brain is *not* involved.

- Alpha (α) cells in the pancreas detect a fall in blood glucose concentration and respond by secreting the hormone glucagon.

- Glucagon activates enzymes in the liver that convert glycogen to glucose.

- Adrenaline, secreted by the adrenal glands, has a similar effect on liver cells as glucagon.

- The action of glucagon and adrenaline is known as the second-messenger model since both hormones stimulate the action of cAMP, which in turn stimulates the action of the enzymes that break down glycogen.

- Beta (β) cells in the pancreas detect a rise in blood glucose concentration and respond by secreting the hormone insulin.

- Insulin increases the rate of uptake of glucose by body cells by stimulating the movement of carrier protein molecules from the cytoplasm to the cell surface membrane. These carrier protein molecules move glucose into the cells by facilitated diffusion.

- Insulin also activates enzymes in the liver that catalyse condensation reactions that convert glucose to glycogen.

- The actions of both insulin and glucagon result in negative feedback, bringing the blood glucose concentration back to normal.

- When control of blood glucose fails, diabetes can develop.

- Type I diabetes occurs when cells in the islets of Langerhans are damaged and no longer produce insulin.

- Type II diabetes is a result of cells in the body becoming resistant to the effects of insulin.

hsw

Making human insulin

The search for ways of producing human insulin began when doctors realised that insulin from animals was not always effective and that it caused side effects in many patients. Insulin is a protein. It has two peptide chains joined together by sulfur bridges. It is a small molecule and was one of the first proteins to be described in terms of its primary structure. The sequence of amino acids that make up insulin was worked out in the early 1950s. However, although its structure was known, repeated attempts to synthesise insulin gave very poor yields because it was difficult to get the sulfur bridges to form between the two peptide chains to make the functional molecule.

The breakthrough came when a protein called proinsulin was discovered in the pancreas. This longer protein is the molecule from which insulin is formed in the body.

Proinsulin is a relatively simple molecule to synthesise because it is a single chain consisting of 86 amino acids. If the 35 amino acids in the C peptide are then chopped off (see Fig. 4), the remaining parts of the molecule form insulin using the correct sulfur bridges. This mirrors exactly what happens in the beta cells of the pancreas.

Fig. 4   Human insulin

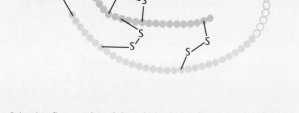

C peptide
A peptide
B peptide

Scientists first produced the whole proinsulin molecule. They then removed 35 amino acids, the C peptide as shown. This left two separate peptides, the A and B peptides joined by sulfur bridges.

10 What is a peptide?

11 What type of enzyme will remove the C peptide from proinsulin?

12 Apart from length, what other differences are there likely to be between the three peptides?

13 How many nucleotides are needed to code for proinsulin?

13.3 Temperature regulation in mammals

During some illnesses caused by microbes, the body temperature rises as high as 40 °C, causing weakness or exhaustion. Such a temperature is not in itself a serious threat to health as long as it does not persist. However, serious fevers, in which body temperature rises to 42 °C, even for a short time, can result in convulsions and death. Although infection by microbes can be the cause of a severe fever, ironically, many of these infective microbes are extremely

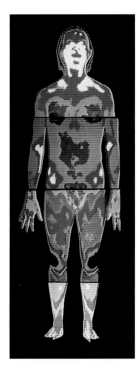

A thermogram of the body showing different skin temperatures in different regions of the body.

susceptible to a rise in body temperature. A fever can therefore be regarded as one of the body's defence mechanisms against microbes.

Body temperature regulation is another good example of homeostasis. Mammals try to keep a high, constant body temperature so that the enzyme-controlled reactions in their bodies always proceed at the optimum temperature. In normal conditions, even when we are exercising, the core temperature of humans does not vary by more than 1 °C, and it does not exceed 37.8 °C. How does the body manage such tight control?

Much of the energy released by respiration in mammals is transferred as thermal energy. It is this thermal energy that keeps body temperature above that of the surroundings. When the surrounding air is warmer than ideal body temperature, a mammal must lose some of its heat to the surroundings.

The photograph is a thermogram of the body showing infra-red radiation. The deeper the colour, the more intense the radiation. When the surrounding air is cooler than the mammal, the body loses heat, mainly by radiation and convection, and it must conserve heat, or increase its generation of heat, to maintain its body temperature.

The body's thermostat

The nervous system plays a major part in the homeostatic control of body temperature in mammals. The **hypothalamus**, a region at the base of the brain (Fig. 5), acts a bit like the thermostat in a central heating system, but with an important difference. A room thermostat simply switches the circulating pump on when the room temperature falls, and switches it off when the room temperature reaches the thermostat setting. It can control the transfer of energy as heat into a room, but it can do nothing to cool the room on a hot day. In contrast, the hypothalamus is able to detect an increase or a decrease in body temperature, and respond to both. It has two regions:

Heat loss centre – a group of cells at the front of the hypothalamus, both switches on sweating and increases blood flow to the skin surface if the body temperature rises.

Heat gain centre – a group of cells at the rear of the hypothalamus, initiates responses to prevent the body from cooling down too much.

What happens when the body needs to cool down?

Inside the hypothalamus, both sets of receptor cells constantly monitor the temperature of the blood. An increase in blood temperature stimulates cells in the heat loss centre. This centre is known as a **coordinator** because it coordinates what happens next. Nerve impulses are sent from the heat loss centre along neurones to effectors. Effectors bring about a change or response. In this case the effectors are the sweat glands in the skin and arterioles that supply the skin capillaries. When the nerve impulses arrive, sweat glands increase their production of sweat and the arterioles increase the supply of blood to capillaries near to the surface of the skin. Sweating and vasodilation, the two main processes that help the body cool down are both negative feedback mechanisms since they help to restore normal body temperature.

Sweating

Sweating cools the body as the sweat evaporates. Sweat is produced by the body's numerous sweat glands – the skin of the average adult contains between two and five million sweat glands. That's between 150 and 300 per square centimetre. With this number of glands, even on a cool day you produce at least one litre of sweat.

The efficiency of sweating as a cooling mechanism depends on the humidity of the surroundings. If there is very little water vapour in the surrounding air (humidity is low), sweat

Fig. 5   The hypothalamus

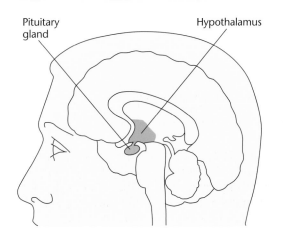

Pituitary gland

Hypothalamus

Dancing brings sweat glands into action.

False-colour scanning electron micrograph of the opening of a sweat gland in the surface of human skin.

will evaporate more easily and the cooling will be effective. In dry air, humans can tolerate air temperatures of 65 °C for several hours. If, however, the air is saturated with water vapour (humidity is high), the sweat cannot evaporate. In these conditions, sweat simply drips off the body and temperatures of only 35 °C can cause the human body to overheat. Some mammals are adapted for hot conditions, but humidity and high temperature can lead to heat stress in even the best-adapted animal.

Vasodilation

When you exercise on a hot day your skin becomes flushed. This is caused by an increase in the rate at which blood flows through the capillaries in the surface layers of the skin. This transfers more thermal energy from the blood to the surface layers of the skin, which increases the skin temperature and thus the rate of radiant energy loss from the skin. (*Remember not to make the all-too-common error of stating that blood vessels move towards the surface of the skin!*)

The increase in the rate of blood supply to the capillaries in the surface layers of the skin is brought about by changes to two sets of arterioles in the skin (see Fig. 6). The arterioles that supply the capillaries **dilate** (become wider), while the 'bypass' arterioles, which divert blood away from the capillaries, constrict (become narrower). As blood flows through the surface capillaries, it becomes cooler as the effect of radiation is greater. As this cooler blood returns to the core of the body, thermal energy is transferred to it from the tissues by conduction, cooling the tissues.

Fig. 6 Vasodilation

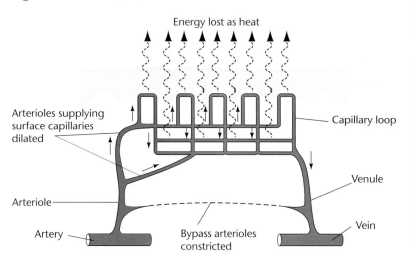

Energy lost as heat

Arterioles supplying surface capillaries dilated

Capillary loop

Arteriole

Venule

Artery

Bypass arterioles constricted

Vein

14

a How does the structure of arterioles enable them to constrict?

b What brings about the dilation of arterioles?

c Draw a flow chart to show the sequence of events as the body heats to beyond its normal body temperature.

A dog's tongue acts as its 'radiator'.

Special techniques for cooling down

Panting

Some fur-covered mammals like dogs and cats do not possess sweat glands. Fur reduces the rate of evaporation of sweat to such an extent that sweating produces no cooling effect. The sweat glands have therefore been lost during evolution. So what do these animals do to cool down? You have probably noticed dogs panting. To lose thermal energy, they breathe rapidly but shallowly. The rapid passage of air cools the blood in the blood vessels surrounding the mouth and lungs. The correct name for this process is thermal panting. A dog may pant up to 300 times a minute on a hot day.

hsw

how science works

Football can be a dangerous game

This statement is true – but not for the reasons you might think! The highest cause of sports-related death among American high school football players is not physical injury, but heat stress! About six players die each year from this and many more are made seriously ill.

The problem is mainly due to the padding used to prevent physical injuries. The padding around the chest and shoulders has a mass of about 6 kg. But most types of padding do not allow sweat to evaporate. A typical footballer produces about 5 kJ of heat energy per minute during a game. Without all the padding the footballer would lose about 4.5 kJ of heat energy per minute by the evaporation of sweat. However, if the sweat can't evaporate, the core body temperature may rise by about 1 °C every five minutes. If precautions are not taken the footballer will soon feel heat-stress symptoms.

To alleviate this problem, scientists have developed a special quilted fabric with which to make the padding for footballers. This fabric consists of three layers:

- an inner, porous, good thermal-conducting layer
- a sandwiched layer of water-absorbing fibres
- an outer layer of breathable material.

15 Explain how this new material will help the footballers to keep cool.

16 Suggest how the beefy physique of American footballers contributes to heat stress.

The tiny kangaroo rat (photograph, left) and the huge wallowing hippopotamus (photograph, right) have very different strategies for keeping cool, but both involve getting wet.

Getting wet

Sweat cools the body when it evaporates so it makes sense for animals that don't sweat to find another way to make use of evaporation.

The kangaroo rat lives in hot deserts where water is scarce. These rats have short fur and no sweat glands. To cool themselves down on a hot day they produce saliva that they then lick all over their body. This evaporates, taking thermal energy from the rat's body and produces a cooling effect.

By comparison, the hippopotamus is a giant, but it also lives in a hot environment, in tropical areas of the world. Although it produces sweat, it has a small surface area to volume ratio and must find extra ways of keeping cool. One of its favourite activities is to spend most of the day wallowing in a pool of water, only coming out at night to graze. This has a dual effect: during the day, when the environment is at its hottest, the hippopotamus transfers thermal energy to the water by conduction. Since water has a higher specific heat capacity than air, this process is much more efficient than trying to cool down in air. At night, when the hippopotamus emerges from its pool, the water evaporates from the surface of its skin, providing an adequate cooling effect in the lower temperatures of the night.

What happens when the body needs to keep warm?

So far we have only considered what happens when mammals overheat. Most small mammals have the opposite problem: they have a large surface area to volume ratio and can lose heat too rapidly. They need mechanisms to conserve heat and to increase core body temperature when necessary. This is particularly important for small mammals such

as the water vole that spend their lives swimming in cool water.

Peripheral cold receptors in the skin detect when the temperature of the surroundings is lower than body temperature and send nerve impulses along neurones to the heat gain centre in the hypothalamus. This then coordinates the body's response to the cold conditions by inhibiting the activity of the heat loss centre and by sending impulses along neurones to a range of effector mechanisms across the body. These bring about:

- vasoconstriction
- hair erection
- shivering
- an increase in metabolic activity in brown fat
- the release of hormones.

> **17** Detecting a drop in the temperature of blood flowing through the hypothalamus might be too late for the body to correct. Explain why.

Vasoconstriction

Reducing the amount of heat lost through the skin is achieved by reducing the amount of blood that flows through capillaries near the surface, as shown in Fig. 7. When detector cells in the heat gain centre are stimulated, they send impulses to the arterioles in the skin. Their action has the opposite effect to impulses coming from the heat loss centre. Impulses from

Fig. 7   Vasoconstriction

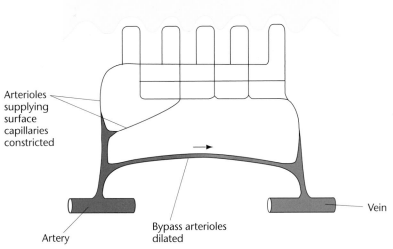

Arterioles supplying surface capillaries constricted

Artery

Bypass arterioles dilated

Vein

the heat gain centre force the arterioles that supply the capillaries to constrict, and they cause the 'bypass' arterioles that divert blood away from the capillaries to dilate. This reduces the rate of blood flow through the surface capillaries and consequently reduces heat loss by radiation. In pale-skinned people, the skin becomes much whiter, or even bluish, as the blood is no longer visible in the surface layers of the skin – hence the expression 'to turn blue with cold'.

Hair erection

Hair erection is another good strategy for conserving body heat – particularly in long-haired mammals. In the arctic fox, the insulation provided by the air trapped in its fur is so efficient that the animal does not begin to

The arctic fox is happy at temperatures as low as –40 °C, thanks to its furry insulation.

increase heat production until the external temperature drops to –40 °C.

Each hair in mammalian skin has a muscle attached to a point near its base. When a nerve impulse arrives from the heat gain centre of the hypothalamus, the hair is pulled upright. This process is known as **hair erection** (Fig. 8).

Fig. 8   Hair erection

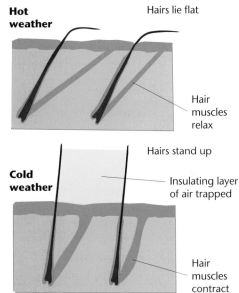

Hot weather — Hairs lie flat — Hair muscles relax

Cold weather — Hairs stand up — Insulating layer of air trapped — Hair muscles contract

You have probably noticed this on your own skin – a 'goose pimple' is a part of the skin where a hair has been pulled upright. The effect of all the hairs in the skin being pulled upright is to trap a layer of air next to the skin. Air is a poor conductor of heat, so the layer of trapped air acts as an insulator, reducing energy transfer by radiation and convection. This method of conserving energy may be important in fur-covered mammals; however, it is of little value to humans because we have so little hair.

Shivering

When we get cold, the brain sends impulses to the skeletal muscles that cause them to contract spasmodically. We call this **shivering**. Shivering muscles have a higher rate of respiration to provide the energy for muscle contraction. Since respiration in muscle cells is very inefficient – up to 80% of the energy released by the oxidation of glucose is transferred as thermal energy – lots of extra heat is released. This is transferred to the blood passing through the muscles, and then to the tissues and organs.

Brown fat

Some mammals, including some humans, have layers of cells containing 'brown fat', mainly under the skin at the back of the neck. These cells have many more mitochondria than white fat cells. When impulses arrive from the hypothalamus, the cells containing brown fat rapidly oxidise fatty acids to release energy as heat. The position of this tissue means that it acts rather like an electric blanket for the blood supply to the brain. Human babies have an almost complete covering of brown fat, but most of this disappears as we grow up. Babies use brown fat rather than shivering to generate thermal energy when they are cold. Many hibernating mammals can switch on this 'electric blanket' during particularly cold weather to prevent frostbite or freezing to death.

18 Why can brown fat cells generate much more thermal energy than white fat cells?

Hormones

Several hormones are important in temperature regulation. One of the hormones is adrenaline. When adrenaline reaches body tissues it causes an almost immediate, but short-lived, increase in the rate of respiration in the cells. The energy released in respiration is transferred as heat to the blood passing through these tissues. Another is thyroxine, a hormone produced by the thyroid gland. This has the same effect on the rate of respiration as adrenaline, but the effect is slower and more prolonged.

19 Explain why adrenaline causes an increase in the respiration rate of cells.

Ectotherms and endotherms

Birds and mammals maintain a constant high temperature by a high metabolic rate; they are called **endotherms** to indicate that the source of heat is internal. Reptiles choose a warm environment when they are active, and this supplies the necessary heat; they are called **ectotherms**, indicating that the heat source is external to the animal. A familiar sight in tropical countries is to see reptiles basking on rocks in the Sun. Heat from the Sun warms the reptiles until their metabolic rate is high enough for them to become active. If their body temperature rises too high, reptiles seek out shade. Many crocodiles shuttle between land and water during the day to try and maintain a constant body temperature. Ectotherms are not cold-blooded animals – all but the largest have body temperatures that depend mainly on the environment.

An iguana basking in the Sun.

As a result, an active reptile may use less than a tenth as much metabolic energy as an endotherm. Even when at rest the metabolic rate of a reptile is only 10%–20% of that of birds and mammals of similar size. In non-active periods of the day the body temperature can also drop, further reducing the metabolic rate.

Most of the energy used for muscular activity in reptiles is obtained from anaerobic metabolism, rather than aerobic as in endotherms. Glycogen is immediately available within their muscles to facilitate bursts of activity. But in many cases the animals would be completely exhausted by three to five minutes of maximum activity and could require several hours to completely regenerate their energy stores.

The graphs in Fig. 9 summarise the effects of temperature on ectotherms and endotherms.

20 Explain the shape of the curves on the graph for oxygen consumption by ectotherms and endotherms.

Fig. 9   The effect of external temperature on ectotherms and endotherms

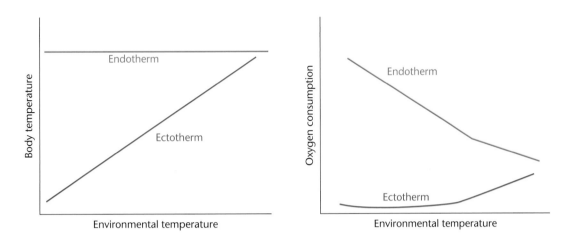

hsw

how science works

Temperature regulation in the crocodile

Scientists in Australia investigated the hypothesis that crocodiles would become warmer with increasing mass and show increased stability in body temperature. The graph in Fig. 10 summarises their results.

Daytime air temperatures averaged about 35 °C in summer and 30 °C in winter.

Fig. 10   Temperature regulation in crocodiles

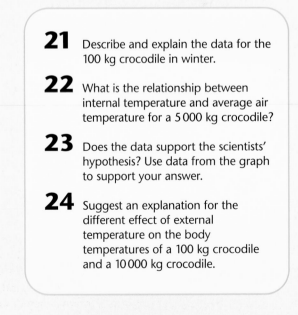

21 Describe and explain the data for the 100 kg crocodile in winter.

22 What is the relationship between internal temperature and average air temperature for a 5 000 kg crocodile?

23 Does the data support the scientists' hypothesis? Use data from the graph to support your answer.

24 Suggest an explanation for the different effect of external temperature on the body temperatures of a 100 kg crocodile and a 10 000 kg crocodile.

key facts

- The hypothalamus monitors and controls body temperature.

- The principal receptors involved are temperature receptors in the hypothalamus and in the surface layers of the skin. The skin receptors are particularly important in detecting a fall in external temperature.

- Heat loss mechanisms include:
  ○ sweating – the evaporation of sweat cools the body
  ○ vasodilation – the arterioles that supply the capillaries dilate, while the 'bypass' arterioles that divert blood away from the capillaries constrict.

- Heat gain mechanisms include:
  ○ vasoconstriction – the arterioles that supply the capillaries constrict, while the 'bypass' arterioles that divert blood away from the capillaries dilate
  ○ contraction of the hair erector muscles – the erect hairs trap an insulating layer of air
  ○ shivering – skeletal muscles contract, releasing thermal energy
  ○ high metabolic rate in the brown fat tissue.

- The adrenal and thyroid glands produce hormones that increase the rate of metabolism.

- Remember that blood vessels do NOT move up and down in the skin. Always make clear whether you are referring to arterioles or to capillaries. Vasodilation and vasoconstriction refer to arterioles – NOT capillaries.

- Endotherms maintain a constant high temperature by a high metabolic rate.

- The body temperature of an ectotherm is largely dependent on the external temperature. Ecotherms have behaviour patterns to obtain optimal body temperatures.

1 In an investigation, a locust was given alternating supplies of atmospheric air and pure carbon dioxide. The rate of pumping movements of the insect's abdomen was measured.
The graph shows the results.

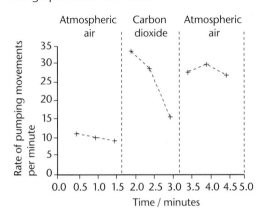

a Explain what caused
  i the rise in the rate of abdominal pumping movements between 1.5 and 2.0 minutes, (1)
  ii the fall in the rate of abdominal pumping movements between 2.0 and 3.0 minutes. (2)

b The rate of abdominal pumping movements increases between 3.0 and 3.5 minutes.
Suggest the advantage of this change to the locust. (1)

Total 4

AQA, A, June 2006, Unit 6, Question 2

2 The echidna is an Australian mammal. In winter, its body temperature falls to a temperature similar to that of its environment and it hibernates. However, during the period of hibernation, it becomes active every few weeks and at these times its temperature rises to a level similar to its summer temperature. The graph shows how the echidna's temperature varies in the summer and in the winter.

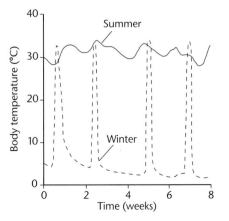

a Explain how the fall in body temperature to that of the environment helps the echidna to survive the winter. (2)

b Explain how a higher body temperature is of benefit to an active echidna. (2)

Total 4

AQA, A, January 2005, Unit 6, Question 3

3
a **i** What is meant by *homeostasis*? (1)
  ii Giving **one** example, explain why homeostasis is important in mammals. (2)

b **i** Cross-channel swimmers experience a large decrease in external temperature when they enter the water. Describe the processes involved in thermoregulation in response to this large decrease in external temperature. (7)
  ii A person swimming in cold water may not be able to maintain their core body temperature and begins to suffer from hypothermia. Explain why a tall, thin swimmer is more likely to suffer from hypothermia than a short, stout swimmer of the same body mass. (2)

c Cross-channel swimmers may suffer from muscle fatigue during which the contraction mechanism is disrupted. One factor thought to contribute to muscle fatigue is a decrease in the availability of calcium ions within muscle fibres. Explain how a decrease in the availability of calcium ions could disrupt the contraction mechanism in muscles. (3)

Total 15

AQA, B, June 2006, Unit 4, Question 8

4 Many diabetics inject insulin, because their pancreas has stopped producing it. Attempts have been made to transplant pancreatic cells from human embryos into diabetics but these foreign cells are often destroyed as a result of antibodies produced by the diabetic's immune system.
The diagram shows a new type of transplant which has been tested in rats.

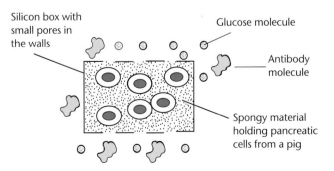

a i Explain why this transplant is not destroyed by the rat but can respond to changes in the rat's blood glucose concentration. (2)

ii Suggest why there might be controversy if this transplant was used in humans. (1)

b Explain how the cells in the transplant control the blood glucose concentration of the rat. (3)

Total 6

AQA, B, June 2003, Unit 4, Question 2

5

a One effect of getting into a cold shower is a reduction in the amount of blood flowing through the capillaries near the surface of the skin. Explain how the cold water causes this response. (4)

b i When exercising at 30 °C, the body is more likely to overheat in humid conditions than in dry conditions. Explain why. (2)

ii Strenuous exercise leads to exhaustion more quickly in hot conditions than in cool conditions. One reason for this is a reduced blood supply to the muscles, which means that they receive less oxygen. Suggest an explanation for the reduced blood supply to the muscles. (2)

Total 8

AQA, B, June 2004, Unit 4, Question 3

6 Read the following passage.

Diabetes

Diabetes mellitus is a group of disorders that all lead to an increase in blood glucose concentration (hyperglycaemia). The two major types of diabetes mellitus are type I and type II. In type I diabetes there is a deficiency of insulin. Type I diabetes is also called insulin-dependent diabetes mellitus because regular injections of insulin are essential. It most commonly develops in people younger than age twenty.

Type II diabetes most often occurs in people who are over forty and overweight. Clinical symptoms may be mild, and the high glucose concentrations in the blood can be controlled by diet and exercise. Some type II diabetics secrete low amounts of insulin but others have a sufficient amount or even a surplus of insulin in the blood. For these people, diabetes arises not from a shortage of insulin but because target cells become less responsive to it. Type II diabetes is therefore called non-insulin-dependent diabetes mellitus.

a Describe how blood glucose concentration is controlled by hormones in an individual who is **not** affected by diabetes. (6)

b Suggest how diet and exercise can maintain low glucose concentrations in the blood of type II diabetes. (3)

c Glucose starts to appear in the urine when the blood glucose concentration exceeds about 180 mg dm^{-3}. Explain how the kidney normally prevents glucose appearing in urine. (3)

Total 12

AQA, B, June 2000, Unit 3, Question 8

Treating hypothermia

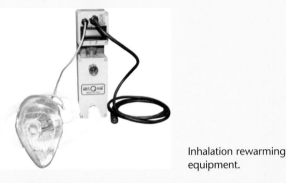

Inhalation rewarming equipment.

Hypothermia is a fall in core body temperature that occurs when the body remains in a cold environment for too long. Hypothermia can affect anyone, from a child who falls through thin ice into cold water, to a super-fit climbing enthusiast stranded on a mountain during bad weather, to a frail elderly woman whose house is too cold because she can't afford heating bills.

Fig. 11 Effects of different treatments for hypothermia

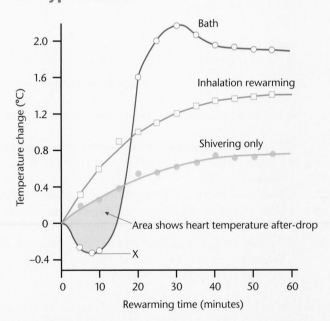

Treating hypothermia is difficult and people have been known to survive long periods of cold, even though they appeared dead when discovered. The basic principles in treating hypothermia are:

- prevention of further heat loss
- rewarming as soon as is safely possible at a 'successful' rate, i.e. a rate that will not itself produce further damage
- rewarming the core before the shell, in an attempt to avoid inducing lethal side effects during rewarming.

Hypothermia causes several reactions within the body as it tries to protect itself and retain its heat. The most important of these is vasoconstriction, which halts blood flow to the extremities in order to conserve heat in the critical core area of the body.

When core temperature falls below 36 °C, the major source of heat production is shivering. This maintains peripheral vasoconstriction, which minimises the severity of vascular collapse during rewarming. Induction of vasodilation in these patients may precipitate rewarming shock.

Rapid shunting of cold blood from the periphery to the core as the direct result of vasodilation may cause the core temperature to drop. This phenomenon of a drop in temperature after initiation of therapy is termed 'core temperature after-drop'. Prevention of vasodilation is the reason why it is imperative that the patient's extremities should not be rewarmed before the core. If vasodilation occurs, cold blood returning to the heart may be enough to cause ventricular fibrillation (abnormal heartbeat).

The patient must also be handled very gently and not be allowed to exercise, as muscular action can pump cold blood to the heart. One method of treating hypothermia is inhalation rewarming, getting the patient to breathe warmed air. This can be done with equipment such as RES-Q-AIR, shown in the photograph on this page.

Another method, only to be used if the patient has relatively mild hypothermia, is to place the patient in a warm bath. The graph shows the effects of these two treatments on a patient with mild hypothermia, compared with simply leaving the patient to shiver.

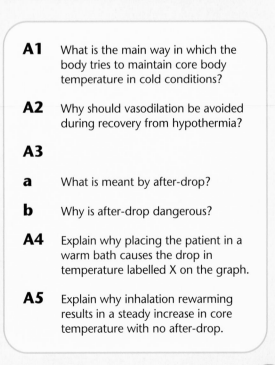

A1 What is the main way in which the body tries to maintain core body temperature in cold conditions?

A2 Why should vasodilation be avoided during recovery from hypothermia?

A3

a What is meant by after-drop?

b Why is after-drop dangerous?

A4 Explain why placing the patient in a warm bath causes the drop in temperature labelled X on the graph.

A5 Explain why inhalation rewarming results in a steady increase in core temperature with no after-drop.

14 DNA and protein synthesis

DNA vaccine offers painless jabs

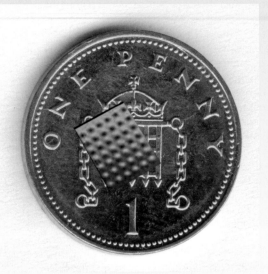

A tiny plate containing up to 400 needles is used and tests have found them to be completely painless.

Squeamish patients who cannot stand the sight of needles could be offered pain-free injections, thanks to the work of scientists at the Welsh School of Pharmacy at Cardiff University. The new micro-needles are long enough to penetrate the skin but not to reach pain receptors. They were designed to introduce a DNA vaccination directly into skin cells.

The micro-needles, manufactured by the Tyndall National Institute in Ireland, measure up to 300 microns (0.3 mm) across, and are barely visible to the naked eye. A tiny plate containing up to 400 needles is used and tests have found that these injections are completely painless.

James Birchall, head of the gene delivery research group, said: "Think of the bed-of-nails effect – the forces are spread over a wide surface area." Conventional needles go in too deep for the method of 'genetic vaccination' his team are developing. The micro-needles work by creating temporary channels in the skin to allow the vaccine to reach 'immune-responsive' skin layers.

They are currently made of silicon, but biodegradable needles that dissolve in the skin are a future possibility. Dr Birchall said DNA vaccines delivered via a micro-needle could have several advantages over standard vaccines.

"They are likely to be cheaper and easier to make. The micro-needle system might also be developed as a patch for self-application, avoiding the need for a clinician," said Dr Birchall. "There is also reduced risk of transmission of blood-borne pathogens by inappropriate re-use of needles. These are all particular advantages for the developing world. Pain-free vaccination could also be useful for childhood vaccines in developed countries." The new needles are not likely to be seen in hospitals or surgeries for at least five years.

The next stage of the research will investigate delivery of vaccines such as hepatitis B and influenza. The vaccinations work by introducing DNA directly into the skin cells. This DNA initiates the production of proteins that are recognised as antigens by the body. In response, the B lymphocytes produce antibodies that are effective against the pathogen.

In this chapter you will learn about the link between DNA codes and protein manufacture.

Back to basics

In the AS book you learned the structure and function of DNA.

- DNA molecules consist of two **polynucleotide** strands linked together.
- The sequence of bases in the nucleotides enables the DNA to store information.
- The **triplet sequences** of bases on the **sense strand** of the DNA molecule that code for different amino acids are called **codons**.
- Since there are 64 possible triplet sequences and only 20 amino acids to code for, the DNA code is **degenerate** (i.e. each amino acid has more than one codon). However, the reverse is not true, there are *no* codons that specify more than one amino acid.
- **Exons** are the functional part of the DNA molecule.
- **Introns** are regions of 'junk DNA'.

14.1 Transcription

Ribonucleic acid (RNA)
DNA carries the coded information to make polypeptides and proteins. But how can one small section of DNA on one chromosome in a cell nucleus make enough of a protein to supply the whole cell? Imagine trying to make enough copies of a best-selling compact disc from a single original master disc. Clearly it is much more efficient to make several copies of the DNA master template and then use these copies to produce the quantity of protein that a cell needs. And this is exactly what the cell does. Copies, or 'imprints', of the DNA code are produced. These copies are **messenger**

ribonucleic acid molecules **(mRNA)** (Fig. 1). They pass from the nucleus into the cytoplasm and are then used as guides to manufacture the protein encoded in their sequence of bases.

Fig. 1   Messenger RNA

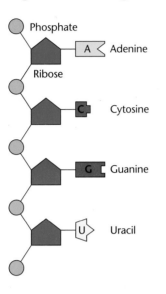

RNA molecules are well suited to their function. They use a similar four-base system as DNA, enabling the genetic code to be copied from DNA to mRNA. The bases are exposed on a single strand of mRNA, and this strand can be used to assemble amino acids. The molecules of mRNA are small enough to pass through pores in the nuclear membrane. Unlike DNA, RNA molecules are quite short-lived; this enables the cell to change protein production to suit its needs.

The structure of RNA is similar to that of a single strand of DNA, except that the sugar **ribose** replaces **deoxyribose**, and the base **thymine** is replaced by another base called **uracil**. Uracil and thymine molecules are similar in size and shape, and uracil pairs readily with adenine.

The process of copying the code in DNA to form mRNA is called **transcription** (Fig. 2). Transcription starts when an enzyme catalyses a reaction that makes the DNA of a gene untwist. Another enzyme, **RNA polymerase**, then assembles the RNA nucleotides along one side of the DNA molecule. This active side is called the **sense strand**. As the RNA polymerase moves along the sense strand, it produces a single-stranded molecule of mRNA. mRNA carries coded information in the same way as DNA; the order

Fig. 2   Transcription

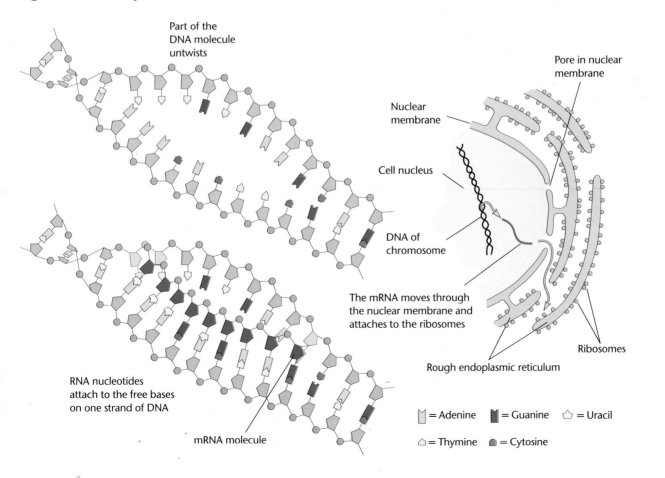

of bases on an mRNA molecule is a 'mirror image' of those on the sense strand of DNA, except that uracil bases are used in place of thymine.

The sections of DNA sense strand and mRNA shown in Fig. 2 have these bases:

DNA   A A A C A C T T C
mRNA   U U U G U G A A G

The three mRNA codons are therefore UUU, GUG and AAG.

The mRNA detaches from the DNA and passes out of the nucleus through the nuclear pores and into the endoplasmic reticulum. The mRNA attaches to the **ribosomes**, which are also made of RNA. They have a specially shaped 'pocket' that is rather like the active site of enzyme molecules. The mRNA molecule fits into this pocket and the process of translation begins.

RNA splicing

In most eukaryotic cells the mRNA contains both exons and introns. This form of mRNA is called **pre-mRNA**. Before it leaves the nucleus this pre-mRNA is treated to remove the introns, a process known as splicing. The resulting mRNA is called **mature mRNA**.

The steps of pre-mRNA splicing are shown in Fig. 3:

- **snRNPs** (small nuclear ribonucleoprotein particles) are formed. Each snRPN particle consists of an snRNA (small nuclear RNA – a RNA molecule consisting of about 150 nucleotides) and a protein molecule, joined to form a complex. This complex binds onto the pre-mRNA to form the spliceosome.

Fig. 3   RNA splicing

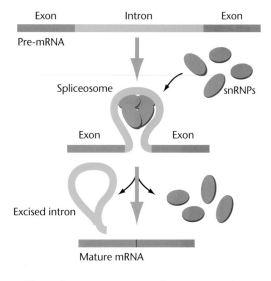

- The spliceosome causes the intron to form a loop shape.
- The intron is excised, and the exons are then spliced together.
- The resulting mature mRNA may then exit the nucleus.

---

1 List four differences between molecules of DNA and mRNA.

2 What will be the order of nucleotides in the mRNA molecule produced by this section of a strand of DNA?

  A C G A T T G T G C A C G A G

14.2 Translation

Once the DNA code has been transcribed and the mRNA copies have passed out of the nucleus to the ribosomes, the code on the mRNA is used to assemble the amino acids of the protein in the correct order. This process is called **translation** (Fig. 4).

Transfer RNA (**tRNA**) plays a key role in this process. Molecules of free tRNA are present as tRNA–amino acid complexes in the endoplasmic reticulum, near to the ribosomes. Each of the 20 amino acids has a specific tRNA molecule. The tRNA molecule is arranged in a cloverleaf shape with three bases sticking out from one of the 'leaves'. The distance between these bases and the amino acid at the other end of the molecule is the same in all tRNA molecules.

When one end of the mRNA strand attaches to the ribosome, a tRNA molecule that matches the first codon on the mRNA binds to the ribosome, carrying its amino acid with it. The triplet of bases on the tRNA that binds to the codon on the mRNA is called the **anticodon**. A second tRNA molecule, also carrying its amino acid, then binds to the next codon on the mRNA. So, two tRNA molecules bind to the ribosome at once.

Fig. 4 Translation

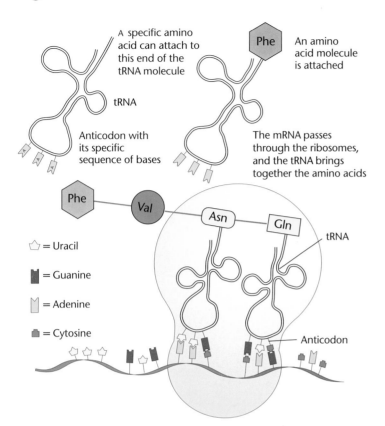

A specific amino acid can attach to this end of the tRNA molecule

tRNA

Anticodon with its specific sequence of bases

Phe
An amino acid molecule is attached

The mRNA passes through the ribosomes, and the tRNA brings together the amino acids

⬠ = Uracil

⬛ = Guanine

⬜ = Adenine

⬛ = Cytosine

tRNA

Anticodon

moved across the ribosome, the first tRNA molecule, minus its amino acid, falls off and the second tRNA molecule moves across to take its place. The next codon becomes available to bind a tRNA molecule with the next amino acid, which is then added to the growing polypeptide chain.

The order of codons on the mRNA molecule determines which tRNA molecules bind, and the tRNA molecules determine which amino acids are brought together. The whole system ensures that the amino acids are assembled in the correct sequence to make the polypeptide chain encoded by the original gene on the DNA molecule.

A ribosome can translate any piece of mRNA. This means that a group of 40 ribosomes could work on 40 different mRNA molecules to produce 40 different proteins. Alternatively, the 40 ribosomes could work on 40 copies of the same piece of mRNA to produce large amounts of the relevant polypeptide very quickly.

Usually the polypeptide that is released from the ribosome after translation needs to be processed by enzymes to make it fully functional. Some proteins are made from more than one polypeptide; others need to be combined with polysaccharides or metal ions before they can work properly. The amount of a particular protein needed by a cell varies; it does not need to make large amounts of the same protein all the time. The cell controls which genes are switched on and which are quiet. This process is incredibly complex and we are only just beginning to understand how some of the control systems work.

When they are both in place, the amino acids at the far end of the molecules are very close together. The amino acids are then joined together by a peptide bond. Energy from ATP is needed for this reaction to occur. The mRNA is

---

3 Give two similarities and two differences between the structure of mRNA and that of tRNA.

4 Copy and complete Table 1, showing the codes at each stage of the process in assembling the first seven amino acids of an insulin molecule.

5 A polypeptide consists of 145 amino acids; 14 different amino acids are contained in its structure.

a How many base pairs must there be in the gene that codes for this polypeptide?

b How many nucleotides are there in the mRNA that is transcribed from this gene?

c How many different types of tRNA are needed for the synthesis of this polypeptide?

Table 1

	Amino acid						
	Phe	**Val**	**Asn**	**Gln**	**His**	**Leu**	**Cys**
DNA code in gene	AAA	CAC	TTG	GTC	GTG	GAG	ACG
Codon in mRNA	UUU	GUG					
Anticodon of tRNA	AAA	CAC					

The genetic code

Table 2   mRNA codons for the amino acids

First base	G	A	C	U	Third base
G	GGG glycine	GAG glutamic acid	GCG alanine	GUG valine	G
	GGA glycine	GAA glutamic acid	GCA alanine	GUA valine	A
	GGC glycine	GAC aspartic acid	GCC alanine	GUC valine	C
	GGU glycine	GAU aspartic acid	GCU alanine	GUU valine	U
A	AGG arginine	AAG lysine	ACG threonine	AUG methionine	G
	AGA arginine	AAA lysine	ACA threonine	AUA isoleucine	A
	AGC serine	AAC asparagine	ACC threonine	AUC isoleucine	C
	AGU serine	AAU asparagine	ACU threonine	AUU isoleucine	U
C	CGG arginine	CAG glutamine	CCG proline	CUG leucine	G
	CGA arginine	CAA glutamine	CCA proline	CUA leucine	A
	CGC arginine	CAC histidine	CCC proline	CUC leucine	C
	CGU arginine	CAU histidine	CCU proline	CUU leucine	U
U	UGG tryptophan	UAG stop	UCG serine	UUG leucine	G
	UGA stop	UAA stop	UCA serine	UUA leucine	A
	UGC cysteine	UAC tyrosine	UCC serine	UUC phenylalanine	C
	UGU cysteine	UAU tyrosine	UCU serine	UUU phenylalanine	U

The table above shows which amino acids are encoded by the mRNA codons. There are three particularly important things that you should notice about the code.

- Often it is only the first two bases of the triplet that are specific for a particular amino acid, and any third base will do. This also reduces the chance that a change in the bases will alter the function of the polypeptide.
- There are three 'stop' codes. These indicate the end of a section of mRNA, after which point translation stops.

- The codon for methionine, AUG, is also used as a start code. This means that polypeptides normally start with a methionine group when they are freshly translated. It is often removed in the processing stage that converts the polypeptide into a functional protein.

6 Look at Table 2.

a A section of mRNA has the order of bases: AAG CGC UCU GCA. What will be the order of amino acids in the polypeptide it codes for?

b What are the corresponding DNA codons on the gene that produced this section of mRNA?

c Which anticodons on the tRNA molecules attach to this section of mRNA?

7 The first stages in deciphering the genetic code involved making synthetic mRNA. The polypeptides they produced were then analysed.

a The researchers made mRNA in which all the bases were uracil. The polypeptide produced consisted entirely of the amino acid, phenylalanine. Explain why.

b What amino acids would the polypeptide contain if the bases on the mRNA were all adenine?

c The researchers then produced mRNA in which the bases uracil and cytosine alternated: UCUCUCUC. The polypeptide produced contained equal amounts of two amino acids. Which two? Explain your answer.

Table 3   Characteristics of nucleic acids

Feature	DNA molecule	mRNA molecule	tRNA molecule
Number of strands	2	1	1
Length	Longest	Shorter	Shortest
Situated	In nucleus	Made in nucleus, then attached to ribosome	In cytoplasm
Bases	A T C G	A U C G	A U C G
Attachment			Specific amino acid

● The characteristics of nucleic acids are listed in Table 3.

● The DNA of a gene is not used to make a polypeptide in the nucleus because this would be too slow a process. Instead, RNA copies of the gene's code are made.

● Many RNA copies of the coded information contained in a stretch of DNA can be made. This enables polypeptide products to be produced rapidly.

● One strand of the gene's DNA is used to make the copies of mRNA, which have a matching code. This process is transcription.

● In many eukaryotes, mRNA is transcribed as pro-mRNA that contains both exons and introns.

● In a process called splicing, introns are removed from pro-mRNA and the exons are spliced to form mature mRNA.

● The mature mRNA passes out of the nucleus and attaches to ribosomes in the endoplasmic reticulum.

● The endoplasmic reticulum has a plentiful supply of tRNA molecules that are attached to specific amino acids. The tRNA molecules have anticodons that recognise and bind to the corresponding mRNA codon.

● As the mRNA moves through a ribosome, the amino acids carried by the tRNA are combined in the correct sequence to form the polypeptide. This process is translation.

● The polypeptides formed can then be used to make a specific protein, for example, an enzyme, a membrane protein or a structural protein.

14.3 Mutation

From time to time, errors occur during DNA replication. For example, one nucleotide in a strand may be replaced by another, or extra nucleotides may be added. As a result, the sequence of bases in the DNA is changed. A change in the order of bases in a gene is called a **gene mutation**. Gene mutations can result from a change of just one base.

There are three basic types of gene mutation:

• **Addition** – an extra nucleotide is inserted, so an extra base is added to the sequence.
• **Deletion** – a nucleotide is removed.

• **Substitution** – a nucleotide is replaced by a nucleotide with a different base.

These sentences illustrate the three types:

Original – THE OLD MEN SAW THE LAD

Addition – THE COLD MEN SAW THE LAD

Deletion – THE OLD MEN SAW THE AD

Substitution – THE OLD HEN SAW THE LAD

Other errors include the **inversion** of a sequence (THE OLD MEN WAS THE LAD), or **duplication** (THE OLD OLD MEN SAW

THE LAD). Sometimes errors involve several nucleotides and thus a significant chunk of the 'message'. The effect of the mutation depends on how much the code is disrupted. A single substitution will only affect one codon, whereas an addition or deletion may affect all the codons beyond the error.

---

8 One strand of DNA has the following sequence of nucleotide bases:

CATCATAGATGAGAC

a Which type of mutation could have produced each of the following mistakes during replication of the original DNA sequence?

CATCGTAGATGAGAC
CATCATAAGATGAGACC
CATCAGAGATGAGAC
CATATAGATGAGAC

b Use the genetic code in the table on page 244 to work out the amino acid sequence that the original code and each of the mutations would code for. Don't forget that the table shows the mRNA codons, not DNA.

c Describe the effect that each of these mutations would have on the polypeptide produced by the gene.

The consequences of mutation

The change in the code caused by a mutation may mean that a different polypeptide, and hence a different protein, is produced by a gene. This protein may not have the same properties as the original, and often does not work in the same way. A mutation can produce a different form of the gene, and so a new allele.

When the mutated DNA replicates, the new form is copied, so the mutation passes on to other cells. If a mutation occurs in an ovary or testis as the gametes are being produced, the new allele may be passed on to offspring, and may spread to many individuals. Often an allele cannot spread because its effects are too damaging; for example, if the protein the original allele coded for is vital and the mutation cannot produce it, then the organism will not develop.

An albino thrush.

The albino thrush in the photograph has a gene mutation which means that it cannot make black pigment. It probably has a poor chance of survival because it is so conspicuous. Sometimes the absence of the correct protein may be either harmless or at least not too serious a problem. Occasionally mutations can increase survival chances; such mutant alleles provide the genetic variation that permits natural selection and evolution.

Gene mutations occur naturally at random. As we get older, more and more cells will contain gene mutations. Mutations in body cells cannot be passed on to offspring. Mutations that occur during development may cause abnormal growth of the parts formed from the cell with the mutation, as you can see in the photograph of a horse chestnut with a patch of leaves without chlorophyll.

A horse chestnut tree showing a patch of leaves that cannot make chlorophyll.

Mutations and cancer

When mutations occur in the genes that control cell division, unchecked irregular growth takes place and a tumour develops. The frequency of mutations that lead to cancer is increased by certain **mutagenic agents**. Mutagens may cause DNA molecules to break, or change a small section of DNA chemically. Breaks in a DNA molecule in a cell are mended by an enzyme, **ligase**, which joins the broken ends, but in this process it is possible for a nucleotide to be deleted or for some other defect to occur. High-energy radiation, including X-rays, gamma rays and ultraviolet light, are mutagens, as are high-energy radioactive and ionised particles. X-rays and gamma rays can penetrate deep into the body and may cause mutations in any tissue. Damage is especially serious in tissues where cell division is rapid, such as the bone marrow where blood cells are made. The effect is cumulative, so many small doses have the same effect as one large dose. Radioactive substances such as uranium and plutonium release particles with higher energy levels than radiation, so they can have an even greater mutagenic effect. Atomic particles do not penetrate tissues in the same way as radiation, but absorbing radioactive substances into the body in food or breath is very dangerous, because they continue to decay and emit particles. Many chemicals, especially organic compounds such as those that occur in tobacco tar, cause mutations. All new drugs and pesticides must be tested to see if they are likely to be mutagenic.

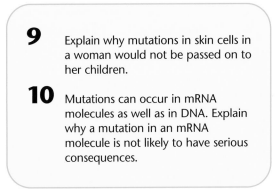

9 Explain why mutations in skin cells in a woman would not be passed on to her children.

10 Mutations can occur in mRNA molecules as well as in DNA. Explain why a mutation in an mRNA molecule is not likely to have serious consequences.

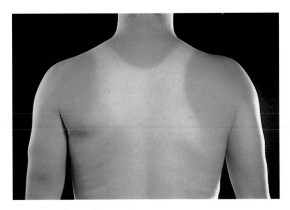

This person is probably experiencing the discomfort of severe sunburn. This will pass quickly, but repeated exposure to strong ultraviolet light causes an increase in the mutation rate in skin cells. Skin cancer is then much more likely to develop. Thinning of the ozone layer, which normally acts as a shield, is allowing more ultraviolet rays to reach the Earth's surface, especially in regions nearer the poles. This is increasing the incidence of skin cancer, especially in white-skinned people in several countries.

Genes and cancer

Normally, cell division is tightly controlled. The growth process is controlled by genes, called **proto-oncogenes**. Only when specific proto-oncogenes are switched on by a growth factor does a cell grow and divide. For much of the time the activity of tumour-suppressor genes inhibits the proto-oncogenes. Sometimes, however, something goes wrong with this 'switching-off' mechanism. A mutation of the relevant tumour-suppressor gene can also allow a proto-oncogene to keep cell division going. It stimulates continuous cell division long after it has ceased to be necessary for normal body function.

11 Cancer research scientists are discovering more genes that can affect cell growth and cancer development. For example, a gene called p53 has been found to activate a tumour-suppressor gene. This gene is often mutated in cells taken from a colon cancer tumour. Draw a flow diagram to show how mutation of the p53 gene could cause colon cancer to develop.

Fig. 5

A skin tumour behind the ear of an elderly man (top) and a coloured chest X-ray (right) showing a cancerous tumour in the left lung.

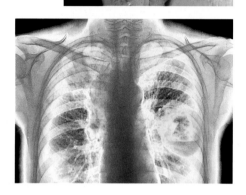

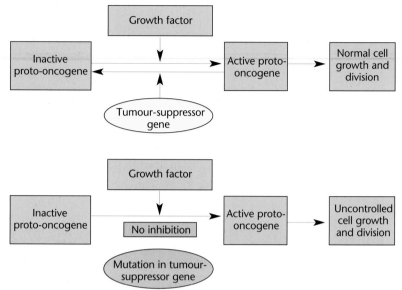

---

12 In a metabolic pathway a series of reactions takes place. Each reaction is catalysed by a different enzyme. Look at the following pathway:

Enzyme A   Enzyme B   Enzyme C
Substance W → Substance X → Substance Y → Substance Z

A mutation of the gene that codes for an enzyme may result in the protein produced having a different tertiary structure so that it cannot function. Suppose that the gene for Enzyme B mutates, and no Enzyme B is produced.

a Explain why production of Substance Z stops.

b Explain why Substance X accumulates.

c Explain what would happen if Substance Y were then supplied.

---

key facts

● A gene mutation occurs when there is a change in the sequence of bases in the DNA of a gene. Bases may be added, deleted or substituted. Segments of DNA may be inverted or duplicated.

● A mutation produces a change in the DNA codons and is likely to result in a polypeptide with a different amino acid sequence.

● New alleles arise from mutations in existing alleles.

● Mutations in reproductive cells can be passed on to following generations, but mutations in body cells will only affect the tissues in which they occur.

● Mutations occur naturally at random, but the rate of mutation is increased by mutagens such as radiation and some organic chemicals.

● The rate of cell division is controlled by proto-oncogenes that stimulate cell division and tumour-suppressor genes that slow cell division.

● A mutated proto-oncogene, called an oncogene, stimulates cells to divide too quickly. A mutated tumour-suppressor gene is inactivated, allowing the rate of cell division to increase.

14.4 Control of gene expression

What makes your fingers grow to a certain size and then stop? How is it that some cells in your fingers make bone, while others produce skin, nerves, muscles, blood vessels and so on? As the skin on your fingertips wears away, cells underneath produce new layers, following exactly the same pattern of fingerprints. How do these cells 'know' what to do? And when to stop doing it?

Cells produced by mitosis may grow and become specialised, or they may have a relatively short growth period before dividing again. Some, for example, most nerve cells, may never divide again. They nevertheless remain active and their genes continue to produce the proteins necessary for their function as nerve cells. Until recently it was thought that no new brain cells are made in humans after about the age of 16 years, and that as cells die they are never replaced. However, even with a loss of several thousand per day, there are still plenty to spare. More recently this idea has been challenged, and there is evidence that even specialised nerve cells may be stimulated to divide again.

Cells in embryos and those in tissues that have a high cell turnover, such as the skin, gut lining and bone marrow, have quite a short interphase. During the first part of interphase the genes are actively involved in transcription, and growth occurs. New organelles are formed, and some of these, such as mitochondria and chloroplasts, contain their own small sections of DNA that enable these organelles to reproduce independently.

Totipotent cells

The explanation of how cells specialise begins with normal sexual reproduction and the development of an embryo. First, a sperm and an egg, each with 23 chromosomes, fuse to form a fertilised egg with a full complement of 46 chromosomes – 23 from the father and 23 from the mother. This fertilised egg has the potential to form any and every type of human cell; the fertilised egg is **totipotent** – a cell with total potential. The fertilised egg cell divides and each new cell divides again. This process is repeated many times.

After many such divisions, these totipotent cells begin to develop into a specific type of cell – a liver cell or a skin cell, for example. The kind of cell is determined by the gene or genes that are turned on in that particular cell. For example, if the genes that control liver formation are activated, the totipotent cell becomes a liver cell. But, if the genes that control muscle formation are turned on, that same totipotent cell can become a muscle cell. When becoming specialised, totipotent cells translate only the relevant parts of their DNA.

Under normal conditions, once the gene (or genes) that control muscle formation is activated, that cell loses the ability to become anything other than a muscle cell – it could never become a liver cell or a brain cell. In other words, the cell can never become totipotent again – at least that is what scientists once thought until the remarkable experiment that resulted in the birth of the sheep Dolly.

Dolly was grown from a single udder cell from a six-year old sheep. Before Dolly's birth it was thought that once a cell had been specialised in an organ such as the heart or udder, it was impossible to reprogramme it to go through the same full cycle of development as a fertilised egg.

At first it was thought that Dolly might be genetically damaged and would already be an 'old' ewe when born. It is still not certain whether

Dolly, the world's first cloned sheep with her lamb Bonnie.

Dolly's DNA had more than average disruption as a result of mutation, but in 1998 she gave birth to a perfectly normal lamb. However, Dolly died in 2003, at a much younger age than is normal for a ewe. Premature death has also been seen in clones of other species.

Although widely publicised as a scientific success story, Dolly's birth was achieved only after many failures. Would it be acceptable for deformed or handicapped babies to be born in the pursuit of a perfect clone? What would be the legal status of a child born by cloning one individual, and would such children have problems accepting their own identity? Would cloning only be available to the super-rich who can afford the expense? These issues need debate and legislation and the future is uncertain. The only thing that seems sure is that there is no possibility of reversing the scientific understanding and technological advances that have made cloning a reality.

Micropropagation

Micropropagation involves the growth of **plantlets** from single cells, pieces of tissue or organs, using sterile laboratory techniques (Fig. 6).

Plantlets are very small plants that are produced by asexual reproduction. In mature plants many cells remain totipotent. These cells have the ability to give rise to an identical genetic copy of the parent plant. In the case of the oil palm in the photograph below, the clones are derived from tiny pieces of leaf tissue.

A plantation of oil palms.

This plantation of oil palms is 15 years old. The oil palm is an important tropical crop. The oil has many uses, for example, in foods such as margarine and in soap. Traditional breeding programmes to produce plantations take several years to generate enough plants with the desired characteristics. Oil palms are now bred successfully by micropropagation, and plantations developed using this technique have been grown in Malaysia since 1977. Oil palm plantlets are obtained by growing small pieces of leaf tissue, so large numbers of plantlets can be obtained rapidly and in a small space. The technique is particularly useful for obtaining clones from hybrid plants that are naturally sterile and unable to reproduce.

Growth rooms with a constant environment are used for raising cultured plant material such as oil palm plantlets. The conditions are kept sterile, which means the plantlets are kept free from disease. Environmental conditions (for example, temperature) are kept at the best levels for healthy growth.

Fig. 6   Micropropagation

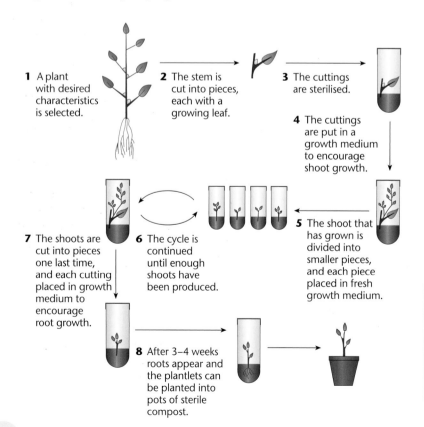

1 A plant with desired characteristics is selected.

2 The stem is cut into pieces, each with a growing leaf.

3 The cuttings are sterilised.

4 The cuttings are put in a growth medium to encourage shoot growth.

5 The shoot that has grown is divided into smaller pieces, and each piece placed in fresh growth medium.

6 The cycle is continued until enough shoots have been produced.

7 The shoots are cut into pieces one last time, and each cutting placed in growth medium to encourage root growth.

8 After 3–4 weeks roots appear and the plantlets can be planted into pots of sterile compost.

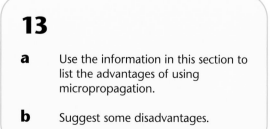

13

a   Use the information in this section to list the advantages of using micropropagation.

b   Suggest some disadvantages.

A constant environment room.

Stem cells

In mature mammals only a few cells remain totipotent. These cells are called **stem cells**. Stem cells share the following general characteristics:

- the ability to differentiate into specialised cells
- the ability to regenerate an infinite number of times
- the ability to relocate and differentiate where needed.

There are three main classes of stems cells, listed below and shown in Fig. 7:

- **Totipotent cells** – after fertilisation (union of sperm and egg), the zygote created is a totipotent cell. This one totipotent cell divides into multiple totipotent cells for up to five days after fertilisation (three or four cellular divisions).
- **Pluripotent cells** – after about five days, these totipotent cells begin to differentiate, or

specialise, and form a hollow ball of cells called a **blastocyst**. The blastocyst has an outer layer of cells that eventually form the placenta, and a cluster of cells inside the hollow sphere called the **inner cell mass**. The cells of the inner cell mass are pluripotent cells, meaning that they each have the potential to create every cell of the body but not the placenta. Pluripotent cells can be isolated from embryos. The use of these cells in stem cell research creates moral and ethical dilemmas that are still being debated in most countries.

- **Multipotent cells** – pluripotent cells soon undergo further specialisation into multipotent cells, usually referred to as **adult stem cells**. These cells can give rise to a limited number of other particular types of cells. For example, haematopoietic cells (blood cells) in the bone marrow are multipotent and give rise to the various types of blood cells, including red cells, white cells and platelets.

Fig. 8 overleaf shows the main sources of stem cells for research.

Multipotent cells are found in both developing fetuses and fully developed human beings. There are certain limitations to using multipotent cells, however. Scientists have not identified multipotent cells for every type of mature body cell. Unlike pluripotent cells, multipotent cells are often in minute quantities and their numbers usually decrease with age.

Multipotent cells from a specific patient may take time to mature in culture in order to produce adequate numbers for treatment. They often contain DNA damage due to ageing, toxins, and random DNA mutation during replication. Spontaneous mutations are more likely to show up in older multipotent cells than in younger pluripotent cells. Research on the early stages of cell specialisation may not be possible with multipotent cells because they are further along the specialisation pathway. Thus, study of both pluripotent and multipotent stem cells is vital to fully understand cell specialisation and potentially develop new treatments or even cures for diseases.

Fig. 7   Stem cells

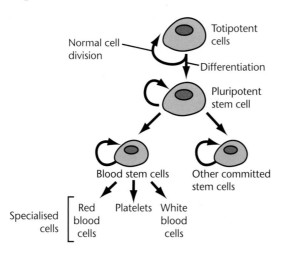

Normal cell division
Totipotent cells
Differentiation
Pluripotent stem cell
Blood stem cells
Other committed stem cells
Specialised cells
Red blood cells
Platelets
White blood cells

14 Suggest why using cells from embryos in research creates ethical and moral dilemmas.

Fig. 8   Obtaining stem cells for research

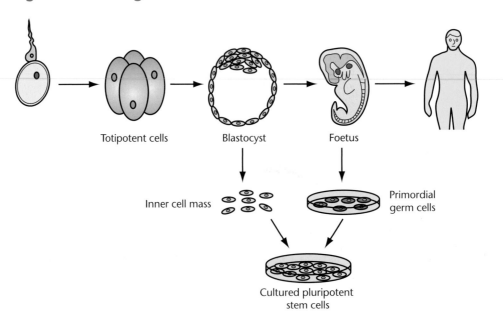

Totipotent cells   Blastocyst   Foetus

Inner cell mass   Primordial germ cells

Cultured pluripotent stem cells

how science works

hsw

Using stem cells to treat human disorders at the clinic – hope or hearsay?

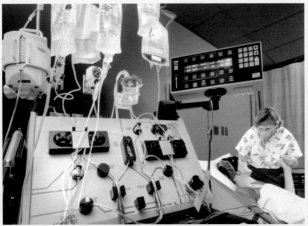

Preparing to collect human stem cells.

Research has shown that stem cells can grow into specialised cells that are able to repair damage by dividing, restoring and replenishing at cellular level. In this way, stem cells can offer new ways to treat human disorders. There is an increasing belief that injecting stem cells into patients with illnesses, ranging from Alzheimer's or Parkinson's disease to autism and multiple sclerosis (MS), is the cure-all for human disorders and will ultimately make its way into clinics. Is this hope or just hearsay?

Indications that the hope could become a reality, are the proposal of clinical trials, which involve injecting heart attack patients with stem cells; and research showing the possibility of using stem cells to enable blind mice to see.

Although scientists are at the stage where some stem cell research is moving from using animals, towards the clinic, it is agreed that the progress in stem cell research should not be exaggerated.

Efforts are being made to inch stem cells into the clinic, but some of these efforts are progressing more than others. For years stem cells have been used successfully to grow skin for grafting onto victims of burns; and they have been used in bone marrow transplants to treat leukaemia patients. Stem cell treatment has been used for type-2 diabetes and emphysema patients. Research is also being done with the view to using stem cells to repair eyes, cartilage and even the spinal cord.

Skin that has been produced from stem cells.

But, experts agree that while much publicity is being given to the good that stem cells can perform, no attention is being drawn to the problems that still need to be overcome. Consensus is that stem cell research has a long way to go. There are also issues to resolve: treatment for many conditions is not yet freely available; the cost is prohibitive; and there are issues of ethics and legality.

With ethical issues and legal implications in mind, many scientists believe that it will be some time before the somatic (adult or foetal) stem cells are made use of in the clinic, and a decade or more before embryonic stem cells are used.

There is concern about what is known as 'stem cell cowboys', who are said to charge huge sums of money for the promise of curing a myriad of diseases. It is believed that these people inject what stem cell experts believe are 'dubious' cells into people suffering from diseases such as multiple sclerosis. Some of these people pursue couples who are about to give birth and charge thousands of pounds to store their baby's umbilical cord cells. Leading medical research laboratory doctors regard these activities as unethical. There are still basic problems as to how stem cells (whether they are adult, foetal or embryonic) will react once they are in the body.

While it is considered certain that stem cell transplantation will be a therapy of the future, the focus should be on the basic mechanisms of stem cell differentiation – how stem cells make the decision to move from their stem cell state into a more committed phenotype. It should be stressed that the medical profession does not yet know nearly enough about how to control what stem cells do in the test tube, the animal and the patient. Knowledge is still so limited that most doctors would not yet dream of injecting stem cells randomly into patients.

In spite of the fact that other areas of stem cell transplant advances may be further off than hoped, a leading UK centre for stem cell research is working on brain stem cells, and is making progress. A doctor at the centre explains that it would be unrealistic to build up patients' hopes that the cells could be used as a means of repair in transplants, but the brain stem cell research will provide a highly useful tool that will promote understanding of the basic biology of disease.

The creation of stem cells from an animal or human with a certain disease or form of cancer allows scientists to study what is happening in the cells as they change into other cell types and to track what happens during the process of the cells' development and growth. Knowledge of how a disease works could help to reveal potential treatment or even a cure, and could be used for, for example:

Drug screening – access to human neural material offers scientists the prospect of seeing how they can, for example, affect cell behaviour and test toxicity.

Cancer research (one of the more probable success stories) – for many tumours such as brain tumours, or for breast cancers, there seems to be a subpopulation of cells within the tumour that can be called cancer stem cells – targeting and understanding these is a major area of study.

For now stem cell research is progressing slowly but surely, and scientists agree that the area of stem cell research is both promising and exciting. Initially, however, before stem cell treatment is regularly used in the clinic, it may be that the largest advances will be in forming an understanding of the workings of these specialised cells, deducing how they can be controlled and harnessed, and using them to explore our basic biology.

15 Use the information above to evaluate the use of stem cells in treating human disorders.

key facts

- Totipotent cells can mature into any body cell.

- During development, totipotent cells translate only part of their DNA, resulting in cell specialisation.

- In mature plant cells many cells remain totipotent. In micropropagation, these cells can develop into organs or whole plants.

- Only a few totipotent cells, called stem cells, remain totipotent in mammals.

- Stem cells are used to treat some human disorders, but research into, and use of, stem cells raises moral and ethical issues.

Regulation of gene expression

By switching genes off when they are not needed, cells can prevent resources from being wasted. So a typical human cell normally expresses 3%–5% of its genes at any given time. Cancer results when genes do not turn off properly. Cancer cells have lost their ability to regulate mitosis, resulting in uncontrolled cell division.

Gene expression in eukaryotes is controlled by a variety of mechanisms that range from those that prevent transcription to those that prevent expression after the protein has been produced. The various mechanisms can be placed into four categories, illustrated in Fig. 9:

- **Transcriptional** – these mechanisms prevent transcription and thereby prevent the synthesis of mRNA.
- **Post-transcriptional** – these mechanisms control or regulate mRNA after it has been produced.
- **Translational** – these mechanisms prevent translation; they often involve protein factors needed for translation.
- **Post-translational** – these mechanisms act after the protein has been produced.

Transcription factors

Proteins called **transcription factors** function by moving in from the cytoplasm and binding to the DNA. These transcription factors are necessary for RNA polymerase to attach to the DNA chain. Transcription begins when the factors create a loop in the DNA (Fig. 10).

Fig. 10   Transcription factors

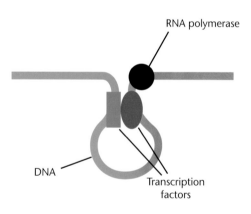

Fig. 9   Regulation of gene expression

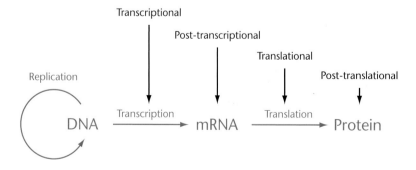

Hundreds of different transcription factors have been discovered, each of which recognises and binds with a specific nucleotide sequence in the DNA. A specific combination of transcription factors is necessary to activate a gene. Transcription factors are regulated by signals produced from other molecules. For example, hormones activate transcription factors and thus enable transcription. Hormones therefore activate certain genes. One example is the involvement of oestrogen in the development of breast cancer. Cells in some types of breast cancer have oestrogen receptors in their cell membranes. The binding of oestrogen with these receptors results in transcription switching on genes for cell growth and division. This results in rapid division of the cells forming the tumour.

16 Tamoxifen is a drug that is used to treat breast cancer in women whose breast cancer cells contain oestrogen receptors.

Suggest how Tamoxifen has its effect.

Small interfering RNA

RNA interference (RNAi) is currently a hot topic in science. Over the last few years, the number of papers published on the subject has escalated. Why the fuss about RNAi? Visualise that you could identify the role of a gene in a disease by switching it off easily, in almost any organism, during only one day. Then imagine that you could take this tool and treat diseases such as cancer by switching off the causative genes. This is what RNAi promises.

During the 1990s, two scientists named Guo and Kemphus, conducted investigations into the function of the par-1 gene in a nematode worm. They studied the effect of blocking the production of the *par-1* protein using antisense – a technique in which a small synthesised piece of RNA, which is complementary to a specific sequence in *par-1* mRNA attaches and stops the mRNA from being translated into the protein Fig. 11).

Fig. 11   Formation of double-stranded mRNA

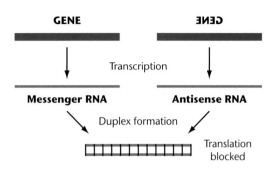

17

a   What is meant by 'antisense'?

b   Explain how the small piece of synthesised RNA prevented translation.

They injected an antisense RNA into the worm, and the result was what they expected: all the embryos died. But Guo and Kemphus were puzzled because injecting the sense strand (the same sequence as the *par-1* mRNA) – a standard negative control for the experiment – also resulted in the death of the embryos.

18 What is meant by 'standard negative control'?

Fig. 12   Dicer and the RISC

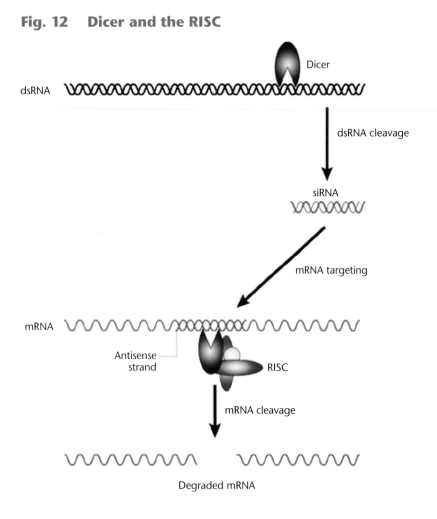

dsRNA

Dicer

dsRNA cleavage

siRNA

mRNA targeting

mRNA

Antisense strand

RISC

mRNA cleavage

Degraded mRNA

Source: McManus and Sharp. Nature Rev Genet 2002; 3: 737–47. Reproduced with permission from Macmillan Magazines Ltd.

Two more scientists, Fire and Mello, set up an experiment that was to become a significant moment in RNAi research. In order to see if there would be an additive effect, they injected sense and antisense RNAs into a nematode worm. However, they found that this double-stranded RNA (dsRNA) mixture was a great deal more potent than the sum of its parts. The target gene was silenced 10 times more efficiently than either strand alone – an effect that became known as RNA interference.

The fundamental process of RNAi is to chop dsRNA into smaller pieces of a defined length – using the suitably named enzyme known as Dicer (Fig. 12). Dicer chops dsRNA into small interfering RNAs (siRNAs) to a length of 21–23 nucleotides. Dicer delivers these siRNAs to a group of proteins called the RNA-induced silencing complex (RISC), which uses the antisense strand of the siRNA to bind to and degrade the corresponding mRNA, which results in gene silencing.

In molecular biology, a minor revolution has been caused by the ease with which genes can be silenced using RNAi. A group at the University of Cambridge has created a library of more than 16 000 dsRNAs. Worms were fed on these clones, thus determining the function of 1 722 genes, most of which were previously unknown. Using this technique, a group at Cold Spring Harbor, New York, in the USA, is attempting to determine the functions of every gene in the human genome.

key facts

● Transcription of target genes is stimulated only when specific transcriptional factors move from the cytoplasm into the nucleus.

● Transcriptional factors prevent transcription and thereby prevent the synthesis of mRNA.

● Cells in some types of breast cancer have oestrogen receptors in their cell membranes. The binding of oestrogen with these receptors results in transcription switching on genes for cell growth and division. This results in rapid division of the cells forming the tumour.

● Small interfering RNA (siRNA) are short pieces of double-stranded RNA that interfere with the expression of specific genes by degrading mRNA.

examination questions

1 The table shows the sequence of bases on part of the coding strand of DNA.

Base sequence on coding strand of DNA	C	G	T	T	A	C
Base sequence of mRNA						

a Complete the table to show the base sequence of the mRNA transcribed from this DNA strand. (2)

b A piece of mRNA is 660 nucleotides long but the DNA coding strand from which it was transcribed is 870 nucleotides long.
  i Explain this difference in the number of nucleotides. (1)
  ii What is the maximum number of amino acids in the protein translated from this piece of mRNA? Explain your answer. (2)

c Give **two** differences between the structure of mRNA and the structure of tRNA. (2)

Total 7

AQA, A, January 2006, Unit 2, Question 6

2

a The table shows the mRNA codons for some amino acids.

Codon	Amino acid
CUA	Leucine
GUC	Valine
ACG	Threonine
UGC	Cysteine
GCU	Alanine
AGU	Serine

  i Give the DNA sequence coding for cysteine. (1)
  ii Name the amino acid coded by the tRNA anticodon UCA. (1)

b A particular gene is 562 base-pairs long. However, the resulting mRNA is only 441 nucleotides long. Explain this difference. (1)

c Tetracycline binds to bacterial ribosomes. This is shown in the diagram.

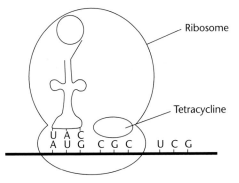

Protein synthesis in bacteria is similar to that in eukaryotic cells. Explain how tetracycline stops protein synthesis. (2)

Total 5

AQA, A, June 2005, Unit 2, Question 7

3

a **Table 1** shows some of the events which take place in protein synthesis.

A	tRNA molecules bring specific amino acids to the mRNA molecule
B	mRNA nucleotides join with the exposed DNA bases and form a molecule of mRNA
C	The two strands of a DNA molecule separate
D	Peptide bonds form between the amino acids
E	The mRNA molecule leaves the nucleus
F	A ribosome attaches to the mRNA molecule

  i Write the letters in the correct order to show the sequence of events during protein synthesis, starting with the earliest. (2)
  ii In which part of a cell does **C** take place? (1)
  iii Which of **A – F** are involved in translation? (1)

b **Table 2** shows some mRNA codons and the amino acids for which they code.

mRNA codon	Amino acid
GUU	Valine
CUU	Leucine
GCC	Alanine
AUU	Isoleucine
ACC	Threonine

  i A tRNA molecule has the anticodon UAA. Which amino acid does the tRNA molecule carry? (1)

257

ii Give the DNA base sequence that codes for threonine. (1)

Total 6

AQA, A, June 2004, Unit 2, Question 1

4

a **Figure 1** shows the exposed bases (anticodons) of two tRNA molecules involved in the synthesis of a protein.

AGC

UUC

Complete the boxes to show the sequence of bases found along the corresponding section of the coding DNA strand. (2)

b Describe the role of tRNA in the process of translation (3)

c **Figure 2** shows the sequence of bases in a section of DNA coding for a polypeptide of seven amino acids.

Figure 2

TACAAGGTCGTCTTTGTCAAG

The polypeptide was hydrolysed. It contained four different amino acids. The number of each type obtained is shown in the table.

Amino acid	Number present
Phe	2
Met	1
Lys	1
Gln	3

Use the base sequence shown in **Figure 2** to work out the order of amino acids starting with met in the polypeptide. (2)

Total 7

AQA, B, January 2006, Unit 2, Question 5

5

a The mRNA codon for the amino acid tyrosine is UAU.
i Give the DNA triplet for tyrosine. (1)
ii Give the tRNA anticodon for tyrosine. (1)

b Give **two** ways in which the structure of a molecule of tRNA differs from the structure of a molecule of mRNA. (2)

c One group of antibiotics, the aminoglycosides, prevent the growth of bacteria by allowing any tRNA molecule to bind to any codon on a mRNA molecule.
i Name the site of action of aminoglycosides in a bacterial cell. (1)
ii Use the information provided to explain how aminoglycosides prevent the growth of bacteria. (2)

Total 7

AQA, B, June 2005, Unit 2, Question 3

6

a Explain what is meant by:
i a totipotent cell
ii a luripotent cell
iii an adult stem cell (3)

b **i** Describe one way in which stem cells can be obtained for research. (3)
ii Describe and explain one way in which stem cells could be used to treat human disease. (3)

Total 9

7

a Describe four different ways in which gene expression may be regulated. (4)

b Explain how small interfering RNA is used in gene silencing. (4)

Total 8

Using siRNA to treat disease

Taking the DNA sequence of a gene and designing dsRNA that can specifically and effectively silence a disease-related gene is analogous to monoclonal antibody production.

> **A1** Explain why dsRNA technology is analogous to monoclonal antibody production.

HIV

Research has shown that siRNA interference can inhibit HIV replication effectively in culture. HIV infection can also be blocked by targeting either viral or human genes (for example, *CD4*, the principal receptor for HIV) that are involved in the HIV life cycle. However, these results have been achieved by transferring the siRNAs into cells; getting the siRNAs to function *in vivo* is likely to be a more difficult task although early attempts have shown that this can decrease replication of HIV considerably.

> **A2**
>
> **a** HIV enters human white blood cells and reproduces inside them. Explain how siRNA interference affects the HIV life cycle.
>
> **b** Suggest why getting siRNAs to function *in vivo* is likely to prove difficult.

Hepatitis

This has provided the first convincing evidence that RNAi could be an effective therapy for diseases in live animals. Early studies showed that RNA silencing was prominent in the liver, which made this organ an attractive target for RNAi therapy. Many immune-related liver diseases are characterised by apoptosis, a form of cell death in which a programmed sequence of events leads to the elimination of liver cells without releasing harmful substances into the surrounding tissues. A protein called Fas plays an important role in apoptosis. Scientists injected siRNA-targeting-Fas into mice suffering from autoimmune hepatitis. This decreased Fas mRNA and protein levels in liver cells and protected the cells against liver injury from apoptosis.

> **A3**
>
> **a** Why is the liver chosen for research into RNAi?
>
> **b** Explain why aptopsis in liver cells is important for health.
>
> **c** Suggest what is meant by 'autoimmune hepatitis'.
>
> **d** Explain how siRNA-targeting-Fas helped mice suffering from autoimmune hepatitis.

Cancer

Scientists have used RNAi to silence expression of p53 – the 'guardian of the genome', which protects against tumour-associated DNA damage. They have done this by introducing p53-targeting siRNAs into stem cells and looking at the effect in mice. The siRNAs produced a wide range of tumours, ranging from benign to malignant tumours. The work with these modified stem cells gives hope that this could treat diseases in which stem cells can be modified ex vivo and then re-introduced into the affected individual.

Researchers at the charity Cancer Research UK have recently announced that they intend to generate a large library of human cells, each containing a silenced gene. They want to silence 300–8000 cancer genes. Their aim is to uncover all the genes that become over-expressed in human cancers and to find out precisely what needs to be taken away from a cancerous cell in order to make it normal again.

> **A4**
>
> **a** Suggest what is meant by 'guardian of the genome'.
>
> **b** Explain why the results of the introduction of p53-targeting siRNAs into stem cells gives hope for disease treatment.
>
> **c** Suggest what is meant by 'genes that become over-expressed'.
>
> **A5** Evaluate the use of RNAi in the treatment of disease.

15 Gene technology

The use of genetic engineering in food production has provoked strong feelings in many people.

Today, the blood sugar levels of someone with diabetes mellitus can be controlled by regular carefully measured injections of insulin. Until the early 1980s all insulin for injection came from cattle and pigs. Although it controls blood glucose perfectly well, animal insulin is not exactly the same as human insulin and some people are allergic to it. The development of genetic engineering techniques made it possible to use microbes to manufacture an exact copy of human insulin.

The gene for human insulin was one of the first to be inserted into microbes to manufacture drugs for human use. Other substances now made by genetic engineering include human growth hormone and Factor VIII. Growth hormone is used to treat children who do not grow properly because of a pituitary gland disorder. Factor VIII is the blood-clotting factor that is used to treat haemophilia. Making these proteins using microbes instead of extracting them from animal or human organs or human blood reduces the possibility of contamination by, for example, viruses that cause AIDS or hepatitis. It also increases drug yield and so makes manufacturing cheaper.

Genes can also be inserted into crop plants to improve their qualities. For example, genes have been added to tomato plants to slow down the ripening process so that the fruit will stay fresh for longer. Genes have been transferred to soya bean and maize crops in the USA to make them more resistant to insect pests. Other suggestions for the future include adding genes to plants to enable them to make plastics.

However, there is growing concern about the possible consequences of genetic engineering, especially in crop plants. Might genetically modified foods be a danger to health? Might genes transferred to crops or microbes spread to other organisms and create environmental havoc? Might the widespread use of pest- and herbicide-resistant crops devastate wildlife? Some of the arguments of opponents of genetic engineering may appear emotive and unscientific at times, but the caution they demand may be wise. Indeed, many of the researchers in the field urge a careful assessment of the use of the technology. There are serious issues that need to be considered calmly and carefully. In this chapter we look first at how genetic engineers transfer genes from one organism to another, and then we consider some of the possible benefits and dangers of this novel technology.

15.1 Manipulating genes

Isolating the DNA

Fig. 1 shows the stages in isolating a gene from an organism and inserting it into a microbe.

To remove DNA from a cell, the cell membrane needs to be disrupted and the nucleus broken open. The method used depends on the type of cell. In eukaryotes, the cell surface membrane and the nuclear membrane both need to be broken open. In plants, the cell wall must also be disrupted. In prokaryotes, the cell wall needs to be broken, but the absence of a nuclear membrane makes the second stage easier. One common way to disrupt a cell uses a detergent called **sodium dodecyl sulfate** (SDS). This breaks down cell membranes and cell walls. Once the DNA is free, the surrounding proteins are removed using digestive enzymes.

1

a   Suggest how the detergent breaks down the cell membranes.

b   What type of digestive enzyme could be used to remove the proteins in the chromosomes of a human cell?

Cutting up the DNA

Once the DNA has been isolated from the rest of the cell, the part of the DNA molecule that contains the required gene has to be cut out and the rest of the DNA discarded. This is important because genetic engineering must be as precise as possible; only known genes should be transferred to the donor organism.

Genetic engineers isolate genes by using enzymes that cut across DNA molecules at particular positions. These are called **restriction endonuclease enzymes**. Several different restriction enzymes occur naturally in bacteria. Their function is to chop up and destroy the DNA of any viruses that infect the bacterial cell. Each enzyme cuts across the double-stranded DNA molecule at a different point in a nucleotide sequence. For example, one enzyme, known as *EcoRI* cuts the strands only at the sequence shown in Fig. 2.

The names of restriction enzymes seem strange when you first come across them but they are actually quite logical. *Eco*RI was the first restriction enzyme found in the R strain of the bacterium *Escherichia coli*.

You will see that this enzyme does not slice straight across a DNA molecule. It separates the strands over a stretch of four bases, leaving each part of the broken DNA molecule with a short single-stranded tail. These tails are called **sticky ends** because they are easy to join with other sticky ends to make complete DNA molecules.

Fig. 1   Transferring a gene

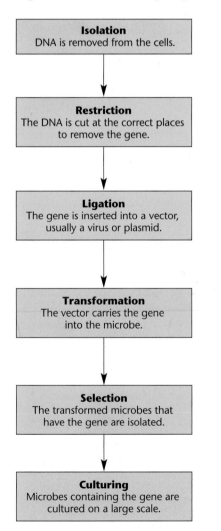

Isolation
DNA is removed from the cells.

↓

Restriction
The DNA is cut at the correct places to remove the gene.

↓

Ligation
The gene is inserted into a vector, usually a virus or plasmid.

↓

Transformation
The vector carries the gene into the microbe.

↓

Selection
The transformed microbes that have the gene are isolated.

↓

Culturing
Microbes containing the gene are cultured on a large scale.

2   Use your knowledge of enzymes to explain why *Eco*RI only cuts DNA at one particular position.

Fig. 2   The action of an endonuclease

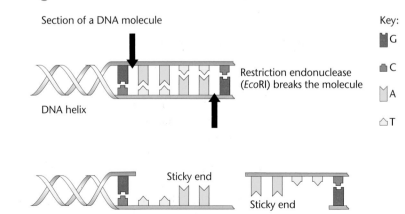

Section of a DNA molecule

Restriction endonuclease (*Eco*RI) breaks the molecule

DNA helix

Sticky end

Sticky end

Key:
G
C
A
T

3

a One of the sticky ends produced by cleavage with *Eco*RI in Fig. 2 has nucleotides with the bases:

A T T G
T A A C T T A A

Which bases would attach to the sticky end to make a new DNA molecule?

b A new DNA molecule can only be made by joining this sticky end with a section of DNA with these bases on a sticky end. Use your knowledge of DNA structure to explain why.

hsw

Restriction endonucleases

The table below shows the base sequences of DNA that are cut by four different restriction endonucleases.

Restriction endonuclease	Cutting points
*Bam*I	C↓CTAGG
	G GATC↑C
*Eco*RII	C↓GGACCG
	G CCTGG↑C
*Hind*III	T↓TCGAA
	AAGCT↑T
*Pst*I	G↓ACGTC
	CTGCA↑G

4 Draw diagrams to show the sticky ends produced when each of the restriction endonucleases cuts a DNA molecule.

5 A section of a DNA molecule has the following sequence of bases:

T C C G G A C C G A C G T C G G T T C G A A T C
A G G C C T G G C T G C A G C C A A G C T T A G

This DNA is treated with a mixture of all four enzymes in the table. How many DNA fragments will be produced? Draw the fragments produced and name the enzymes involved at each cut.

Fig. 3

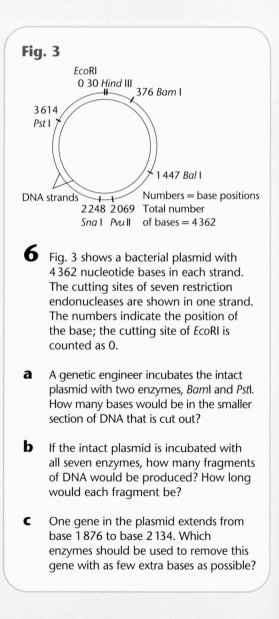

Fig. 3

EcoRI
0 30 *Hind* III
376 *Bam* I
3 614 *Pst* I
1 447 *Bal* I
DNA strands
2 248 2 069
Sna I *Pvu* II
Numbers = base positions
Total number of bases = 4 362

6 Fig. 3 shows a bacterial plasmid with 4 362 nucleotide bases in each strand. The cutting sites of seven restriction endonucleases are shown in one strand. The numbers indicate the position of the base; the cutting site of *Eco*RI is counted as 0.

a A genetic engineer incubates the intact plasmid with two enzymes, *Bam*I and *Pst*I. How many bases would be in the smaller section of DNA that is cut out?

b If the intact plasmid is incubated with all seven enzymes, how many fragments of DNA would be produced? How long would each fragment be?

c One gene in the plasmid extends from base 1 876 to base 2 134. Which enzymes should be used to remove this gene with as few extra bases as possible?

Extracting human genes

It is not easy to isolate a specific gene from human cells. The first difficulty is finding the gene, although this will be easier now that the human genome has been fully mapped. Genetic engineers use **genetic probes**. A genetic probe is a marker that can reveal the position of a gene in the human genome.

Fig. 4   Extracting human genes

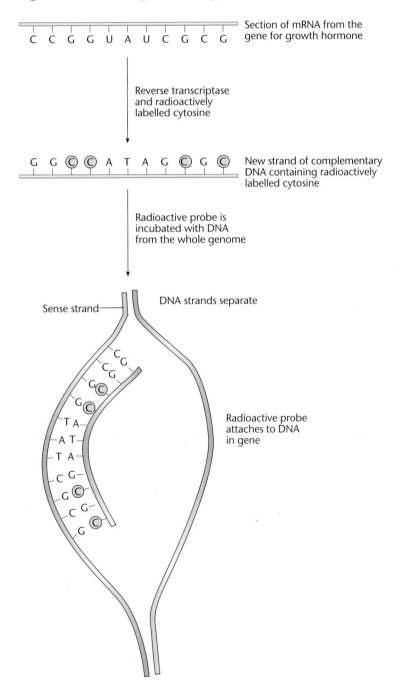

The first step in making such a probe is to find cells that actively produce the gene product. Human growth hormone, for example, is synthesised in the anterior lobe of the pituitary gland. The gene for growth hormone is actively expressed in these cells, and their cytoplasm contains messenger RNA (mRNA) for growth hormone. This mRNA can be extracted and used to make a complementary strand of DNA using an enzyme called **reverse transcriptase**. This enzyme reverses the usual process by which a DNA template is used to make an mRNA copy. It is obtained from viruses that have RNA instead of DNA as their genetic material. Such viruses, including the AIDS virus, HIV, use the enzyme to replicate inside their host cells. Genetic engineers use the enzyme in association with radioactive nucleotides to produce a single strand of radioactive DNA called **complementary DNA (cDNA)**.

Radioactive DNA made from the mRNA produced originally by the growth hormone gene is cultured with DNA from the whole human genome that has been split into its individual strands. The DNA probe attaches to the matching strand of the section of DNA that carries the gene for growth hormone. When the position of the radioactivity is identified, the growth hormone gene has been located (Fig. 4).

Double-stranded DNA can be obtained from cDNA as shown in Fig. 5. This double-stranded DNA can then be spliced into bacterial plasmids which, when inserted into bacteria, will initiate production of the appropriate protein.

An alternative method is to use the extracted mRNA to synthesise an artificial gene. The order of nucleotide bases in the mRNA can be determined, and from this the order of bases in the DNA of the gene can be worked out. DNA that has the nucleotides in the correct order can

Fig. 5   Producing cDNA

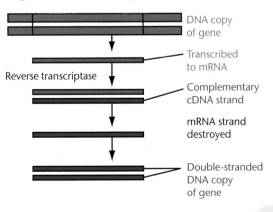

then be made in the laboratory. This method has been used to synthesise the gene for human insulin, which is an unusually short protein with only 51 amino acids in the active hormone. Recombinant gene technology has then been used to incorporate this artificial gene into the bacteria that produce insulin.

> **7** Explain how the order of bases in DNA can be worked out from the order in the corresponding mRNA.
>
> **8** How many bases are in the section of single-stranded DNA that corresponds to the 51 amino acids in the active human insulin molecule?

Getting the gene into bacterial DNA

The next stage is to insert the isolated gene into a **vector**. A vector is a piece of DNA that can take the gene into the chosen microbe. A common vector is a small circular molecule of DNA, called a **plasmid**. Plasmids occur naturally in bacteria in addition to the larger molecule of chromosomal DNA. Genetic engineers find plasmids very useful because these loops of DNA can replicate independently from the bacterial chromosome.

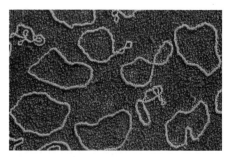

Transmission electron micrograph showing pBR322 plasmids from *Escherichia coli*.

The same restriction enzyme used to cut out the gene from the donor DNA is used to cut open the plasmid DNA. This creates a broken loop of DNA with sticky ends that match those on the donor gene. The donor gene can then be inserted into the plasmid loop using the enzyme **ligase**. Ligase catalyses the **ligation** reaction that joins two sections of DNA (Fig. 6).

In practice, the DNA from the donor organism and the plasmids from the bacterial recipient are incubated with the same restriction endonuclease in separate tubes for 2–3 hours to

Fig. 6 Inserting the gene into the plasmid

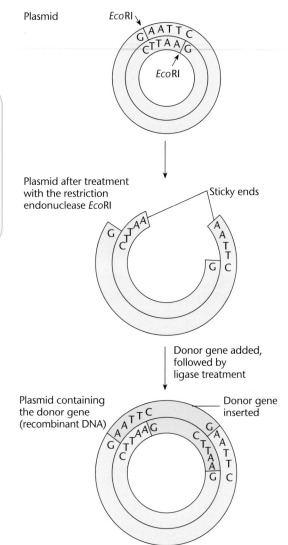

create identical sticky ends. The tubes are then heated to denature the restriction endonuclease. The contents of the tubes are mixed and ligase is added. The sticky ends of the donor DNA join with the corresponding sticky ends in the plasmids. Hydrogen bonds form between complementary bases and the ligase joins the sugar–phosphate backbone. The new DNA is called **recombinant DNA**.

> **9** Explain why it is important to denature the restriction endonuclease before mixing the contents of tubes containing donor DNA and bacterial plasmids.

Getting the gene into the bacterium

Plasmids containing the donor gene must now be transferred into the microbe. A culture of the intended bacterial recipients is placed in cold calcium chloride solution for about 30 minutes. This makes the cell membranes of the bacteria more permeable. Plasmids with the recombinant DNA are added to the culture and the mixture is warmed up for a short time. This shock treatment causes some of the bacteria to take up plasmids. Those bacteria that do contain plasmids with recombinant DNA are said to have undergone **transformation** (Fig. 7).

The transformation process is not very efficient and only quite a small proportion of bacteria in the culture will be transformed. The genetic engineer wants to grow only transformed cells, so the next step involves identifying and isolating these. One commonly used technique is to insert a **marker gene** into the plasmids, in addition to the donor gene. A marker gene may make the bacteria resistant to a particular antibiotic. If the culture containing the transformed bacteria is grown on a medium that contains that antibiotic, bacteria with plasmids that have the antibiotic resistance marker gene (and the recombinant DNA) will survive and grow better than those that do not.

Replica plating can be then be used to produce several cultures. Fig. 8 shows how a single replica plate is made. The process is repeated several times to eliminate colonies containing non-resistant bacteria that start to grow before the antibiotic takes effect. Replica plating also increases the supply of bacteria that have the added gene.

Fig. 7   Getting the gene into bacteria

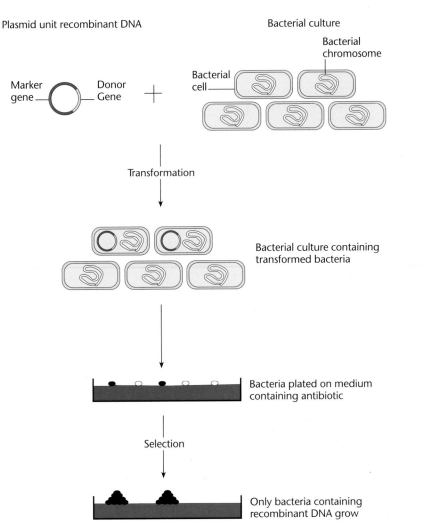

Fig. 8   Replica plating

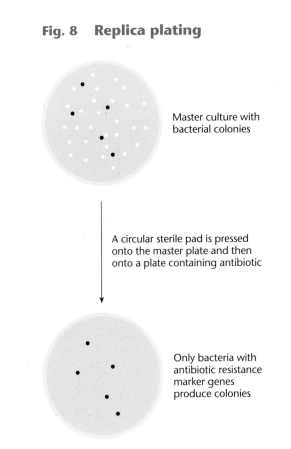

Genetic engineering and food

Fig. 9   Blocking the production of polygalacturonase

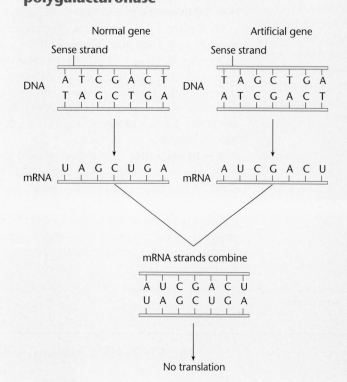

Fruit sold in supermarkets is often picked well before it is ripe and transported long distances. It is then ripened artificially just before being sold. It doesn't have the same taste as freshly ripened fruit but it can be displayed for longer as it does not become soft too quickly.

As tomatoes ripen they produce **pectinase**, an enzyme that breaks down the pectin that normally holds the cell walls together. As the cells separate, the fruit goes soft and squishy and rots. Genetically modified tomatoes have been given an additional artificial gene. The bases on the sense strand of the DNA of this artificial gene are exactly opposite to those on the sense strand of the gene that codes for pectinase. The two strands of mRNA are therefore attracted to each other and bind together to form a double strand. This prevents the normal mRNA joining onto the ribosomes and being translated to make pectinase.

10

a   Draw the mRNA that would be produced by the sense strand of the section of normal gene shown in the diagram.

b   Draw the mRNA that would be produced by the sense strand of the section of artificial gene shown in the diagram.

c   Explain why the two strands bind together to form a double strand of mRNA.

d   Explain how the artificial gene stops the tomatoes from going soft.

Ripe tomatoes, ready for eating.

Other parts of the ripening process are not affected, so the flavour of a ripe tomato still develops. However, since the modified tomatoes do not produce pectinase, this happens without the tomato going soft. This means that it is possible for growers to leave the tomatoes to ripen naturally on the plant; they can be sure that the fruit will remain in good condition for several days longer than traditional tomato varieties, allowing plenty of time for transport to the shop and a few days of display on the shelves.

key facts

- In genetic engineering genes are removed from one organism and inserted into another. Genes that code for useful substances, such as hormones, enzymes and antibiotics, can be transferred into microorganisms, which then produce large quantities of these substances.

- A gene is isolated from the DNA of the donor organism using a restriction endonuclease enzyme. This cuts out the relevant section of the organism's DNA, leaving sticky ends that will enable the gene to be inserted into a small circular piece of bacterial DNA called a plasmid.

- Plasmids are often used as vectors to incorporate the selected gene into bacterial cells. Plasmids occur naturally in bacterial cells and replicate independently of the main bacterial DNA.

- Human genes can be copied by using the enzyme reverse transcriptase to convert mRNA to cDNA.

- The same restriction endonuclease is used to cut the plasmid. This leaves complementary sticky ends to which the selected gene can be attached by another enzyme, ligase.

- The plasmids are then introduced into the bacteria, and transformed cells are selected and cloned.

- Genetic markers in the plasmids, such as genes that confer antibiotic resistance, enable genetic engineers to identify bacteria that have successfully taken up the selected gene.

- This method of cloning genes is known as *in vitro* cloning since the genes are transferred into host cells 'in glassware'.

15.2 Investigating DNA

In the early days of DNA research, sequencing a nucleotide fragment or analysing a sample of DNA was a slow process. It was technically difficult and expensive to obtain a small fragment of DNA in large enough quantities to make analysis possible. It could take months to sequence a short section of DNA, and it was impossible to extract sufficient DNA from spots of blood at the scene of a crime. The whole field of DNA analysis has been revolutionised by the development of a technique that can make a billion copies of a strand of DNA in a few hours.

The polymerase chain reaction

The **polymerase chain reaction** (PCR) is like a nuclear chain reaction in that it proceeds at an ever-increasing rate. PCR can amplify tiny amounts of DNA into quantities large enough for scientific analysis. It is the basis of **genetic fingerprinting**, a technique that allows forensic scientists to identify a criminal from a microscopic blood spot or single hair left behind at the scene of a crime. Even dried blood or semen stains that are several years old may contain enough DNA for PCR, enabling an individual to be identified. PCR is the technique that was used in the book *Jurassic Park* by Michael Crichton to obtain the DNA from which the dinosaurs were recreated. Its more realistic uses include amplifying DNA from:

- samples of tissue from extinct animals, such as the Tasmanian wolf, to establish their closest relatives
- buried human bodies where some soft tissue has been preserved; these studies are helping us to understand the migrations of early human populations.

PCR uses the enzyme DNA polymerase. This enzyme occurs naturally in cells – it catalyses the replication of DNA in the nucleus. The first stage in PCR is to heat DNA until the two strands of the molecule separate. DNA nucleotides are added and the DNA polymerase attaches them to each strand, just as it does in normal replication. The process can then be repeated endlessly. The two new double-stranded molecules make a total of four, then eight, 16 and so on. PCR can now be managed by machines so that each cycle of replication takes only a matter of minutes. Fig. 10 shows the stages in a single PCR cycle.

Fig. 10   A single cycle of the polymerase chain reaction

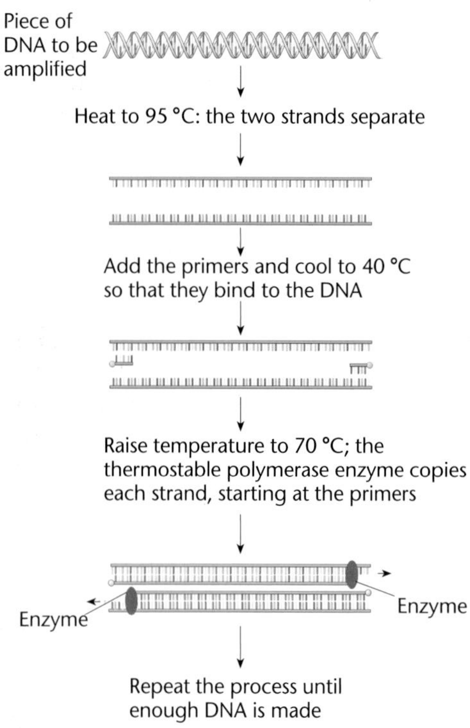

Piece of DNA to be amplified

Heat to 95 °C: the two strands separate

Add the primers and cool to 40 °C so that they bind to the DNA

Raise temperature to 70 °C; the thermostable polymerase enzyme copies each strand, starting at the primers

Enzyme   Enzyme

Repeat the process until enough DNA is made

Analysing samples of DNA

Once a large enough sample of DNA has been generated by PCR, it can be analysed and compared with samples from known sources. This allows us to compare the DNA from different species, or from individuals from the same species, or even from members of the same family. Forensic scientists often have to compare the DNA from samples found at a crime scene with the DNA from suspects. In every case, the first stage in the process of analysis is to cut the DNA into short lengths with restriction endonuclease enzymes. This produces DNA fragments of varying lengths, because the sequences of bases where the enzymes cut occur at irregular intervals along the DNA molecules.

The DNA fragments are then separated by electrophoresis (Fig. 11). The mixture of DNA fragments is placed at one end of a long piece of agar gel in a trough containing a dilute solution of ionic salts. Electrodes are placed in the solution at either end and a voltage is applied. The phosphate groups in the fragments of DNA give them a negative charge, so they are attracted through the gel towards the positive electrode. The smaller fragments move more rapidly than the larger ones, so the different-sized fragments are separated in much the same way as in chromatography.

11 Look at Fig. 10. How many different types of DNA nucleotide must be added to the mixture in the final stage? Why?

12 Suggest why using DNA polymerase from bacteria living in hot springs is essential.

13 How many cycles of the process would be needed to produce a million copies of a DNA molecule from one DNA molecule?

14 Explain why a restriction enzyme cuts a DNA molecule at specific positions.

15 PCR is used to amplify a section of DNA. When this amplified DNA is treated with a restriction enzyme, 18 DNA fragments of different lengths are always detected by electrophoresis, no matter how many different samples are run. Explain why the samples have a fixed number of DNA fragments.

16 Phosphate groups give the DNA a negative charge. Explain why DNA fragments contain phosphate groups.

Fig. 11   Electrophoresis

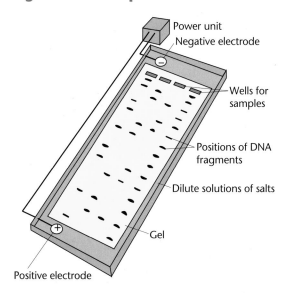

Power unit
Negative electrode
Wells for samples
Positions of DNA fragments
Dilute solutions of salts
Gel
Positive electrode

The pattern of fragments is a sort of 'fingerprint' but to be able to compare one sample with another, the fingerprint must be made visible. The pattern of DNA fragments in the gel is first transferred to a nitrocellulose sheet, which binds DNA. This nylon-based sheet

Fig. 12   Analysing DNA samples

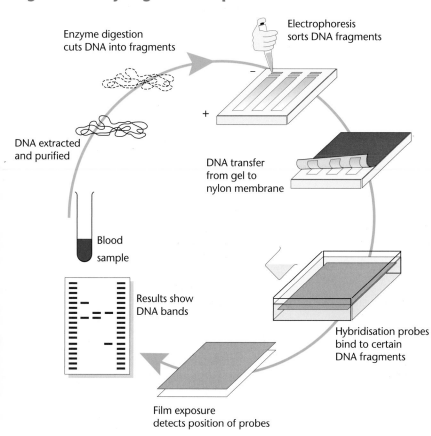

Enzyme digestion cuts DNA into fragments

Electrophoresis sorts DNA fragments

DNA transfer from gel to nylon membrane

DNA extracted and purified

Blood sample

Results show DNA bands

Hybridisation probes bind to certain DNA fragments

Film exposure detects position of probes

is placed on top of the gel and then a stack of paper towels is used to press it down onto the gel (see Fig. 12). The gel and nitrocellulose sheet are left in place for several hours, usually overnight. The DNA fragments in the gel transfer to the nitrocellulose sheet to form an 'imprint' of the pattern of the fragments in the original gel.

A probe labelled with radioactivity or a fluorescent dye is then used to reveal the positions of the bands on the sheet containing the DNA sequence of interest. A general probe can be used to make all the bands show up or a specific probe can be used. A specific probe is a piece of single-stranded DNA that is complementary to the base sequence of the specific stretch of DNA. The nitrocellulose sheet is incubated with the probe in a sealed plastic bag containing a buffer solution. The sheet is removed and placed next to an unexposed piece of photographic film. The radioactivity in the probe causes a band to show on the film. If a fluorescent dye is used in the probe, the sheet is illuminated with a UV lamp to show the bands.

DNA sequencing

The order of nucleotide bases in a fragment of DNA can be found using several methods. Dideoxy sequencing (also called chain termination or the Sanger method) uses DNA polymerase and bases labelled radioactively or with fluorescent dyes to synthesise DNA chains of different lengths. The DNA sample is divided into four separate sequencing reactions containing all four of the standard **deoxynucleotides** (dATP, dGTP, dCTP and dTTP) and the **DNA polymerase**. To each reaction is added only one of the four (ddATP, ddGTP, ddCTP, or ddTTP). These **dideoxynucleotides** lack the OH group required for the formation of a bond between two nucleotides during DNA strand elongation. Incorporation of a dideoxynucleotide into the elongating DNA strand therefore terminates DNA strand extension, resulting in various DNA fragments of varying length. The resulting fragments in each tube are run on a gel that separates them according to their size. The shorter fragments move faster and appear at the bottom of the gel. An autoradiograph is produced by placing a photographic film next to the resulting gel for several hours as shown. The sequence is read from bottom to top. The gel autoradiogram in Fig. 13 shows the result of a sequencing gel carried out for a strand of DNA containing only 12 bases.

Fig. 13   DNA sequencing

1. Sequencing reactions loaded onto polyacrylamide gel for fragment separation

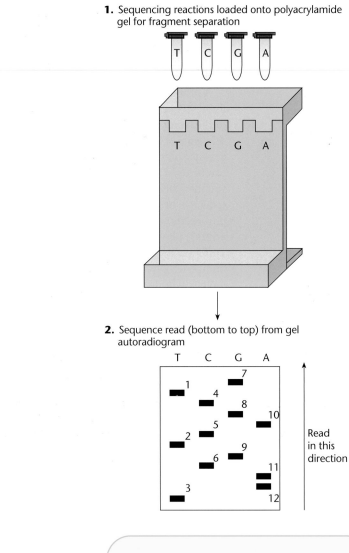

2. Sequence read (bottom to top) from gel autoradiogram

Read in this direction

Restriction mapping

Another way to locate gene sequences is **restriction mapping**. This enables us to locate and isolate DNA fragments for further study and manipulation. The location of different restriction enzyme sites relative to one another is determined by digesting DNA with different restriction enzymes. These enzymes are used both alone and in different combinations. The digested DNA is separated by gel electrophoresis and the fragments stained with a reagent such as **ethidium bromide** and viewed under UV light. The sizes of the fragments produced by digestion are used to build a 'map' of the sites at which the restriction enzymes act. For circular DNA molecules (such as plasmids) restriction enzymes that cut at a single site will generate a linear molecule, which will run as a single band in gel electrophoresis.

Fig. 14 illustrates the principle of restriction mapping. Diagram a shows the effects of two enzymes, I and II, acting singly. The two enzymes cut the DNA at different sites. Enzyme I cuts the DNA into fragments with relative sizes of 0.6 and 0.4. Enzyme II cuts the DNA into fragments with relative sizes of 0.9 and 0.1 respectively. Diagram b shows the effect of a **double digest**, using the two enzymes together. Enzyme I cuts the DNA into fragments as before, and Enzyme II cuts a fragment with relative length of 0.1 from either the 0.6 fragment or the 0.4 fragment.

17

a   Look at the autoradiograph in Fig. 13. Which of the patches at positions labelled 1 to 12 contain sections of DNA with only one nucleotide? Give a reason for your answer.

b   How many cytosine bases did the original DNA strand contain?

c   What was the order of the 12 bases in the original DNA strand?

d   The original DNA was a section of the sense strand. What would be the sequence of the mRNA strand that it would transcribe?

Fig. 14   Restriction mapping

a   Cutting DNA with single enzymes

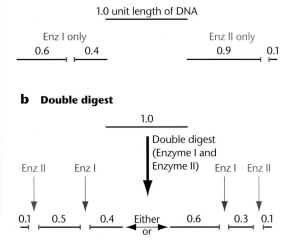

b   Double digest

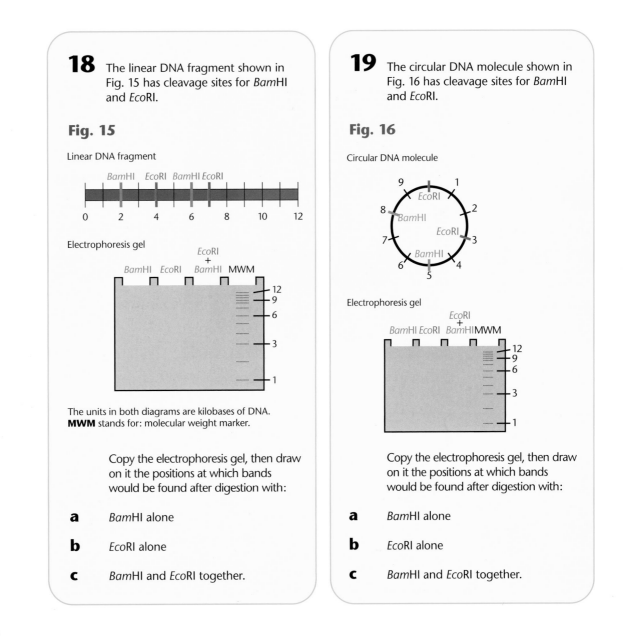

18 The linear DNA fragment shown in Fig. 15 has cleavage sites for *Bam*HI and *Eco*RI.

Fig. 15

Linear DNA fragment

Electrophoresis gel

The units in both diagrams are kilobases of DNA.
MWM stands for: molecular weight marker.

Copy the electrophoresis gel, then draw on it the positions at which bands would be found after digestion with:

a   *Bam*HI alone

b   *Eco*RI alone

c   *Bam*HI and *Eco*RI together.

19 The circular DNA molecule shown in Fig. 16 has cleavage sites for *Bam*HI and *Eco*RI.

Fig. 16

Circular DNA molecule

Electrophoresis gel

Copy the electrophoresis gel, then draw on it the positions at which bands would be found after digestion with:

a   *Bam*HI alone

b   *Eco*RI alone

c   *Bam*HI and *Eco*RI together.

key facts

- DNA can be replicated artificially by the polymerase chain reaction (PCR). The enzyme DNA polymerase is used to make new double-stranded DNA by synthesising a new complementary strand to a pre-existing strand, just as in natural replication.

- The PCR method is known as an *in vitro* method since the process takes place outside living cells 'in glassware'.

- PCR is an amplification reaction that is self-sustaining, which makes it possible to synthesise large numbers of copies of very small samples of DNA.

- DNA fragments can be separated by gel electrophoresis. A voltage is applied to the gel and the negatively charged DNA fragments move towards the positive electrode. Smaller fragments move faster than larger ones.

- The bands of DNA can be seen if radioactive nucleotides are used in the PCR. The pattern of bands in the gel can be made visible by placing the gel next to a sheet of unexposed photographic film overnight. The radioactive bands cause the film to turn black.

- In restriction mapping, the locations of different restriction enzyme sites relative to one another is determined by digesting DNA with different restriction enzymes. These enzymes are used both alone and in different combinations.

- The digested DNA is separated by gel electrophoresis and the fragments stained with a reagent such as ethidium bromide and viewed under UV light.

15.3 Genes and medicine

Tracey, the world's first transgenic sheep, with her two lambs.

In 1993, Tracey was probably the most famous sheep in the world. She was the star performer in a family of sheep produced in the laboratories of Pharmaceutical Proteins in Edinburgh. When she was just an embryo, scientists had transferred human genes into her cells. When she grew to adulthood, Tracey was able to make milk containing 35 grams per litre of a substance called human alpha-1-antitrypsin (AAT). AAT is used to treat the lung disease, emphysema, and can also be useful for some children who have cystic fibrosis. It can be extracted from human blood, but only in very small amounts, so the 35 g per litre available in Tracey's milk was a positively bountiful supply. A single flock of about 2 000 ewes like Tracey could satisfy the needs of every hospital in the world.

Apart from her ability to secrete AAT into her milk, Tracey was a completely normal ewe and she gave birth to two healthy lambs. Moreover, one of these lambs inherited the capacity to manufacture AAT, showing that the transferred gene could be passed on from one generation to the next. This is a great advantage compared with having to go through the complex procedures necessary to introduce the gene into every individual embryo in order to obtain extra sources of AAT.

In this section we explore how genetic engineering can be used to produce animals that are sources of medicines used to treat human diseases. We also look at the area of **gene therapy**. This is the field of medicine that aims to treat people with genetic disorders by giving them a copy of a healthy gene to overcome the problems produced by their mutated gene. We also consider some of the implications of this technology for the future.

Genetically modified animals

Animals that have had genes from another species transferred into them are called **transgenic organisms**. Transgenic mammals are particularly useful because they can express human genes. It is not always possible for bacteria to express human genes. Prokaryotic bacterial cells do not have the ribosomes and other cellular machinery necessary to produce complex mammalian proteins.

The AAT gene that was given transferred into Tracey codes for a protein that consists of 394 amino acids. It is actually a **glycoprotein**, a protein with an attached sugar group. The normal function of the AAT glycoprotein is to inhibit an enzyme called elastase that is produced by some types of white blood cells. Elastase, if not inhibited, breaks down elastic tissue in the lungs, causing emphysema. In emphysema, the walls of the alveoli disintegrate so that much less oxygen is absorbed into the blood and fluid leaks into the airspaces. The fluid disrupts the normal functioning of the lungs and can encourage infections that could make a weakened patient even worse. Emphysema can be caused by a number of different factors; having the Z allele for AAT is quite rare. The much more common causes include smoking and working with fine dust particles. Many coal miners and quarry workers suffer from emphysema as a result of their jobs.

There are many different alleles of the AAT gene. About 3% of people in the UK have the Z allele. The Z allele codes for a version of AAT that differs from the normal form by only one amino acid at one position. However, this defective AAT cannot function and patients with Z alleles frequently develop emphysema. Affected people can be treated with an aerosol spray containing AAT. When inhaled, this stops the breakdown of the alveoli and can help patients' breathing immensely. However, it does not provide a permanent cure for the disease. Patients must continue the treatment for the rest of their lives.

20

a The DNA of the Z allele differs from the normal allele by just one nucleotide base. What type of mutation could have caused this difference?

b How many bases does the gene for the complete AAT protein contain?

c Suggest why the AAT protein produced by the Z allele does not inhibit elastase.

d Explain why treatment with an aerosol containing AAT does not provide a permanent cure for emphysema.

e Could treatment with AAT help someone whose emphysema is caused by smoking? Give reasons for your answer.

Fig. 17 Making Tracey

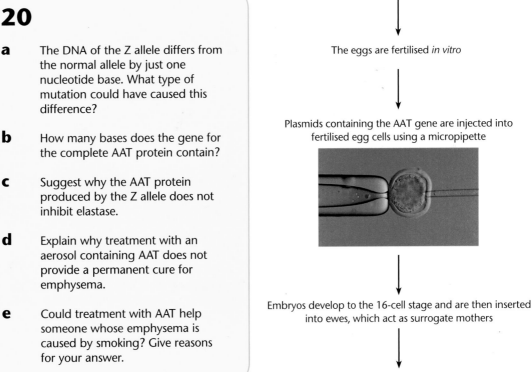

Mature eggs are removed from the sheep's ovary

The eggs are fertilised *in vitro*

Plasmids containing the AAT gene are injected into fertilised egg cells using a micropipette

Embryos develop to the 16-cell stage and are then inserted into ewes, which act as surrogate mothers

Many embryos were inserted before a successful pregnancy was established; Tracey was the world's first transgenic lamb

Making Tracey

How was Tracey modified to produce the AAT protein in her milk? The human gene was identified and isolated from a culture of human cells using the techniques described earlier in this chapter. The gene was then combined with a promoter sequence that allows the gene to be expressed in sheep mammary glands. The protein can be collected easily and in large amounts from the animal's milk, without causing distress to the animal.

Mature egg cells were removed from the ovary of a sheep and fertilised *in vitro* (Fig. 17).

The AAT gene and its promoter sequence were injected into the nucleus of the fertilised egg cells. Once the zygote had divided to form a small embryo, this was placed in the uterus of a sheep surrogate mother. Not all attempts at establishing a pregnancy were successful but Tracey demonstrated that the technique worked. Since then other transgenic animals have been produced with genes that manufacture other useful products in their milk, such as the blood-clotting protein Factor VIII, which is used to treat haemophilia.

key facts

- Animals can be genetically modified to produce substances that are useful for treating human diseases. Animals that have had a gene from another species transferred into them are called transgenic organisms.

- The human gene that codes for the required protein is isolated from human cells and is then injected into a fertilised egg from the animal.

- The tiny embryos that develop are placed in the wombs of surrogate mothers, which later give birth to young that carry the human gene in their cells.

- Transgenic sheep are used to produce alpha-1-antitrypsin (AAT), a protein that is used to treat emphysema in people who have a mutation in their AAT gene. The sheep secrete the AAT in their milk.

Gene therapy

Using genetically modified animals to produce substances that can be used to treat people who are unable to make the substances themselves is a great step forward. But another approach is to insert the relevant gene into the human organ where the substance is needed. In the case of AAT, this would mean putting healthy AAT genes directly into the cells within lung tissue of a patient so that they can produce the substance for themselves. This type of procedure is called **gene therapy**.

Gene therapy is being developed to treat the genetic disease **cystic fibrosis**. Most people have a gene that produces a protein called the cystic fibrosis transmembrane regulator (CFTR). This has a complex molecule consisting of 1 480 amino acids, and it is one of the essential channel proteins in cell membranes. The function of this particular channel protein is to transport chloride ions through the cell membrane.

Sally, the girl in the photograph, has cystic fibrosis, because she has a mutation in both alleles of her CFTR gene. The effect of this mutation is that one of the 1 480 amino acids in her CFTR protein is missing. This affects the three-dimensional shape of the channel protein so that it cannot transport chloride ions through a membrane. The mutation affects all the cells in Sally's body, but it is a particular problem in her lungs. The epithelial cells lining the airways in the lungs normally pass chloride ions out and absorb sodium ions (Fig. 18). However, cells with CFTR produced by the mutant genes retain chloride ions, and the increased concentration of ions in the cells causes them to retain water, which normally leaks out of the cells into the thin layer of mucus that lines the airways.

With less water the mucus layer becomes more sticky and thicker and it is not easily moved out of the lungs by the beating cilia that line the airways. For this reason, Sally has to be helped to expel the excess mucus from her lungs. If her airways become blocked with mucus, she will get breathless because her lungs cannot absorb enough oxygen. The sticky mucus also traps bacteria, so patients with cystic fibrosis are prone to chest infections. These can lead to pneumonia and bronchitis. Although these can be treated by antibiotics, repeated infections can make a child weak and can cause scarring of the delicate lung tissue.

The accumulation of mucus also affects the digestive system, as the duct linking the pancreas to the duodenum tends to become blocked, preventing pancreatic juices from reaching food in the gut.

The physiotherapist is helping to clear excess mucus from Sally's lungs, a process that has to be carried out two or three times every day.

21

a Suggest how an increased concentration of ions in the cells reduces the amount of water that they lose.

b Explain how blockage of the pancreatic duct might affect digestion.

Getting healthy CFTR genes into cells

Gene therapy for cystic fibrosis involves inserting functional CFTR genes into epithelial cells of the lungs to replace the defective genes. Two techniques are being tested:

- wrapping CFTR genes in lipids that can be absorbed through the cell surface membranes
- inserting CFTR genes into harmless viruses that are then allowed to 'infect' the cells.

The first method uses tiny lipid droplets called **liposomes**. Liposomes can fuse with the phosopholipid membranes that make up a cell surface membrane and so carry genes into target cells. CFTR genes are isolated from healthy human tissues, cloned and inserted into plasmids using recombinant DNA technology. The plasmids are then inserted into liposomes. The next hurdle is to get the liposomes into the area of the body where the genes are required. Aerosol sprays are used to get the liposomes into the lungs, in much the same way as people with asthma use inhalers. In the lungs, the liposomes fuse with the surface membrane of the epithelial cells, and the plasmids are transported across the membrane into the cells. However, the technique has its difficulties. One problem is developing an aerosol that can produce a fine enough spray to get the liposomes through the very narrow bronchioles of the lungs and into the alveoli. Another is that only a small proportion of the genes that are absorbed into cells are actually expressed. These problems must be overcome before gene therapy for cystic fibrosis can become widely available.

Fig. 18   Transport of ions by CFTR

a Epithelial cell in a healthy person

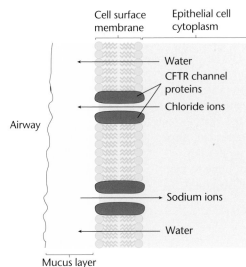

Cell surface membrane — Epithelial cell cytoplasm — Water — CFTR channel proteins — Chloride ions — Airway — Sodium ions — Water — Mucus layer

b Epithelial cell in a person with cystic fibrosis

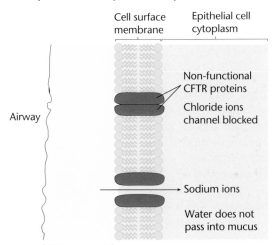

Cell surface membrane — Epithelial cell cytoplasm — Non-functional CFTR proteins — Chloride ions channel blocked — Airway — Sodium ions — Water does not pass into mucus — Thick and sticky mucus layer

22 In one clinical trial, doctors used a gene probe to test for the presence of CFTR mRNA to find out if CFTR genes absorbed into cells were being expressed. Explain how this would confirm that the genes were working.

Adenovirus vectors

In the second method, viruses called **adenoviruses** are used. Adenoviruses reproduce themselves by injecting their DNA into host cells. The viral DNA uses the cells enzymes and ribosomes to replicate and produce copies, which then reconstruct the rest of the virus before being released from the cell (see Fig. 19).

Adenoviruses infect the cells of the airways in the lungs, and they are adapted for replication in the epithelial cells. They normally cause colds and other respiratory diseases. The adenoviruses used for gene therapy have been modified so that they infect cells but do not cause illness. The modification involves disabling the genes that allow the virus to replicate. Modified adenoviruses are cultured in epithelial cells grown in the laboratory and are exposed to plasmids that have the normal CFTR gene. The CFTR gene is incorporated into the viral DNA. The adenoviruses containing the CFTR gene are then extracted from the epithelial cells, purified and sprayed into the lungs. Here they infect the epithelial cells, taking the CFTR gene into the cells. The CFTR channel protein is synthesised as normal, but as the viruses cannot replicate, they do not damage the cells.

Fig. 19 Adenovirus replication

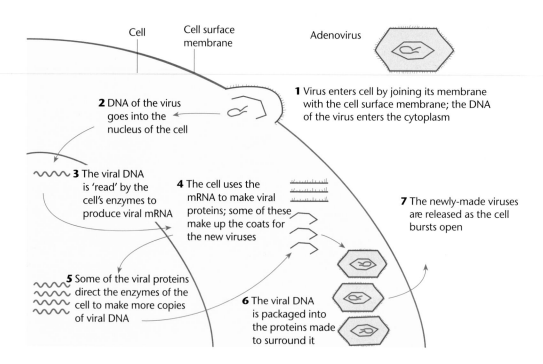

Adenoviruses are well adapted for entry into lung cells, so they are highly efficient vehicles for getting the CFTR gene into target cells. However, there are some concerns about possible dangers of infection by adenoviruses. Infection could occur if the genes that allow the virus to replicate have not been completely inactivated. Also, patients treated repeatedly may develop antibodies that make them immune to the viruses and so resistant to the treatment.

> **23** Use the information above to compare the advantages and disadvantages of using adenovirus vectors rather than liposomes to get healthy CFTR genes into lung epithelial cells.

Perfecting gene therapy techniques

Neither of the gene therapy techniques described provides a permanent cure for cystic fibrosis, but both have the potential to alleviate its most distressing symptoms and improve patients' quality of life. Before using either method on people, research workers had to carry out careful testing to make sure that gene therapy was safe. The first tests, carried out on laboratory animals such as mice, had the following objectives:

- to find the best ways to get the gene into cells
- to check that the gene was expressed in the target cells
- to detect any ill effects of the therapy itself.

Following encouraging results in animals, the next step was to start testing the gene delivery systems in people. Although mammals are similar in many respects, each species is different and scientists needed to check that the techniques that worked in mice also work in humans. This involved human volunteers, who, for example, had small doses of liposomes containing the recombinant gene applied to the lining of the nose. The nose was chosen because the epithelial cells here are much easier to study than cells deep in the lungs. The next stages, which are on-going, involve clinical trials with cystic fibrosis patients. Many trials need to be analysed before gene therapy for cystic fibrosis can become an officially approved treatment, available to all cystic fibrosis patients. All these very necessary precautions take time, but progress in the basic science underlying gene therapy continues at a rapid pace.

> **24** Explain why gene therapy does not provide a permanent cure for cystic fibrosis.

how science works

Gene therapy for SCID

This child is playing inside a biological isolation garment, which is a type of space suit that has been customised for patients who are highly vulnerable to infection. A biological isolation garment is often used by children suffering from SCID.

One of the first examples of gene therapy in humans was in the treatment of a rare inherited condition called Severe Combined Immunodeficiency (**SCID**). Children who have this condition have no effective immune system, so they are susceptible to infections that can develop into life-threatening illnesses. The only treatment for such children has been to be protected in an isolation 'bubble' to prevent exposure to germs.

The child with SCID has inherited defective alleles and as a result cannot make an enzyme called adenosine deaminase (ADA). Without this enzyme, toxins build up and destroy the white blood cells called T lymphocytes, which normally recognise infectious microbes that enter the body. The aim of gene therapy for SCID is therefore to replace the child's T lymphocytes with cells containing the normal ADA gene. The method used is shown in Fig. 20.

25

a What is the advantage of inserting the ADA gene into developing lymphocytes from the bone marrow, instead of mature lymphocytes in the blood?

b Suggest why the treatment has to be repeated at regular intervals.

c Suppose a child that is given healthy ADA genes in this way eventually has his or her own children. Could the healthy ADA genes be passed on to the children? Explain your answer.

Fig. 20   Gene therapy for SCID

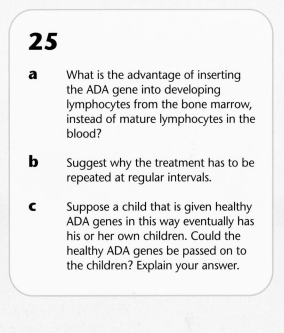

A restriction endonuclease is used to isolate the normal ADA gene from the DNA of healthy cells

The normal gene is inserted into a virus

The virus is incubated with a cell culture; the virus injects its genetic material into the cell, which makes many copies of the virus with the ADA gene

The viruses with the ADA gene are mixed with the patient's white blood cells; the viruses inject their genetic material into the nuclei of the white blood cells

Developing white blood cells are removed from the bone marrow of the patient with ADA deficiency

Nucleus

The white blood cells are cloned in the laboratory

White blood cells, which now contain normal ADA genes, are transfused back into the patient; in the body of the patient these young cells will divide to produce a supply of cells with ADA genes

Genetic testing

Genetic testing can be used to find out whether a person's cells carry normal CFTR genes or defective, mutant CFTR genes. This is unnecessary for people who have cystic fibrosis as they must have the mutant genes. Genetic testing can be useful for a couple who are planning to have a child; if the test reveals that they are both carriers of one mutant CFTR allele, there is a one in four chance that they will have a child that has cystic fibrosis. If they decide to go ahead with a pregnancy, the fetus can be tested. Fetal cells are obtained by amniocentesis or chorionic villus sampling, two antenatal screening tests that are performed routinely. Amniocentesis is performed between weeks 15 and 22 of the woman's pregnancy. A small amount of the fluid that surrounds the fetus is removed. This contains cells that have come from the fetus. Chorionic villus sampling is done earlier, at about weeks 10 to 12 of the pregnancy. This method involves removing a small piece of placenta, the organ which links the fetus with the mother and which has the same genetic make-up as the fetus.

Once the parents know the results, they can decide whether or not to go ahead with the pregnancy. This is never an easy decision, but it does provide the family with some choice about their future and the future of their child.

The future of gene therapy

Progress in genetic research is proceeding at a great rate. It may, for example, soon be possible to:

- incorporate a healthy gene into one of the chromosomes before putting it into a cell; the chromosome would be replicated as the cell divides, thus providing a longer-lasting treatment
- treat rapidly dividing cancer cells with genes that kill them
- provide protection from viral infection of cells by inserting genes that interfere with the replication of the virus.

More effective ways of getting genes into cells will make the treatment of genetic diseases such as cystic fibrosis more effective. Liposomes or adenoviruses get the CFTR gene into target cells where it is expressed, but treatment has to be repeated at regular intervals because copies of the gene are not passed from cell to cell during

26 About 1 in 2000 babies in the UK is born with cystic fibrosis. Do you think genetic screening of fetuses could be used to eliminate cystic fibrosis? Explain your answer.

27 In theory, genetic screening of a fetus could be used for each of the following:

- to detect Tay-Sachs disease, a genetic disease for which there is no treatment, and which causes serious brain damage, paralysis and death in early childhood
- to detect a genetic condition that results in deafness
- to detect the presence of a gene that increases the chances of developing breast cancer in later life
- to choose the hair or eye colour of a child
- to choose the sex of a child.

a In which of these situations, if any, do you think it is acceptable to screen the genetic make-up of the fetus?

b How would you justify where to draw the line?

mitosis. Epithelial cells in the lining of the respiratory tract divide often, so this is a serious limitation. However, if the healthy gene were inserted into a chromosome, it would be replicated each time the cell divided. One significant problem is that it might insert at a point in the DNA that could damage other genes.

28 Explain how inserting a gene into one of the cell's chromosomes might cause damage to other genes.

Gene therapy might also become an important weapon in the war against cancer and disease. There are genes that cause cell death. If cancer cells could be targeted reliably, so that only actively dividing cells took up or expressed these genes, cancerous tumours could be destroyed.

Reproduction of viruses in cells could also be prevented by inserting genes that produce mRNA that is complementary to the viral RNA. This would inactivate the viruses in much the same way as the ripening enzyme is inactivated in genetically modified tomatoes.

29 Use the information about inactivating the ripening enzyme in tomatoes earlier in the chapter, and the information on adenovirus replication, to explain how cells could be protected from viral infection by gene therapy.

s&c

stretch and challenge

Helping haemophiliacs

Haemophilia is caused by a small defect in a single human gene. When this particular defect occurs, as in classic or type A haemophilia, the body lacks an important protein that helps to form fibrin, a protein that is essential for the process of blood clotting. This protein is called Factor VIII. About 85% of all haemophiliacs are missing the gene that instructs the body's cells to produce Factor VIII. Haemophilia B is a less common type of genetic disease, caused by a deficiency in another necessary protein, called Factor IX. In each type of bleeding disorder, proteins are either missing or deficient and thus fibrin is not able to form. Approximately two out of every 10 000 males are affected by either type A or type B haemophilia. Fortunately, type A haemophiliacs can be treated with transfusions of concentrated Factor VIII protein. This has increased the life expectancy of some haemophiliacs. The protein concentrate can be prepared by combining volumes of blood donated from many humans with normal clotting blood. However, this method can spread viral diseases that were present in the original donated blood. Many haemophiliacs are chronically infected with these viruses, which include HIV, the virus that causes AIDS, and hepatitis viruses. In fact, in the early to mid-1980s, over 90% of the people with severe haemophilia contracted HIV through contaminated human blood plasma. Only recently has a process been developed where certain viruses, including HIV, can be detected in the blood supply. Cow and pig blood have also been concentrated and used as a therapy, since these animals have higher concentrations of Factor VIII than humans. Transfusion reactions and other problems have caused a decline in the use of animal-derived Factor VIII.

The human Factor VIII and IX genes were isolated in the early 1980s. This discovery led to the development of recombinant factor concentrates in the late 1980s. The human Factor VIII or IX gene is isolated through genetic engineering. This gene contains the code that has instructions for the cell on how to make human Factor VIII or IX. The gene is inserted into non-human cells, such as baby hamster kidney cells or Chinese hamster ovary cells. These cells are grown in a cell culture. Human plasma protein solution is added to the cell culture to make Factor VIII. This is not necessary in the preparation of Factor IX. The Factor VIII or IX is separated from the cell culture and purified. Human albumin is then added to stabilise the final Factor VIII product. Factor IX is a much more stable protein and does not require the addition of albumin to stabilise it. Researchers are working on several ways to improve genetically engineered Factor VIII concentrates.

- Factor VIII concentrates currently in clinical trials are made without the addition of albumin as a stabiliser. While albumin has never been associated with the transmission of a blood-borne pathogen, it is believed that a recombinant product without human proteins provides an extra margin of safety.

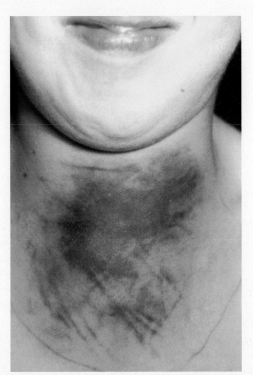

Haemorrhage in the neck of a haemophiliac.

- Researchers are looking for cell lines that will be better at expressing or manufacturing Factor VIII in order to reduce the cost of production. The product would be more widely available at an affordable price. More people would have access to prophylactic therapy.
- One of the major problems with factor replacement therapy, including recombinant therapy, is that 10%–20% of Factor VIII hemophiliacs develop antibodies to the Factor VIII molecule infused. Their bodies' immune systems reject the infusion as a foreign agent. In some patients, this inhibitor is powerful and long lasting. Treatment becomes very complicated and extremely expensive. Researchers are hoping they can change Factor VIII to make it less immunogenic. In other words, Factor VIII would no longer trigger the body's immune system into mounting a defence. Scientists would use genetic engineering to modify those parts of the Factor VIII molecule that are responsible for the immune response.
- Factor VIII is an extremely unstable protein with a short half-life. This means that the interaction of Factor VIII with other particles in the blood reduces the specific activity of Factor VIII and shortens its life. Researchers are hoping to modify the structure of Factor VIII to change this. If Factor VIII were more active, and could last longer in the bloodstream, smaller doses would be needed. This would make Factor VIII cheaper and more widely available.

30 Summarise the advantages of genetically engineered Factor VIII over Factor VIII obtained from blood products.

31 Summarise the disadvantages of using genetically engineered Factor VIII.

32 Suggest why human plasma is needed to produce genetically engineered Factor VIII.

33 What is meant by 'cell lines better at expressing Factor VIII'?

34

a Explain what is meant by 'less immunogenic'.

b What changes would make Factor VIII less immunogenic?

35

a Suggest what is meant by 'Factor VIII has a short half-life.'

How would extending this half-life benefit haemophiliacs?

Difficult choices ahead

The most effective treatment for genetic disorders would be to replace the defective genes completely. This could be done using *in vitro* fertilisation methods similar to those already used to produce transgenic animals. However, the danger of damaging the developing embryo has meant that this technique is not acceptable in humans. Such a genetic change would affect the **germ line**. This means that the genes of the sex cells would be altered and therefore this genetic change would be passed on to future generations.

One proposed method of improving the technique is to create an artificial chromosome. This chromosome would contain only healthy extra genes, together with the promoter genes needed for them to be expressed. Once added to a nucleus, this chromosome should be replicated and take part in mitosis in the normal way. This approach reduces the risk that the extra genes will cause damage to other chromosomes and genes.

Geneticists also need to take into account that most human characteristics are under the control of several genes. Once the full human genome has been decoded and the functions of many more genes are understood, there is the possibility of adding packages of genes to such artificial chromosomes. This might raise the possibility of dealing with more complex genetic disorders in addition to the single gene defects that can be treated at present.

However, some of the possibilities raise difficult moral and ethical questions. Balancing the benefits of this emerging technology against the potential for its abuse will take many years of consultation and consideration by the whole of society, not just the scientists and patients directly involved.

36 The suggestion that extra chromosomes might be inserted into embryonic cells could make it possible for parents to select a range of features that they would like their children to have. Prepare for a group discussion of the potential benefits and dangers of being able to 'design' babies in this way. You might consider some of the following aspects.

- What features might parents select for?
- What might the social consequences be if, for example, wealthy parents were able to favour their children in this way?

- How might the relationship between parents and their children be affected?
- How might parents react if children failed to match up to expectations, for example if enhanced intelligence or sporting prowess was not achieved?
- Are there possible benefits if such techniques became widely available? Could human society be improved in the long term?
- How might the system be abused, for example by totalitarian regimes?

key facts

- Cystic fibrosis is a genetic disorder caused by a mutant allele that produces a defective form of the channel protein called CFTR. This protein normally transports chloride ions out of cells.

- The defective CFTR protein causes chloride ions to build up in cells, resulting in water retention in those cells. This is a particular problem in the lungs and intestines. Water fails to pass into the mucus that lines the airways and gut, causing the mucus to become thick and sticky. In the lungs, this leads to breathing difficulties and the risk of infection; in the gut, the mucus blocks ducts that carry digestive enzymes.

- Some severe genetic disorders can be treated by gene therapy. Healthy genes are cloned and then transferred into target cells in the body to take over the function of defective genes that cause the disorder.

- Two forms of gene therapy are being developed to treat cystic fibrosis. In the first, healthy CFTR genes are inserted into liposomes, which fuse with cell membranes and take the genes into the cells. In the other, harmless viruses are used to insert the CFTR genes into the cells.

- Genetic testing can be used to screen people for genes that produce disease conditions.

- The information obtained may be used in genetic counselling.

1 Plasmids can be used as vectors to insert lengths of foreign DNA into bacteria. The diagram shows how this is achieved.

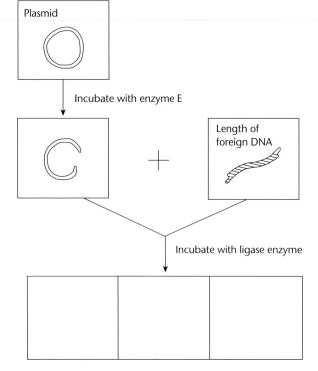

a Name enzyme **E**. (1)

b Cut plasmids and lengths of foreign DNA can join. What features of their ends allows them to join? (2)

c Draw **three** different structures that could be formed by incubating cut plasmids and lengths of foreign DNA with ligase. Use the spaces as provided on your copied diagram. (3)

Total 6

AQA, A, June 2006, Unit 2, Question 4

2 Read the following passage.

DNA tests were used to confirm the identity of deposed Iraqi leader Saddam Hussein, after his capture in December 2003. DNA tests were carried out to prove the suspect was not one of the many alleged 'look alikes' of the former leader. Firstly, the DNA was extracted from the mouth of the captured man using a swab. Great care was taken to check that the swab did not become contaminated with any other DNA. DNA extracted from the swab was then subjected to a standard technique called the polymerase chain reaction (PCR), which takes a couple of hours. Lastly, the sample was 'typed' to give the genetic fingerprint. This was produced within 24 hours of Saddam Hussein's capture. Tests for use in criminal cases often take much longer because samples are very small or contaminated. It appears that Hussein's genetic fingerprint was already stored for comparison. This was obtained from personal items such as his toothbrush. DNA from the toothbrush would have been subjected to PCR before a DNA fingerprint could have been obtained.

Source: adapted from SHAONI BHATTACHARYA, *New Scientist* 15 December, 2003

Use information from the passage and your own knowledge to answer the questions.

a Describe how the technique of genetic fingerprinting is carried out and explain how it can be used to identify a person, such as Saddam Hussein. (6)

b Explain how DNA could be present on a toothbrush (line 19). (2)

c i Explain why the polymerase chain reaction was used on the sample of DNA from the toothbrush (lines 20–21). (2)

ii Explain **one** way in which the polymerase chain reaction differs from DNA replication in a cell. (2)

d Tests for use in criminal cases often take much longer because samples are very small or contaminated (lines 14–16). Explain why it takes longer to obtain a genetic fingerprint if the sample is

i very small (1)

ii contaminated. (2)

Total 15

AQA, A, June 2005, Unit 2, Question 8

3 **Figure 1** shows how the gene for human growth hormone (hGH) can be transferred into a bacterium.

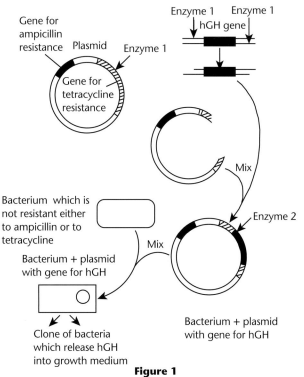

Figure 1

a Name Enzyme **1** and Enzyme **2**. (2)

b After mixing with the plasmid, the bacteria are first grown in Petri dishes of agar containing ampicillin. What is the reason for this? (2)

c **Figure 2** shows how colonies of bacteria can be transferred from a Petri dish of agar containing ampicillin to identical positions on a Petri dish of agar containing tetracycline.

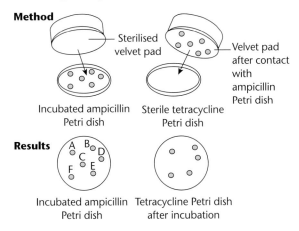

Figure 2

i Give the **letter** of **one** bacterial colony which contained a modified plasmid with the gene for hGH. (1)

ii Explain your answer to part **c i**. (2)

d Describe **one** possible danger of using plasmids which contain genes for antibiotic resistance. (2)

Total 9

(AQA, A, June 2003, Unit 2, Question 7)

4

a The polymerase chain reaction (PCR) can be used to produce large quantities of DNA.
Describe how the PCR is carried out. (6)

b About 20% of the DNA produced by the PCR is copied inaccurately. Suggest and explain why it is not safe to use the PCR to clone the CFTR gene for use in treating cystic fibrosis. (4)

Total 10

AQA, B, June 2006, Unit 2, Question 7

5

a Plasmids are often used as vectors in genetic engineering.

i What is the role of a vector? (1)

ii Describe the role of restriction endonucleases in the formation of plasmids that contain donor DNA. (2)

iii Describe the role of DNA ligase in the production of plasmids containing donor DNA. (1)

b There are many different restriction endonucleases. Each type cuts the DNA of a plasmid at a specific base sequence called a restriction site. The diagram shows the position of four restriction sites, **J**, **K**, **L** and **M**, for four different enzymes on a single plasmid. The distances between these sites is measured in kilobases of DNA.

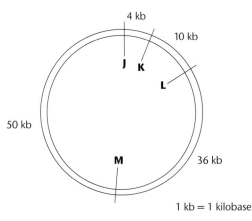

The plasmid was cut using only two restriction endonucleases. The resulting fragments were separated by gel electrophoresis. The positions of the fragments are shown in the chart below.

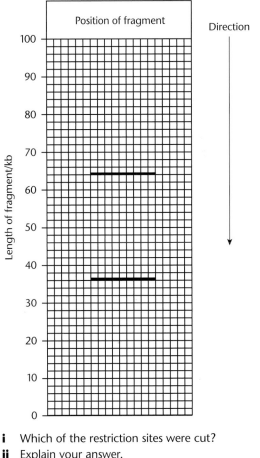

i Which of the restriction sites were cut? (1)

ii Explain your answer. (1)

Total 6

AQA, B, June 2006, Unit 2, Question 6

Genetically modified foods

Rapid developments in genetic engineering are making it possible to alter living organisms in ways that were not dreamt of even a few years ago. In theory, it is possible to transfer genes from almost any organism to any other. Scientists are investigating a myriad of potentially useful applications of recombinant DNA technology, not least ways of increasing crop yields and improving food supplies. However, many people have become suspicious of these developments and are concerned about the possible ways in which the technology could go wrong. Some people think that it's a bad idea to tamper with natural food

sources and they feel that new technology is being imposed without adequate safeguards by companies who just want to make money. Some people are worried that eating genetically modified (GM) foods may in itself be harmful. Others are concerned about the impact of GM crops on the environment.

The arguments for and against

Consider the following summaries of some of the arguments put forward by advocates and opponents of GM foods.

"Foods made from GM crops may be a danger to health."

Anti It is unnatural to swap genes from one species into another. The food we eat contains foreign genes, perhaps from bacteria or even humans, and we don't know what we are eating. The new combinations of genes may have unknown effects, such as making poisonous or carcinogenic substances. Also, when a foreign gene is inserted into a crop plant additional genes are normally added too. These might stimulate other genes in the crop to make harmful products. GM crops modified to produce substances intended to protect them, for example from insect pests, may have dangerous long-term effects.

Pro Genes are not harmful to eat. We eat large amounts of nucleic acids in our food. These are all digested, so we could not somehow incorporate foreign genes from food into our body. All crop plants have been genetically altered by selective breeding, so no food plants are naturally occurring wild plants. Genetic modification speeds up what has been a slow and expensive process of cross-breeding. Genetic modification is precise; only a small number of genes are transferred. Cross-breeding mixes large numbers of genes with unpredictable effects. Food products are extensively tested for safety before being sold, which includes checking for the presence of possibly harmful products.

"Genes might be transferred from GM plants to others, perhaps creating 'superweeds'."

Anti Genes can be passed from one organism to another, and from one species to another. Bacteria regularly exchange genetic material, including plasmids. Genes in a crop that protect it from the effect of weedkillers may be passed to other plants, including weeds, and these might also become resistant to weedkillers. Pollen from GM crops may be distributed by wind or insects for long distances, and natural cross-pollination may pass on the transferred genes to non-GM crops, or possibly produce harmful hybrids. Organic farmers are particularly concerned about cross-contamination.

Pro This is no different from the interbreeding that occurs anyway. Weeds containing genes that confer resistance to a particular herbicide would have no competitive advantage except in environments where the herbicide is being applied. Other weedkillers could still be used, so the weed could not become some sort of uncontrollable monster. Cross-contamination of crops can be avoided by separating organic and GM crops by adequate distances.

"Widespread planting of GM crops will seriously affect wildlife in the environment."

Anti Pesticides produced by genes in crops to protect them will kill not only crop pests but also other useful insects. This will remove pollinating insects such as bees and therefore affect pollination of fruit trees as well as wild flower populations. The loss of wild plants and insect food for other species such as birds will also cut down the diversity of wildlife in the countryside.

Pro The use of resistant crops will reduce the amount of pesticide that farmers need to spray on their crops. Only insects that actually feed on a particular crop will be affected and there will be less environmental damage to neighbouring habitats. Reduced diversity of wildlife is an unavoidable consequence of modern intensive farming practices in which large areas are planted with the same crop, and the use of GM crops would make no difference.

"GM crop seeds are more expensive. This would affect the livelihood of poorer farmers, especially in developing countries."

Anti The use of advanced biotechnology in agriculture will make smaller farms less economical and thus favour large-scale agrobusiness and the companies that develop and patent particular GM crop varieties. The use of 'terminator genes' that prevent germination of the seeds from a crop will hit poorer farmers particularly hard, because traditionally they keep a proportion of their seeds to plant the following season. Restrictive practices by large GM companies, such as making crops resistant to a particular weedkiller that only one company sells, may lead to inflated prices for the weedkiller and excessive profits for the company.

Pro The development of GM crops with improved quality, greater yields and reduced wastage due to pests will be of benefit to all and will increase food supplies for an ever-increasing population. The technology allows crops that flourish in difficult environments, such as saline or very arid soils, to be developed. These will be of great benefit to poorer countries where population expansion is necessitating the use of more marginal land. Competition between companies will maintain costs at realistically low levels.

GM foods: you decide

A1 Use information from the above summaries and from additional research of your own to prepare a presentation either in favour of or against the use of GM foods. You may even change your views about genetically modified foods as a result of this feedback!

285

16 Interpreting data

In this chapter we look at the skills required when interpreting unfamiliar data. These may be the results of actual experiments, or information collected from various investigations.

16.1 Calculations

The calculations in A2 may be more difficult than those that you were required to do in the AS exam. You need to be able to calculate percentage changes and rates. For some calculations, you may need to transform formulae.

Percentage change

It is useful to calculate a percentage change when a difference occurs in one of the variables investigated in an experiment. For example, suppose we were investigating the effect of hormone concentration on the growth of plant roots. It would be almost impossible to find roots that were all the same length. If the roots had different lengths to start with, then simply reporting the actual increase of length of each root would be meaningless. In cases such as this we measure the percentage increase in length of each root. Table 1 shows

Table 1   The effect of hormone concentration on root growth

Concentration of hormone (μM)	Initial length of root (mm)	Final length of root (mm)	Change in length of root (mm)
0.02	16	30	14
0.03	20	36	16

some typical results.

At first sight, it looks as though the roots in the 0.03 μM hormone solution have grown faster than those in the 0.02 μM hormone solution. But the roots are different lengths to start with, so we cannot tell which grew faster simply by looking at the increase in length. Instead, we have to calculate how much each of

them grew per unit of length during the experiment. The best way to do this is to measure the length before the experiment starts, and again at the end, and then calculate the percentage increase in length. We do this by using the equation:

$$\frac{\text{Increase in length}}{\text{Original length}} \times 100$$

For the root in 0.02 μM hormone solution:

$$\frac{14 \text{ mm} \times 100}{16 \text{ mm}} = 87.5\%$$

For the root in 0.03 μM hormone solution:

$$\frac{16 \text{ mm} \times 100}{20 \text{ mm}} = 80.0\%$$

So, in fact, the rate of growth of the shoot in the 0.02 μM hormone solution was the greater.

Do not be surprised if you get results that are greater than 100%. The data for a root placed in 0.01 μM hormone solution are: original length 15 mm, final length 33 mm; increase in length 18 mm. The percentage increase in length for this root is therefore:

$$\frac{18 \text{ mm} \times 100}{15 \text{ mm}} = 120\%$$

Now have a go at Question 1 on the next page. Here is a worked example to show how to answer this. To calculate the percentage, first calculate the actual increase:

$$10.96 \text{ ppm} - 0.23 \text{ ppm} = 10.73 \text{ ppm}$$

Now calculate 10.73 as a percentage of 0.23:

$$\frac{10.73 \times 100}{0.23} = 4\,665\%$$

> **1** Dieldrin is a pesticide that was used to treat wheat before planting. An investigation was carried out to find the effect of treated wheat grain on the dieldrin concentration in the tissues of mice living in wheat fields. Mice were trapped before and after the treated wheat was planted. The results are shown in Table 2.
>
> Calculate the percentage change in mean dieldrin concentration in the tissues of mice.

Table 2   Results of the Dieldrin experiment

Period	Number of mice caught	Number of mice analysed	Mean dieldrin content of mice (ppm)
Before sowing	15	4	0.23
After sowing	18	7	10.96

Significant figures

The actual display on the calculator at the end of the above calculation is 4 665.217 391. Clearly, since the data for the mice is in integers (whole numbers), the answer should be given to the nearest whole number: 4 665.

You should not write down more numbers after the decimal point than the number of figures given in the data. It is a good practice, however, to show the number given by your calculator display, then, round up or down as appropriate.

A good way of checking the magnitude of your answer is to do the calculation again, but using whole numbers where possible.

You could check the above calculation by finding 11 × 100 / 0.25.

The answer is 4 400, so the answer 4 665 is clearly of the right magnitude.

Suitable calculations

When looking at data, we must decide which mathematical technique to use. For example, what is the best technique to use when presented with the data in Table 3?

Table 3   Results of the Tonga surveys

Year	Number of palm trees examined	Number of palm trees damaged by beetles
1971	289	48
1978	302	23

The rhinoceros beetle is a pest that damages coconut palms growing on South Pacific islands. One method of control is to introduce a virus that kills the beetles. The virus was first used on the island of Tonga in 1971. Table 3 shows the results of surveys of rhinoceros beetle damage to palm trees carried out at two sites in 1971 and 1978.

> **2** Look at the results in Table 3. Was the introduction of the virus that kills rhinoceros beetles successful in Tonga?

To approach this question, you need to support your answer with suitable calculations from data in the table. Because the samples are not of equal size, you must calculate the percentage of damaged trees in each of the years.

The percentage of trees damaged in 1971 is:

$$\frac{48 \times 100}{289} = 16.6$$

The percentage of trees damaged in 1978 is:

$$\frac{23 \times 100}{302} = 7.6$$

The percentage of trees damaged fell from 16.6% in 1971 to 7.6% in 1978, a reduction of 9%.

The beetle *Delphastus pusillus* is used in the UK for biological control of whitefly, a common plague on greenhouse tomatoes. The beetle kills larvae as well as adult flies, and so is an effective method of control.

Rate of change

You also need to be able to calculate the rate of change from the gradient of a graph. The same principle applies to working with graphs in biology as it does in physics or chemistry. You divide a reading from the vertical axis by the corresponding reading on the horizontal axis. Fig. 1 shows how the mean concentration of carbon dioxide in the atmosphere changed above a Pacific island between 1960 and 1980.

Fig. 1   Mean carbon dioxide changes over time

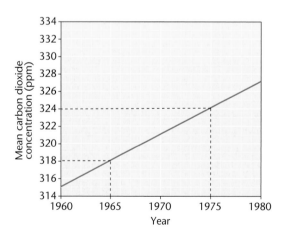

Since this is a 'straight-line' graph we can use it to calculate the rate at which the carbon dioxide concentration rose. It is much easier if you choose whole numbers. For example, the graph crosses intersecting gridlines at 1965 and 1975.

Read off the carbon dioxide concentrations in these two years – 318 ppm and 324 ppm. The carbon dioxide concentration rose by 6 ppm in 10 years, so the rate of increase is:

6 ppm/10 years = 0.6 ppm year^{-1}.

Let's look at another typical question.

3 The graph in Fig. 2 shows the effect of temperature on the mean number of red spider mites eaten by two predatory mites, *Amblyseius fustis* and *Typhlodromus occidentalis*.

Calculate the mean rate of increase in the number of red spider mites eaten per day by *A. fustis*.

Fig. 2   Effect of temperature on predatory mites

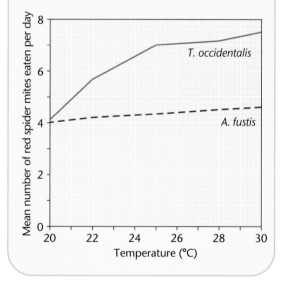

The mean number of mites eaten rises from 4 at 20 °C to 4.6 at 30 °C, which is an increase of 0.6. The rate of increase is therefore:

0.6 mites/10 °C = 0.06 mites °C^{-1}

Remember to refer to the unit when answering this type of question.
Now here is a question to do on your own.

4 Calculate the rate of increase in the mean number of mites eaten by *T. occidentalis* as the temperature rises from 25 °C to 30 °C.

16.2 Transforming formulae

You probably learned the 'magic triangle' method for transforming formulae at GCSE. The 'magic triangle' for the formula $A = B \times C$ is shown in Fig. 3a.

Fig. 3 Magic triangles

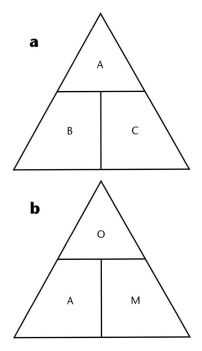

a

b

From the triangle:

$$B = \frac{A}{C}$$

and

$$C = \frac{A}{B}$$

A good example of how a magic triangle formula can be used is the relationship between actual size, observed size and magnification (Fig. 3b):

Observed size (O) = actual size (A) × magnification (M)

Suppose that you are given a photograph on which a mitochondrion has an observed width of 20 mm and you are told that the magnification is ×800 000. You are asked to calculate the actual width in μm.

Using the triangle:

$$A = \frac{O}{M} = \frac{20 \text{ mm}}{800\ 000} = 0.00025 \text{ mm}$$

Divide this by 1000 to convert to μm. The answer is therefore 0.25 μm

The following question, which you can try on your own, requires you to do two separate things: read data and transform a formula.

5 An individual's estimated average requirement for energy (EAR) can be determined by multiplying the basal metabolic rate (BMR) and the physical activity level (PAL):

EAR = BMR × PAL

Table 4 gives the average PAL values for males with different levels of activity; Fig. 4 shows the effect of age on EAR males with low levels of activity. Use the graph to calculate the BMR of a 35-year-old man with a low level of activity. Show your working.

Table 4 Average PAL values

Level of activity	PAL
Low	1.4
Moderate	1.7
High	1.9

Fig. 4 The effect of age on EAR values

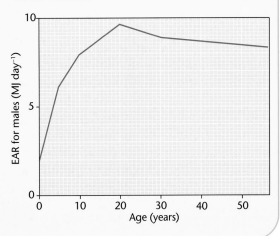

16.3 Patterns in data

Sometimes it is not possible to do an experiment in which one variable, the independent variable, is altered so that the effect of this change on another variable, the dependent variable, can be investigated. For example, in an ecological investigation on predation, it would not be possible to change the number of predators or prey in a habitat. Instead, the scientist needs to measure two or more variables to see if there is any pattern that might relate them. If there is a pattern, the scientist will suggest that the two variables are correlated, or related to each other. In this case the scientist would count the number of predators and prey over a period of time, then look for patterns in population sizes.

Using scattergrams
Data collected from experiments that look to see if two variables are related often show an indirect rather than a direct relationship. Results are usually plotted on a scattergram, as shown in Fig. 5

Fig. 5   Scattergram of weasel predation rate and great tit population density

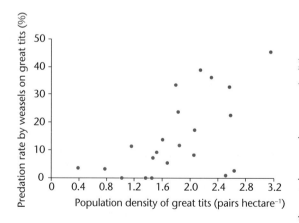

This scattergram shows the results obtained by an ecologist who collected data about the predation rate by weasels on great tits and the population density of great tits. As you can see, the data do not fall conveniently into points on a straight line or a curve, rather it is more scattered – hence the term scattergram.

However, if you look more closely, you will notice that there is a pattern in the data. The greater the population density, the higher the predation rate. But this does not mean that a high population density *causes* a higher predation rate, merely that there is a correlation between the two. What it does mean is that a change in one variable is linked with a change in another variable.

Explaining relationships
Once a correlation is found, we need to think of possible explanations. Further investigations can then help to find whether a suggested explanation is true. The first step, which you may be asked for, is a suggestion, or hypothesis. In this case we might suggest that at high densities, more great tits are killed by weasels, reducing the great tit population, making it more difficult for weasels to locate and then kill great tits. To check this, it would be necessary to find whether weasels really do kill more great tits at high densities by, for example, examining weasel faeces for indigestible remains of great tits.

Sometimes the relationship is negative, as Fig. 6 shows. This graph shows the relationship between the population density of mice and the rate of predation of great tits by weasels.

Fig. 6   A negative correlation

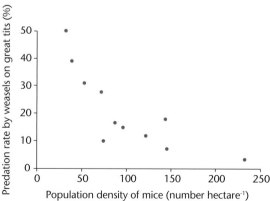

The correlation here is negative: the greater the mouse population, the lower the rate of predation by weasels on great tits. In this case the hypothesis might be that weasels find it easier to catch mice than great tits, so when the mouse population is high, the weasels obtain most of their food from mice rather than great tits. Further investigations, again using faecal examinations, would be done to find what was happening in the real ecosystem.

The results of such investigations often show that food webs are not simply a collection of 'straight' food chains, but have much more complex feeding relationships.

In other investigations, variables may be correlated for some values but not for others. The scattergram in Fig. 7 shows the relationship between the population size of wrens and the number of days that snow covered the ground in the previous winter. You might be asked the following sort of question on these data.

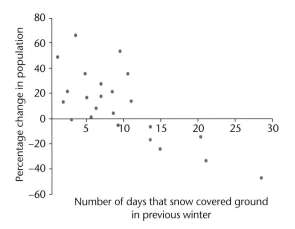

Fig. 7   Scattergram of wren population in snow

When you answer part 'b' of the question, you need to remember that the data are about snow cover and population change. A reasonable hypothesis is that there are fewer birds left to breed in the following summer because some birds starved when their food supply was covered by snow for longer periods.

Looking for patterns

Correlation data are often presented in tables. Again, we look for patterns in the columns of data that might suggest a hypothesis that relates two of the patterns.

Table 5 shows the relationship between the northern limit of tree growth and the resistance to freezing of winter buds.

6

a   Describe the relationship between the number of days that snow covered the ground and the change in population size.

b   Suggest and explain a reason for this relationship.

The answer to part 'a' is that these data show no correlation between number of days of snow and change in the wren population when the number of days of snow is fewer than 13. However, when there are more than 13 days of snow, there is a negative correlation between the number of days of snow and the population size.

---

7   Look at Table 5 and describe the patterns in the data.

Table 5   Tree growth and resistance of buds to freezing

Species	Northern limit (°N latitude)	Mean minimum temperature at northern limit (°C)	Temperature below which the winter buds are damaged by frost (°C)
Oak	32	−6	−7
Magnolia	35	−11	−17
Liquidambar	40	−19	−27
Poplar	48	−33	< −80
Willow	48	−33	< −80
Elm	52	−42	−45
Birch	60	< −46	< −80

There are three columns of data, so we refer to all three. For example:

- The northern limit of the trees is positively correlated with the mean minimum temperature.
- The northern limit of the trees is positively correlated with the temperature at which the buds are damaged by frost.

- The mean temperature at the northern limit is greater than the temperature at which the buds are damaged by frost.

Now let's look at another example. This time the question is for you to answer.

8 Table 6 shows the percentage of woods of different sizes that were inhabited by different species of birds.

a Describe the relationship between the size of wood and the diversity of the birds it contains.

b Use data from the table to explain the different distribution of treecreepers and greater spotted woodpeckers.

Blackbird feeding its young

Table 6 Woods inhabited by different bird species

Species	Food	Typical body mass (g)	Percentage of woods of each size inhabited by named species				
			Size of wood (ha)				
			0.001–0.01	0.01–0.1	0.1–1	1–10	10–100
Blackbird	Wide range of insect and plant food	90	13	34	63	72	91
House sparrow	Weed and grass seeds	30	9	21	44	35	22
Great tit	Insects and nuts	20	0	3	12	16	73
Treecreeper	Insects	10	0	0	2	8	25
Greater spotted woodpecker	Insects	130	0	0	0	1	26

16.4 Experimental investigations

In an experimental investigation the scientist manipulates one variable and then measures the effects of this manipulation on another variable.

If the scientist finds that whenever variable A is changed, then variable B changes, he or she can conclude is that A influences B. Only experimental data can demonstrate conclusively that the change in one variable is caused by the change in the other variable.

Independent variables are those that the experimenter changes, whereas dependent variables are the ones that change as a result of the experiment.

Line graphs

Most data obtained from experimental investigations can be plotted on a line graph. Conventionally the independent variable is plotted on the horizontal axis (also known as the x-axis) and the dependent variable on the vertical axis (the y-axis).

Line graphs include both those where a straight line can be drawn through the data points and those where a curved line is more appropriate. Data points often give curves that change direction abruptly at one point.

The graph in Fig. 8 shows the effect of light intensity on the rate of photosynthesis of a crop plant. In this investigation the experimenter changed the light intensity, so this is the independent variable. The rate of photosynthesis was measured at each light intensity chosen, so this is the dependent variable.

Fig. 8   Effect of light intensity on the rate of photosynthesis

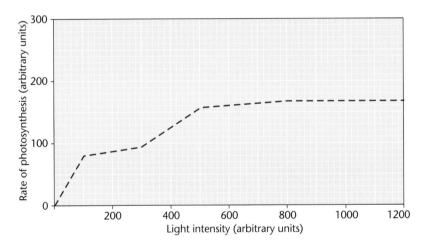

The plotted line for the crop plant levels off at a light intensity of 800 arbitrary units. This means that the rate of photosynthesis remains constant at the rate of 160 units, not that photosynthesis stops.

For light intensities between 0 and 800 units, light intensity is the factor that limits the rate of photosynthesis. At light intensities above this, some other factor such as carbon dioxide concentration is limiting the rate, since increasing light intensity beyond 800 units has no further effect.

Reversing the axes of a line graph

Sometimes it may be better to show data on a graph with axes that are reversed. An example is shown in Fig. 9.

Fig. 9   Loss of mass from a detached leaf

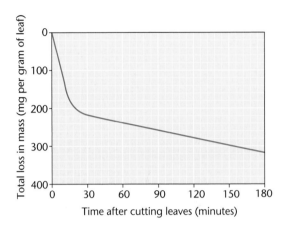

This graph shows the changes in total water loss from a detached leaf over 180 minutes. By having the axis for mass loss going downwards, it shows the actual loss of mass more clearly.

The rate of mass loss in any period can be calculated from the gradient. Thus, in the first 10 minutes after being detached, the rate of mass loss from the leaf is:

100 mg per gram per 10 minutes = 10 mg per gram of leaf per minute

Let's have a look at a question.

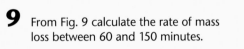

9 From Fig. 9 calculate the rate of mass loss between 60 and 150 minutes.

A good description of the data would be:

'The leaf lost mass at a high constant rate for the first 25 minutes then continued to lose mass at a slower constant rate for the rest of the time.'

Some data give graphs in which the rate of change is exponential for part of the time. This means the increase in rate becomes more and more rapid. The graph in Fig. 10 shows the light absorption of a colony of microbes.

Fig. 10   Light absorption of microbes

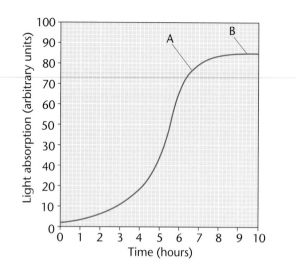

10 Look at Fig. 11. The temperature in different parts of a large grain store varied from 14 °C to 20 °C. The moisture content of the grain was constant at 16%. Beetles of both species were introduced into the store. What would you expect to happen if:

a Species A were a more successful competitor than Species B?

b Species B were a more successful competitor than Species A?

As the number of microbes increases, more light is absorbed. For the first 6.5 hours (A) there is an exponential increase in numbers – the rate of at which numbers increase gets larger – and light absorption increases accordingly. Again, the population does not decrease after 9 hours; it remains at a high, constant level (B).

Using area graphs

Data may also be presented as area graphs.

Two species of beetle are commonly found in stored grain. Fig. 11 shows the range of moisture content and temperature that each species can tolerate. The ecological ranges of the two species overlap. Where their ranges overlap, the beetles will compete, but there will be no competition in the part of the range occupied by only one of the beetles.

Significance

When asked to draw conclusions from data we need to be careful to include only significant differences. This is because living organisms are variable, so minor differences may not be due to changes brought about by the experiment.

A good example is shown by data on plant auxins. Fig. 12 shows the results obtained from experiments that investigated the effect of light on auxin distribution in shoot tips.

Note: the figures in Treatments A to D are mean values derived from several experiments.

11 What conclusions can be drawn by comparing the results for:

a treatments A and B?

b treatments B and C?

c treatments C and D?

The differences between Treatments A and B, and between B and C, are so small that they are likely to be due to variation; they are too small to be significant. So, in question part 'a', light does not affect the total amount of auxin; and in part 'b' equal amounts of auxin diffuse into the two blocks when the two halves of tip are separated. In part 'c', the difference in concentration between Treatments C and D is significant. A reasonable hypothesis would be that the cover glass in Treatment C prevented diffusion of hormone away from the illuminated side of the tip.

Fig 11   Beetles and their tolerance to environmental stress

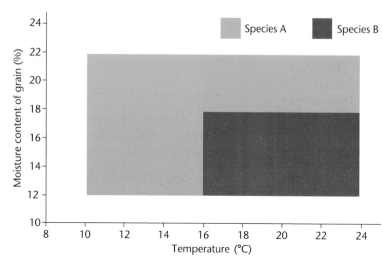

Fig. 12 Auxin experiments

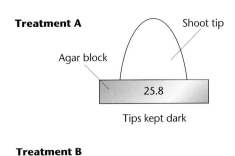

Treatment A

Shoot tip

Agar block

25.8

Tips kept dark

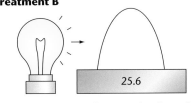

Treatment B

25.6

Tips illuminated unilaterally

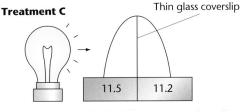

Treatment C

Thin glass coverslip

11.5 | 11.2

Tips completely divided and illuminated unilaterally

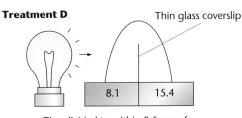

Treatment D

Thin glass coverslip

8.1 | 15.4

Tips divided to within 0.5mm of apex and illuminated unilaterally

16.5 Using statistics

Accuracy and limitations

There is a limit to the accuracy of any measurement made in an experiment. One reason is the limitation of the measuring instrument. A second is the care with which the instrument is used. But in biological experiments there is often a practical limit to the accuracy that it is worth trying to achieve. Although instruments exist which can measure length to a fraction of a micrometre, there would be no point in such accuracy when measuring tail length in mice in an investigation of variation. In fact, with a wriggling mouse, it might be difficult to measure even to the nearest millimetre. This difficulty would be compounded by having to decide exactly where the base of the tail actually starts.

It is therefore important to consider the accuracy that might reasonably be expected from a set of data. Accuracy is often confused with reliability. Consider the data in Table 7.

Taking measurements from several specimens increases the reliability of the results, but it does not make them more accurate. For Plant A, all the results are reasonably similar, which suggests that the value for the mean is likely to be somewhere in the same area. However, if another five leaves were measured, it is highly unlikely that exactly the same mean would be obtained. It is clearly absurd to give a value for the mean that is more precise than the accuracy

Table 7 Comparing loss in mass of leaves from two different types of plants

Leaf	Loss in mass over 24 hours (g)	
	Plant A	**Plant B**
1	1.03	0.28
2	0.96	0.72
3	0.89	0.74
4	1.05	0.69
5	0.94	0.77
Mean	0.968	0.64

of the measurements. Calculators give answers to many places of decimals, but judgement has to be used about the number of significant figures that can sensibly be given in data for means, or other calculations based on manipulating raw data.

The mean for Plant B looks even dodgier. The result for leaf 1 is very different compared to the others, so the mean comes well below all the other results. It may be that this result was a mistaken reading of the balance. On the other hand the anomaly may have been because the leaf was atypical (for example, much smaller, with fewer stomata than normal) or half-dead.

Without information about the original masses from which the losses were calculated it is impossible to guess. Expressing the results as percentage loss rather than as total loss would make comparison more reasonable.

Using statistics to interpret results

The likely accuracy of a mean can be assessed statistically. This test is based on measuring the standard deviation of the results (for details of standard deviation see AS book p. 122). Provided that enough measurements have been made, when the standard deviation is small it is likely that the calculated mean is close to the value that would be obtained if more results were obtained. The results of this test are called the standard error. Standard errors are often shown as a line on a graph, as in Fig. 13.

Fig. 13   Graph of mustard growth against density of planting

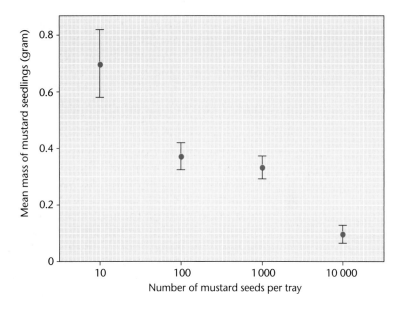

The standard error line indicates that it is 95% certain that the true mean is between its two extreme values. Notice that in Fig. 13 the standard error lines for seed numbers of 100 and 1 000 overlap, so there is no significant difference between them.

When we assess the limitations of an experiment, we consider those aspects which could not be avoided however accurate and careful we were. Dropping or misreading a thermometer is a limitation of the experimenter, not the experiment! Common limitations are the natural variability of living organisms and the impossibility of controlling certain environmental conditions, such as soil texture.

For example, in the experiment for which the results are shown in Table 7, it would be difficult to find leaves that were all exactly the same area in size. This could be allowed for by calculating loss in mass per unit area, but it is possible that leaves of different sizes would have different thickness and different numbers of stomata per unit area. It might also be difficult to seal the broken end of the stalk. The assumption was that loss in mass was due to loss of water, but some loss could be due to respiration or loss of other substances. For an experiment to be valid it must measure what it is aiming to measure. Otherwise, no matter how accurate the measurements are, the conclusions will still be wrong.

12 Fig. 13 shows the results of an experiment to find the effect of planting density on the growth of mustard seedlings.

a Judging from the standard error lines, suggest which planting density produced the greatest variation. Suggest a reason for this variability.

b The seeds were planted in soil in the trays. To measure growth, the plants were cut at the base after 21 days. 50 plants from each tray (all plants from the tray with only 10 seeds) were weighed. Suggest what limitations this method might have.

c Using a sample size of 50 should eliminate variability due to genetic variation between seeds. However, this could be a problem in the tray with only 10 seeds. Suggest how this possible source of error could be reduced.

Probability

How many people in the world have more than the average number of hands? Most people who answer this question without thinking will say 'none' or 'very few', since most of us don't know anybody with three hands. The correct answer, however, is 'nearly everyone'. This is because, although hardly any people have more than two hands, there are unfortunately quite a lot with fewer. There are, for example, those who have lost arms in accidents or were born without them. The arithmetical average, or mean, is therefore less than 2.0, whereas, of course, most people do have exactly 2.0.

This illustrates the importance of being careful when interpreting data. Common sense and statistics don't always go together.

Because living organisms are variable, we need to test whether differences we measure are significant, or just down to chance variation. To do this we use statistical tests. Most statistical tests give us an answer in terms of the probability of our observed differences occurring by chance.

There are several ways of expressing probability:

- 1 in 10
- 1 to 2
- 0.1 (or 1 in 10)
- $\frac{1}{2}$ (1 in 2).

Note that '1 in 2' is not the same as '1 to 2' (1 in 2 is a 0.5 chance and 1 to 2 is a 0.33 chance).

Thus, the probability of a Y sperm fertilising an egg can be expressed in three ways:

- 0.5
- $\frac{1}{2}$
- 1 in 2.

The chi-squared test

Scientific investigations proceed by a process of formulating and testing ideas (hypotheses). You can never prove anything in science, but you can gather support for your ideas. This is where the chi-squared test comes in – it helps scientists to decide how much weight to give to experimental results. (Chi is the Greek letter χ, pronounced 'ky', as in Kylie, and is the mathematical symbol in the formula.)

Suppose that you are trying to find whether using fertiliser makes any difference to the species that grow in a field. You count the number of each of four species in 20 quadrats from fertilised and unfertilised areas and obtain the results shown in Table 8 for the four species.

Table 8 Plant species data

Species	Total number in 20 quadrats	
	Fertilised area	**Unfertilised area**
A	0	182
B	2	51
C	88	124
D	91	102

The chi-squared test can be used when you are counting things, rather than measuring features such as height or mass. It is based on working out how likely it is that the results might have occurred by chance. If you toss a coin 20 times, you could expect that it would fall as heads 10 times and as tails 10 times. However, you could also get 12 heads and eight tails, 15 and five, or other numbers. The further the result is from the expected 10 heads and 10 tails, the rarer it will be. The test formula calculates how likely the actual results are compared with the expected results. In the example above, if the species were randomly distributed between the two areas, you would expect the numbers in each area to be the same.

For the purposes of the test, we assume that any difference between the actual results and the expected results is just due to chance. This is called the **null hypothesis**. The results of the calculation will tell us how rare it would be for the actual results to be due to chance. If the test results show that such results would occur by chance only five times out of 100 or less, we would consider it unlikely that the results were just chance and we would reject the null hypothesis. This means that we are suggesting that we think our research has probably found a significant difference.

Look at the data collected for four plant species in fertilised and unfertilised areas (Table 8).

It seems fairly certain that fertilising the fields affected species A and B, but the results are less clear for species C and D.

Before calculating chi-squared we need a null hypothesis. The null hypothesis in this case is that:

Fertiliser does *not* affect the distribution of the plant species.

In the chi-squared test the observed number in each category is compared with an expected number. The expected number can be calculated according to a theory, as in the case of genetics experiments. It can also calculated by using the means of the observed numbers in the case of environmental investigations.

The formula for the chi-squared test is:

$$\chi^2 = \Sigma \frac{(O - E)^2}{E}$$

Where: χ^2 is the symbol for chi-squared
O is the observed number
E is the expected number.

So, chi-squared test variance is the difference between observed and expected numbers.

If we take the result for species C, the observed numbers are 88 for the fertilised area and 124 for the unfertilised area. The expected number in each area, if the null hypothesis is true, is the mean of the number of plants.

The mean number per quadrat is:

$$\frac{(88 + 124)}{2} = 106$$

We can now put our observed and expected results into a table, as Table 9 shows.

Table 9   Observed and expected results

	Fertilised area	Unfertilised area	Total
O	88	124	212
E	106	106	212
(O–E)	–18	18	
(O–E)²	324	324	
$\frac{(O–E)^2}{E}$	3.06	3.06	6.12

So chi-squared value for plant species C is 6.12. We now need to find if this value for chi-squared confirms or rejects the null hypothesis.

If the value shows that the observed numbers would occur by chance only five times or fewer out of 100 times, we consider it unlikely that the results occur by chance and we reject the null hypothesis. However, we are still not certain that the null hypothesis is not true – only that there is a good chance that it is not true.

We compare the computed result, 6.12, with values in a table of chi-squared values (Table 10).

Table 10   Some values for chi-squared

Degrees of freedom	Probability					
	0.50	0.25	0.10	0.05	0.02	0.01
1	0.45	1.32	2.71	3.84	5.41	6.64
2	1.39	2.77	4.61	5.99	7.82	9.21
3	2.37	4.11	6.25	7.82	9.84	11.34
4	3.36	5.39	7.78	9.49	11.67	13.28

To use Table 10, we have to select the probability we are using and the number of degrees of freedom in the data. Biologists usually choose the 5% probability level. This means a probability of five times in 100, so we select the 0.05 probability column.

The number of degrees of freedom in the data = (number of categories – 1). In this case there are two categories, fertilised and unfertilised. There is therefore (2 – 1) = 1 degree of freedom.

In Table 10, the chi-squared value for a probability of 0.05 and 1 degree of freedom is 3.84. The calculated value of chi-squared is 6.12. Since the calculated value, 6.12, is greater than the table value, 3.84, the null hypothesis is rejected.

This means that the observed results would occur by chance less than five times in 100.

13 Calculate chi-squared for the results for plant species D, referring to Tables 9 and 10.

Another way of calculating expected numbers

In one of his experiments, Mendel crossed true-breeding long-stemmed pea plants with true-breeding short-stemmed pea plants. All the F_1 plants had tall stems. When he self-pollinated the F_1 plants, 787 of the F_2 plants had long stems and 277 had short stems.

There was a total of 1064 plants. If Mendel's hypothesis is correct we would expect a 3 : 1 ratio of tall : short plants (i.e. 798 tall plants and 266 short plants). We can set up a table (Table 11).

Table 11   Mendel's results

	Tall plants	**Short plants**	**Total**
Number observed	787	277	1064
Number expected	798	266	1064

Using a computer or calculator, chi-squared for these data = 0.607. Look back at Table 10 for some values for chi-squared.

In this case there are two categories, long stem and short stem, so the number of degrees of freedom is $(2 - 1) = 1$.

The null hypothesis is that the difference between the observed and the expected results is not significant.

For $\chi^2 = 0.607$ the probability is somewhere between 0.5 and 0.25. This means that there is a probability that the observed results would differ from the expected results by the amounts shown in the data in between 25% and 50% of cases. The null hypothesis is therefore accepted: the difference between the observed and the expected results is not significant.

Using the mean of the observed numbers to calculate χ^2

Cepaea nemoralis is a species of snail that lives in woods and fields. It is prey to birds such as the thrush. The snails show several different colour variations, such as yellow with dark bands, and brown with no bands. Fig. 14 shows two of these variations.

Fig. 14   Snail shell colours

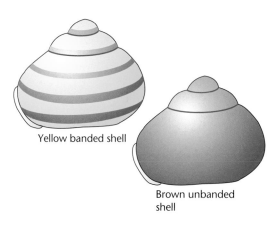

Yellow banded shell

Brown unbanded shell

In an investigation samples of the snail were collected from two sites, one in a beech wood and the other under a hedge. The results are shown in Table 12.

Table 12   Snail data

Collection site	Number of snails collected		
	Yellow banded	**Brown unbanded**	**Total**
Beech wood	32	88	120
Hedges	49	27	76

14

a   Suggest a null hypothesis for the data shown in Table 12.

b   How many degrees of freedom are there in the data?

c   What is the overall mean number of type of snail per location? This is the expected number of snails (*E*).

d   Calculate χ^2 for the data.

e   Look up the probability for this value in Table 10. What do you conclude?

Appendix: The essay

Paper 5 requires you to write an essay. This is to assess your ability to bring together principles from different areas of biology. You will have to choose one of two essay titles.

The instructions for the essay question make clear the skills that must be used to obtain high marks. For example:

> Write an essay on one of the topics below.
> **EITHER**
> **a** The process of diffusion and its importance in living organisms.
> **OR**
> **b** Mutation and its consequences.
>
> - In the answer to this question you should bring together relevant principles and concepts from as many different modules as possible.
> - Your essay will be marked not only for its scientific accuracy, but also for the selection of relevant material.
> - Write the essay in continuous prose.

Maximum marks available for an essay

The maximum number of marks that can be awarded is as follows:

Scientific content	16
Breadth of knowledge	3
Relevance	3
Quality of written communication	3

Although 16 of the 25 marks are for scientific content, high marks can be awarded only for relevant content. You should include material from as many topics as possible from Modules 1–5 so that you cover the whole specification.

It is therefore essential to *plan* the essay. The examination allows 45 minutes for preparing and writing an essay: you should spend about 15 minutes on the planning stage.

Scientific content

The first stage in planning an essay is to 'brainstorm' all the topics from Modules 1–5 that might be relevant to the topic. Let's take one of the titles from the specimen paper, 'Mutation and its consequences', as an example. Brainstorming might produce the following list of topics:

- a definition of mutation
- types of gene mutation (addition, deletion, substitution)
- what causes mutation (spontaneous event, radiation, mutagenic chemicals)
- the effect of mutation on protein synthesis
- a change in base sequence may result in change in amino acid sequence of a polypeptide
- the resulting change in polypeptide may result in change in structure of protein and the way it works
- the possibility of resultant metabolic blocks
- the role of mutation in evolution
- mutation as a source of genotypic variation
- how natural selection operates on mutations, leading to changes in population, changes within a species, the formation of a new species.

It is possible to gain one or two marks for each of the bulleted topics. There might not be time to write about all of them, but if you have to make a selection, be sure that your choices contain topics from every relevant area of the specification. It is not necessary to refer to all the above topics to gain full marks for scientific content. However, it is wise to include at least five or six well-explained points. It is therefore best to select the topics you are most confident with.

Breadth of knowledge

To gain the maximum 3 marks available under this category, you need to show that you can bring together material from different areas of biology.

Relevance

To gain the maximum 3 marks, all the material presented must be clearly relevant to the essay title. For example, in topic b, because mutation may lead to variation, you might be tempted to include other factors relating to variation such as details of meiosis and the random fusion of gametes. Since the essay title is 'mutation' rather than 'factors affecting variation', you would not achieve full marks if you included irrelevant material on meiosis.

Quality of written communication

To gain the maximum 3 marks, all the topics selected must be presented logically in clear, scientific English. You must use technical terminology effectively and accurately.

After selecting your topics, you should write them in a sensible sequence. Think in terms of 'telling a story'. Each major idea should be the basis for a paragraph. Use bullet points and diagrams only if they make your explanations clearer. If you do use bullet points, remember to use punctuation, and use it consistently.

Try making an essay plan for questions 1 and 2.

1 What problems are associated with an increase in size of an organism? How are these problems solved?

2 How is ATP produced and used in living organisms?

For the essay in question 1, you should ensure that you include reference to plants as well as to animals. You should also refer to problems associated with temperature control as well as problems of absorbing and distributing materials such as oxygen.

For the essay in question 2, you should include the production and use of ATP during photosynthesis by plants as well as the obvious topic of respiration.

Answers to in-text questions

Chapter 1

1a Population
1b Habitat
1c Ecosystem
1d Abiotic factor
1e Community
1f Biotic factor

2 Food in soil is a niche exploited by few animals. The first animals to do so were probably quite small, perhaps able to enter natural tunnels and crevices. In the populations of both ancestors there would be differences between forelegs due to genetic variation. Those animals which had forelegs with the largest claws and the most powerful muscles would most likely have been the most successful in obtaining food and surviving. Natural selection would occur, with the genes for advantageous features being passed on through many generations. Because the environmental conditions in soil would be similar in both Australia and Europe, selection pressures resulted in both types of moles having similar features.

3a 0.0625 m²
3b 1.25 m²
3c 0.57 %

4a 25/30 × 100 = 80%
4b 25 %
4c Leguminous plants such as clover have nitrogen-fixing bacteria in their root nodules. These bacteria use gaseous nitrogen to produce organic nitrogen compounds, including amino acids and proteins. When clover plants, or their roots and nodules die, they are decayed by saprophytic bacteria. Then, nitrifying bacteria make nitrates from the ammonium ions produced. The improved supply of nitrate ions enables the grass to grow more quickly.

5 It is much quicker and easier to make decisions using the ACFOR scale. However, this gives less precise data than if you used percentage cover.

6 Oarweed: 0 to 25 cm.
Toothed wrack: 30 cm to 75 cm.
Knotted wrack: 90 cm to 150 cm.
Bladder wrack: 75 cm to 85 cm.
Spiral wrack: 150 cm to 180 cm.
Channelled wrack: 200 cm to 245 cm.

7a 20 × 20 cm.
7b Tor grass, rock rose, ribwort plantain.
7c Some species, such as daisies, have short stems and spreading leaves. They grow well where the grass is short because they get plenty of light, and their leaves are not easily damaged by trampling. To the side of the path, grasses can grow tall and overshadow them.

8 Total number in population = (54 × 63) ÷ 18 = 189.

9 The high concentration of hydrogen ions affects the ionisation of the side chains and other groups of amino acids in the protein; they will hang on to their hydrogen atoms rather than lose them. This can affect the hydrogen bonds that normally form between different parts of the protein molecule, and therefore change its overall shape and behaviour. Enzyme molecules may no longer be able to bind with their substrate.
Heavy metal ions act as non-competitive inhibitors of enzymes. They bind with the enzyme molecule, usually at a site other than its active site. This causes the shape of the whole enzyme molecule to change, so that its active site can no longer bind with the substrate.

10a The oxygen concentration will reduce. Gases such as oxygen are less soluble in warm water than in cold water. Animals may not be able to obtain sufficient oxygen for respiration.
10b Seaweeds photosynthesise, and therefore produce oxygen. Oxygen concentration would rise during daylight hours, since photosynthesis is much faster than respiration. At night, however, there would be a reduction in concentration due to the respiration of the seaweeds.
10c When the pool is exposed, water can evaporate and increase the salinity. Heavy rainfall can dilute the seawater in the pool and decrease the salinity. Changes in salt concentration can cause osmotic effects in organisms, with high salt concentrations causing water loss from cells. Low concentrations may cause water to enter cells, and animals' cells may burst.

11a Water availability, temperature.
11b Temperature, wind speed, depth of soil.
11c Water current, mineral ion availability.
11d Lack of stable attachment surface, wave action, salinity.
11e Acidity of soil, mineral ion availability, high water content of soil resulting in reduced oxygen content.

12 The two populations treated with the worm-killing drug in 1989 (represented by curves b and c) did not crash, while the untreated one (curve a) did. In 1993, the two untreated populations (curves a and b) did crash, while the population that was treated with the drug (curve c) did not.

13 In those populations that were treated with the drug, the numbers still dropped a little in the years when a crash was expected. This can be seen in curve (b) in 1989 and in curve (c) in both 1989 and 1993.

14a

Generation	Number of bacteria
1	20
2	40
3	80
4	160
5	320
6	640
7	1 280
8	2 560
9	5 120
10	10 240

14b

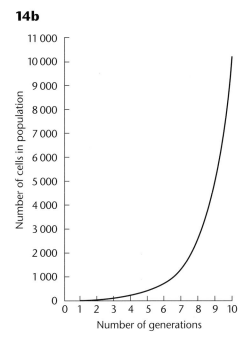

The graph shows an increasingly steep curve. This pattern of population growth is called exponential.

14c The pattern is similar in that there is slow growth at first, followed by a steep increase. The increase is rather irregular, probably because the cells do not all grow and divide at exactly the same rate. The increase in population size levels off after about 15 days.

14d Light and mineral ion availability.

15a There was high mortality among the young rabbits.

15b The population of this colony would decline, and the colony might well die out, because the total number of rabbits at the end of the season (39) was much lower than the number at the start (70).

15c A rabbit colony can produce a large number of young very quickly and so the population could recover rapidly if the population of predators dropped. If the colony were completely wiped out, then the predators might die or move away. Immigration from other colonies could then take place. Also, natural selection might occur, whereby the young that are better at avoiding predators would survive, perhaps by being more cautious or having more acute senses. These more

successful rabbits could pass on their characteristics to their offspring.

16a

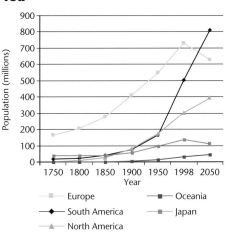

16b To find percentage growth, subtract the population in 2050 from the population in 1750. Divide this by the population in 1750 and multiply by 100. For Europe, percentage growth = $(628 - 163) \times 100 \div 628 = 285\,\%$. If you do this for each area, you will find that the percentage growth in North America is a massive 19 500 %, much greater than any of the others.

16c Most of this huge growth in North America took place between 1750 and 1950, when the country was still accepting large numbers of immigrants. The largest changes more recently have been in South America, where many countries are still undergoing the change towards more development and industrialisation. In recent years, populations in Europe and Japan have actually fallen. This suggests that population growth is greatest in developing countries rather than in developed countries.

17a To determine whether the population fell or grew, subtract the death rate from the birth rate. A positive number means that the population grew, while a negative number means that it fell. The population grew in 2007 in Panama, Tunisia and the United Kingdom.

17b The population decreased in 2007 in Austria and Japan.

17c The birth rate in Japan has been falling steadily since 1990, while the death rate has increased. Between 1990 and 2006, there was increasingly slow population growth, which became population decline in 2007.

17d A very high proportion of the people in Panama and Tunisia are young people, among whom death rates are much lower than in older people. In the developed countries, a very high proportion of people are older, and therefore more of these people die each year.

17e Immigration and emigration.

18a 59 084 000
18b

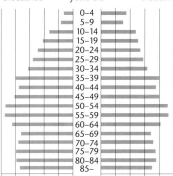

Age structure of UK population in 1999

19a In Great Britain, death rates are highest in later life, in the over 65s. In Botswana, they are highest in people in their 20s and 30s.

19b In Britain, good access to medical treatment, good diet and good hygiene mean that death rates from illness are relatively rare in young people. In Botswana, HIV/AIDS affects a very large proportion of people, and treatment was not yet widespread or effective at the beginning of the 21st century. Many HIV/AIDS patients die while they are still young.

A1a The pyramid suggests that birth rate was high and death rate was also quite high, so the population was probably in stage 2 or stage 3 of the transition.

A1b This pyramid suggests that the size of the population will remain roughly constant as the young people grow up. This is stage 4.

A2 The mean age of the population increased between 1901 and 2001.

A3 The proportion of older people is expected to increase.

A4a The population was growing, so birth rates were greater than death rates. This is probably late stage 2 or early stage 3.

A4b Emigration.

A4c

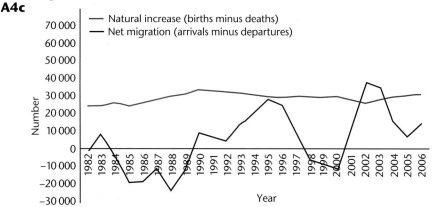

3a The action spectrum for chlorophyll *a* should be the same shape as its absorption spectrum.

3b Having several different pigments allows the plant to absorb a wider range of wavelengths of light, increasing the efficiency of photosynthesis. It is also probable that some of the other pigments absorb wavelengths of light that might otherwise damage the chlorophyll molecules.

4 Seaweeds that live deep in the water will receive only blue light, so they need pigments that will allow them to absorb these wavelengths. Green chlorophyll would not be able to absorb much of this light, whereas red pigments can do so.

5a Chlorophyll *a* absorbs energy from sunlight, which excites an electron so that it leaves the molecule.

5b Water is split to produce protons, electrons and oxygen atoms.

5c The first carrier receives electrons, becoming reduced. It then passes the electrons on to the next carrier, becoming oxidised – the next carrier is reduced. This continues all along the chain.

5d ADP is combined with a phosphate group to produce ATP. This is catalysed by the enzyme ATPases, which is found in the thylakoid membranes. It is an energy-requiring reaction, the energy being derived from the movement of hydrogen ions down their concentration gradient through the ATPases.

5e NADP takes up electrons released from chlorophyll, becoming reduced.

6a Reduced NADP → NADP
NADP is being oxidised, and glycerate 3-phosphate is being reduced.

6b Glycerate 3-phosphate → triose phosphate
Glycerate 3-phosphate is being reduced, and reduced NADP is being oxidised.

7 The NADP is reused in the light-dependent reaction. The ATP and inorganic phosphate are used to make more ATP during photophosphorylation.

8 Energy enters in the form of reduced NADP and ATP. It is incorporated into glucose molecules during the conversion of glycerate 3-phosphate to triose phosphate, which is then converted to glucose.

9 A chloroplast contains chlorophyll for absorption of sunlight. The chlorophyll is arranged on the thyalkoid membranes, which have a large surface area so that they can hold a large quantity of pigment molecules and also absorb large amounts of sunlight. These membranes also hold the molecules making up the electron transport chain, and ATPases, allowing photophosphorylation to occur. The stroma contains all the enzymes required for the light-independent reactions. The envelope (two membranes) surrounding the chloroplast keeps the reactions taking place inside it separate from those taking place in the cytoplasm outside.

10 Y – glycerate 3-phosphate, as this is the first compound to be formed in the light-independent reaction, and will be the first to contain radioactive carbon from the carbon dioxide. X – triose phosphate. This is formed by the reduction of glycerate 3-phosphate, which is why it is not there on the five-second chromatogram.

Chapter 2

1a The most active parts of the body produce most heat. When the body is exercising, the muscles release large amounts of heat.

1b The small intestine, where soluble nutrients are absorbed; the kidneys, where glucose and ions are reabsorbed.

1c Production of proteins from amino acids; production of glycogen from glucose; production of lipids; synthesis of DNA during semi-conservative replication; synthesis of mRNA during transcription.

2a Blue and orange.

2b Because it reflects light from the middle region (green) of the spectrum.

2c Yellow to red – they absorb light from the violet-blue end of the spectrum, and reflect all other colours.

11a The curve falls because ribulose bisphosphate accepts radioactive carbon dioxide to form glycerate 3-phosphate. However, this cannot be reduced to triose phosphate because there is no supply of reduced NADP or ATP from the light-dependent reaction. So no triose phosphate molecules are available for recycling to ribulose bisphosphate.

11b It rises because ribulose bisphosphate accepts radioactive carbon dioxide to form glycerate 3-phosphate, which cannot enter the next step of the reaction because there is no reduced NADP or ATP from the light-dependent reaction. It levels off when all the ribulose bisphosphate has been used up.

12a Light intensity.

12b Either some other factor, such as carbon dioxide, is limiting the rate of photosynthesis, or the increased temperature is decreasing the activity of enzymes involved in photosynthesis.

13 To evaluate the effects of carbon dioxide as a limiting factor, you must compare increases in carbon dioxide when all other factors (light intensity and temperature) are constant. At a chosen light intensity and at 15 °C, an increase in carbon dioxide from 0.03% to 0.13% brings about a large increase in photosynthetic rate. At the same light intensity but at 25 °C, an increase in carbon dioxide from 0.03% to 0.13% brings about an even greater increase in the rate.

A1a The temperature in environment T1 was always greater than that in T2. The difference was smallest in March (about 0.5 °C) and greatest in May and September (just over 1 °C).

A1b T2 had shading, whereas T1 did not. Less light therefore entered the T2 environment, so less long wavelength radiation was produced inside the glasshouse and the air was not warmed as much.

A2 The greatest yield was obtained in the T2 environment, in which the temperature ranged from 32 °C to 39 °C.

A3 Abiotic factors could be:
- carbon dioxide concentration – if this were lower in the open air than in the glasshouse, this would reduce the rate of photosynthesis
- wind speed – this would be greater outside than in the glasshouse, which could cause the plants to lose more water vapour from their leaves, or could cause physical damage to leaves
- humidity – could be higher or lower outside than in the glasshouse, which could affect the rate of water loss from the plant
- water content of soil – could be higher or lower outside than in the glasshouse, which could affect the rate of water uptake by the plant
- soil conditions – could contain more mineral ions in the glasshouse than in the soil.

Biotic factors could be:
- more insects or other pests feeding on the plants, reducing their productivity
- more competition with weeds for light, water or mineral ions
- more diseases such as fungi, thus reducing productivity.

You may be able to think of others. These factors could indeed be very significant, and they do mean that we cannot draw any firm conclusions about the effect of temperature when comparing the results for T1 and T4.

A4 The results could be used by growers in a temperate country, who could ensure that they maintain a temperature in the glasshouse within a range that is shown to produce high yields. However, they should balance costs of heating against expected increases in yields. It may, for example, prove more economical to provide a temperature nearer to the conditions of the T3 environment (between 31 °C and 38 °C) than the

higher temperatures in T4, unless the increase in yield for the higher temperatures outweighs the extra heating costs.

Chapter 3

1 Both have an outer envelope, made up of two membranes. Both have a background material containing enzymes – the stroma in a chloroplast and matrix in the mitochondrion. Both have folded membranes in which the components of the electron transfer chain are embedded – the cristae in a mitochondrion and the thylakoids in a chloroplast. Both have their own DNA.

2 In photosynthesis, triose phosphate is produced by reacting reduced NADP and ATP with glycerate 3-phosphate, which is thereby reduced. In respiration, triose phosphate is oxidised by reduced NAD, and the phosphate group is removed to react with ADP to form ATP. In many ways, therefore, these reactions are the reverse of each other.

3a Per molecule of glucose:
- two molecules of ATP made directly during glycolysis
- two molecules of ATP made directly in Krebs cycle (one per circuit, and one glucose molecule therefore two circuits)
- 30 molecules of ATP from the total of 10 reduced NADs produced in glycolysis, the link reaction and the Krebs cycle
- four molecules of ATP from the two reduced FADs produced in the Krebs cycle making a total of 38.

3b The number of ATP molecules made in oxidative phosphorylation is 34 – the ATP produced from reduced NAD and reduced FAD. If 25% of the energy here is actually lost, then instead of the expected 34 molecules of ATP we will actually get only 25 or 26 molecules. The total yield will then be these 25 or 26, plus the four made in glycolysis and the Krebs cycle, making 29 or 30.

A1 Swimming and running (especially middle distance or long distance) are good aerobic exercises. Weight lifting is not, as it requires short bursts of intense activity.

A2 The myoglobin curve should lie to the left of that for haemoglobin.

A3 The myoglobin would provide an extra store of oxygen in the muscle, which could be used before anaerobic respiration had to begin. This could provide the runner with just a little more energy for muscle contraction, which might just give him or her the edge.

A4 Succinate dehydrogenase is one of the enzymes involved in the Krebs cycle. An increase in its activity means that the Krebs cycle could take place faster, or that more circuits of it could be completed in the same time span. This would provide more ATP and therefore make more energy available to the athlete's muscles.

A5 More myoglobin – greater oxygen store and therefore more aerobic respiration possible before anaerobic respiration is required. More mitochondria – more aerobic respiration (Krebs cycle and oxidative phosphorylation) so a greater rate of ATP formation. More activity of Krebs cycle enzymes – a greater rate of the Krebs cycle reactions and therefore a greater rate of ATP formation. More glycogen stores in the muscle – more glucose can be produced more quickly, so more substrate is available for respiration.

Chapter 4

1 Decomposers feed on all the organisms in the chain, so there would be an arrow from each organism to the decomposers.

2 For example:
a Oak tree → caterpillar → titmouse → weasel

b Other trees and shrubs → decomposers → earthworms → beetles → shrews
c Herbs → decomposers → soil insects → beetles → moles → tawny owls

3 Titmice, parasites, spiders, voles, mice, earthworms, soil insects and mites.

4

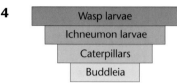

5 Your pyramid should look very strange. If you choose one tree, the bottom layer will contain one organism. Each tree has thousands of aphids, so the next layer up will be wide. The population of blue tits feeding on the aphids will be smaller, so the third layer up will be narrow, but not as narrow as the tree layer. The top layer will be wider as one blue tit could be infested by dozens of mites.

6 Take a sample of grass from a small area, such as a 10-cm square. Make sure that you have collected all parts of the grass, including roots, and cleaned off the soil. Weigh the grass sample and place it in an oven at about 100 °C for several hours. Weigh the sample again. This should be repeated until the weight stays the same, showing that all the water has evaporated. Other samples should be taken for a random sample of 10-cm squares in the meadow and the dry biomass found for each of these. From the mean of the results you can calculate the mean dry biomass per m^2 and from this the total biomass in the meadow.

7a Mammals are homeothermic, and produce a great deal of heat energy to keep their body temperatures constant, even when the environmental temperature is low. This is all lost to the food chain.
7b Plant cells are much less digestible than animal cells, because they are surrounded by a cellulose cell wall.

Few animals are able to digest cellulose. There is therefore more wastage of energy from eating plant-based foods than eating animal-based ones. The faeces of herbivores contain undigested cellulose, which contains energy they have not been able to access.

7c The invertebrate carnivore transfers the greatest percentage of energy to the next trophic level. It loses less heat than either of the mammals, because it does not generate heat to keep its body temperature up. It loses less energy in its faeces than the invertebrate herbivore, because it does not have to digest cellulose. In general, more energy is lost from mammals than insects because they are endothermic. More energy is lost from herbivores than carnivores because more indigestible energy-containing materials are lost in faeces.

8a Abiotic factors are: intensity of sunlight, duration of light, amount of water, temperature, concentration of carbon dioxide, availability of inorganic ions. The other factors are biotic.

8b Tropical forest: a high density of trees and other tropical plants, warm temperatures throughout the year extending the growing season, no shortage of soil nutrients or water, all contribute to the high productivity.
Intensive agriculture: a high density planting of crops during the growing season, application of nitrogenous fertilisers and possibly irrigation, all contribute to the high productivity.
8c Extreme desert and desert scrub: lack of water (and, therefore, lack of plants) and lack of soil nutrients reduce productivity.

9 5.5%

10 23%

11 59.8%

12a Large numbers of chloroplasts in palisade cells; densely packed chlorophyll-containing thyalkoid membranes in chloroplasts; several layers of mesophyll cells.

12b Other pigments (carotenoids) absorb green wavelengths and transfer energy to chlorophyll.

12c Chlorophyll molecules arranged in layers in lamellae, which avoids random shading.

13 The rate of photosynthesis will be less than the maximum for most of the time. Light intensity may only be at its optimum level for short periods during the day. Other factors, such as temperature and carbon dioxide concentration may limit the rate of photosynthesis. Growth rates may be restricted by shortages of water or mineral nutrients.

14a Only a small proportion of the energy in crop plants is transferred to biomass in primary consumers such as cattle. People therefore obtain more energy from a given area of land when they are primary consumers rather than when they are secondary consumers.

14b Animals such as cattle and sheep can feed on plants such as grass that humans are unable to use. On marginal land these may be the only plants that grow. Also, humans use only some parts of crop plants and animals may make efficient use of the rest.

15 150 kg ha^{-1} or slightly less. This concentration results in maximum grain yield. Any less will result in reduced grain yield. Any more will also result in less grain yield and extra expense.

16a Phosphorus. It cannot be nitrogen, as even if we give the plants more nitrogen the yield does not increase (orange curve). However, if they are given more phosphorus the yield does increase (green curve).

16b This has to be guesswork, because the green curve does not go below a nitrogen application of 50 kg ha^{-1}. However, we can be pretty sure that, if the curve follows the trend shown elsewhere, we will get only a low yield. Nitrogen will be the limiting factor – if we give it more nitrogen, it will produce a higher yield (green curve).

17a

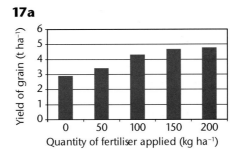

17b The soil in the field must already contain some nitrogen in a form that is available to plants.

17c The increase is less at high nitrogen fertiliser levels than at lower ones. It is probable that at low nitrogen levels, nitrogen is the limiting factor for growth. Once nitrogen levels reach higher levels, then other factors become limiting so adding more nitrogen cannot cause yield to increase accordingly.

17d At 200 kg ha^{-1}, cost of fertiliser is £80 per hectare. Return for grain is £960 per hectare.
Profit = £880 per hectare.
At 150 kg ha^{-1}, cost of fertiliser is £60 per hectare. Return for grain is £940 per hectare.
Profit = £880 per hectare.
At 100 kg ha^{-1}, cost of fertiliser is £40 per hectare. Return for grain is £860 per hectare.
Profit = £820 per hectare.
The farmer will therefore get most profit using either of the two higher values of fertiliser application.

18a Advantages: natural fertilisers may be cheaper, as they may be waste products from other processes, or from animals on the farm. Natural fertilisers such as manure generally contain plant material (such as straw) that can break down in the soil to form humus. This in turn encourages soil organisms such as earthworms, and increases the ability of the soil to hold water as well as increasing aeration and drainage. Natural fertilisers often contain components that break down slowly over time, releasing nutrients gradually rather than all at once.
Disadvantages: natural fertilisers do not contain known amounts of nutrients, so the farmer does not know exactly how much to apply to avoid wastage and leaching.

Natural fertilisers can be smelly, bulky and difficult to handle. If not stored properly, natural fertilisers can cause pollution (for example rain falling onto heaps of manure, washing out nutrients which can pollute nearby water courses).

19 Advantages could include: reapplication of the insecticide is needed less often; it may be cheaper since less insecticide is used; labour costs are reduced. Disadvantages could include: the insecticide is more likely to remain on crops and be ingested by humans or other animals, with possible toxic effects; it could affect useful insects; it is more likely to accumulate in food chains; there is increased chance of resistance developing in pest populations as a result of natural selection.

20 Answers will depend on pesticide choice and information found.

21a 0.04 kg ha^{-1}; 0.52 kg ha^{-1}

21b Natural variation among the budworms could mean that a few individuals have some resistance to the insecticide, so that a larger dose is needed in order to kill them. These would be more likely to survive than the non-resistant ones. They would be more likely to breed, and therefore the alleles conferring resistance would be more likely to be passed on to the next generation. Over time, more and more budworms would carry these resistance genes.

21c There is an increasing likelihood of reaching toxic or lethal doses for non-target species. The rate of bioaccumulation will also increase, with a greater chance of effects on organisms higher up the food chain.

22 The fluctuation may be the result of some kind of density-dependent factor – that acts more strongly on the population when numbers are high than when they are low. This could be interaction between predator and prey, including the control agent.

A1 The greatest fall in peregrine populations was between about 1958 and 1965, following the introduction of dieldrin in 1955. DDT had been introduced earlier, in 1946. When dieldrin and DDT were phased out, in 1967, the population began to rise again. The relationship between the usage of dieldrin and the peregrine population could therefore be a causal one, although the graph only shows there is a correlation, not that one actually causes the other.

A2 The mean concentration of dieldrin in the mice was at least six times greater after the seeds had been treated than before they were treated. There was variation in dieldrin content of the mice, but the difference between the 'before' and 'after' concentrations is much greater than the differences between individual mice after treatment. It can be fairly safely concluded that the mice ate the dieldrin-treated seeds and that some of the dieldrin remained in their bodies.

A3 The numbers of mice analysed were small. Not many were caught, and of these only some were analysed. The considerable variation between the samples after treatment indicates that the results are not especially reliable. It would be useful to know the individual results for each mouse, the means and the standard deviations. Very large standard deviations would make it more difficult to draw firm conclusions.
Trapping and testing more mice both before and after treating the seeds could improve the investigation. Equal numbers of mice should be tested before and after treatment. After seed treatment, it would be useful for trapping to continue for a longer period.
A control should be done in which mice are trapped in exactly the same way, in a similar area, but in which seeds that are not treated with chemicals are sown, rather than treated seeds.
Repeat trials should be done in other areas, using the same techniques.

A4 Answers will depend on the most recent viewpoints of the WWF and WHO.

Chapter 5

1 Glucose.

2 Higher temperatures increase the rate of metabolic reactions, and therefore the rate at which enzymes break down dead matter.

3 Summer, when photosynthesis takes place most rapidly in the higher light intensities, longer days and higher temperatures. Photosynthesis removes carbon dioxide from the air.

4a Arksoy 37%, Harrow 38%, Mandarin 87%, Mukden 36%, Williams 41%.

4b Carbon dioxide concentration is often a limiting factor for photosynthesis, so increasing it results in a greater rate of photosynthesis. This means that the plant is producing organic nutrients at a greater rate, so it has the materials and energy sources to produce more seeds.

4c The varieties may have significant differences among each other, in their physiology and morphology (structure). For example, Mandarin may have enzymes that are better able to make use of the extra carbon dioxide than the other varieties. These differences are genetic, resulting from the different gene pools (collections of alleles) in the different varieties.

5 The winter NAO index slowly increased during the study period, indicating that the weather was becoming warmer. This looks like a genuine change, not just fluctuations, as it has occurred over a 50-year period.

6 It is difficult to pick out a clear pattern. However, there is some indication that peaks in the NAO index tend to occur at the same time as peaks in the aphid numbers (for example, in 1967 and 2001)

and that troughs in one accompany troughs in the other (for example, in 1996). There are exceptions to this pattern, though. If the pattern is a genuine one, it could be that the aphid populations grow larger in warmer years, because there is more food available to them (because of faster plant growth) and because they can fly earlier in the year and therefore have a longer breeding season.

7 The second graph shows that, between 1966 and 2006, the earliest date on which aphids were flying got earlier, by almost 40 days. The gap between first and last captures increased, meaning that the length of the spring migration increased.
The third graph shows a weak but significant correlation between the NAO and the start of the spring migration. In general, the higher the NAO the earlier the migration began. The end of the spring migration does not appear to have been affected. The overall length of time during which the spring migration happened has therefore increased as NAO increased.

8 Measurement of the winter NAO could be used to predict the chance of a major invasion of aphids in spring. If the winter NAO is high, then farmers should begin spraying crops with insecticides earlier than if it is low.

9 In the electron transport system, electrons are transferred from one acceptor to another and then to atmospheric oxygen, producing water. In the absence of oxygen gas, denitrifying bacteria pass the electrons to oxygen in the nitrite or nitrate ions. Energy released in this transfer can be used to synthesise ATP. The oxygen is then available to form water, and nitrogen gas is released.

10 Poorly drained soil becomes waterlogged, and anaerobic denitrifying bacteria can thrive. These reduce the nitrate content of the soil, and hence its fertility. Good drainage will increase the air content of the soil.

11 Nitrogen-fixing bacteria in the root nodules of clover increase the quantity of nitrogen compounds in the soil. These become available for subsequent crops.

12a i During spring and summer, plants and other producers such as algae grow and use nitrates from their water for protein synthesis.

  ii The nitrate concentration rises again as these organisms die and decay slowly during autumn and winter.

  iii The steep rise in ammonium concentration occurs soon after the vegetation dies back in early autumn. Ammonium ions are the first products of saprotrophic breakdown.

  iv The ammonium ions are slowly converted by nitrifying bacteria to nitrates as part of the nitrogen cycle. This process occurs relatively slowly in the lower water temperatures of late autumn and winter.

12b It is unlikely that there is serious pollution. The nitrate concentration falls almost to zero during the summer, suggesting that there is not a large excess of organic material reaching the lake. Ammonium ion concentration stays at a fairly low level during spring and summer, again suggesting that there are not large amounts of organic matter being broken down.

A1 In 2000, agriculture was responsible for approximately one-fifth of the total pollution. This proportion fell gradually, until in 2006 it was just over one-tenth.

A2 A high BOD means that there is a large population of aerobic microorganisms that are using up oxygen from the water. This indicates that nutrients have been added to the water, a sign of pollution.

A3a We want to know how rapidly oxygen is being used up from the water. If there is air in the tube, this supplies extra oxygen, so it will take longer for the oxygen concentration in the water to drop and we will get a smaller difference in oxygen concentration at the end of five days than we should.

A3b We do not want any algae or phytoplankton in the water to be able to photosynthesise, as this would supply extra oxygen and skew the results.

A4 Samples of water could be taken from various points in the watercourse, and their BOD measured. If there is a sample with a high BOD downstream of one with a low BOD, this would suggest that pollutants have entered the water somewhere between them.

A5 Site 1 must be above the source of pollution, as it contains many organisms with a high biotic index (that is, organisms that require a high concentration of dissolved oxygen). Site 2 has very few of these organisms, suggesting that the oxygen concentration here is much lower.

A6 Indicator species can be captured and identified on the spot, and give an instant set of results. There is no need for access to a laboratory. Measuring BOD, however, requires a laboratory and takes five days.
On the other hand, using indicator species might not pick up some types of pollution, if these do not affect the oxygen content of the water. For example, it is possible that some industrial chemicals or heavy metals might cause harm to the community, but not affect the relative proportions of organisms with high or low biotic indices.

A7 Biotic factors such as competition are the likely reason for the absence of bloodworms and rat-tailed maggots from well-oxygenated water.

Chapter 6

1a The grass was no longer mowed.
1b Competition with larger plants, especially for light.

2 Low water content, high wind speed, instability of sand, high salt content.

3 A network of underground stems provides anchorage in the shifting sand. A thick outer cuticle protects from wind-blown sand and reduces water loss. The leaves can roll up, enclosing the stomata and therefore reducing transpiration. Hairs on the leaf surface, and the positioning of the stomata at the base of deep grooves, traps humid air close to the leaf surface and further reduces transpiration.

4 Mainly by wind-blown seeds. These tend to settle in the sheltered areas on the land side of the ridges. Animals or humans may transport some grass seeds.

5 High temperatures in the deposits; lack of water; instability; lack of shelter from wind.

6 It has parachute-like fruits, which help the seeds to disperse far and wide on the wind. It is rapidly growing, so it can complete its life cycle quickly, producing new seeds within a few months of germination. It has woolly, silver-grey foliage – this is caused by a waterproof covering on the leaves, which cuts down transpiration and so allows the plant to live in very dry soils. It produces stolons, which rapidly grow sideways from the plant and then grow into new individuals, allowing it to colonise large areas rapidly. The foliage seems to be unattractive for insects and other animals to eat. It is not very tall, and so can escape the strongest winds.

7 Virtually no plants grew at the site for the first three years following the eruption (due to high temperatures in the deposits; lack of water; instability; lack of shelter from wind). There was then a rapid and fairly steady (but note that the x-axis does not have a linear scale) increase in cover over the next 15 to 16 years (initially wind-dispersed, fast growing species). By 16 years after the eruption, plants covered almost 70% of the ground (species able to take advantage of more stable ground and shelter provided by early colonisers).

8 Species diversity increased gradually from years 1 to 14 after the eruption. This happened as pioneer plants first became established. Over time, their presence altered the abiotic conditions (more nutrients in soil, more water-holding capacity, more protection from wind) allowing other species to grow there. New colonisers would arrive by chance, so the longer the time period the more species would be expected to arrive. Species diversity fell from year 14. This is probably because some plants were now growing strongly and out-competing other species for light or water. Another possibility is that the greater quantity of vegetation would allow more herbivores to live in the area, and they might graze some species in preference to others.

9 Lupines are probably too large to be able to survive in the harsh conditions that the small pearly everlasting able to withstand. They would be more exposed to drying winds. They are also not well adapted for conserving water.

10 The substrate would initially not contain nitrate or ammonium ions. Nitrogen-fixing plants are able to survive in these conditions, because their symbiotic bacteria use nitrogen gas to synthesise ammonium ions. Plants need these to produce amino acids and proteins.

11 Western hemlock is not a nitrogen-fixer, and so needs a source of nitrate or ammonium ions in the soil. Alders are nitrogen-fixers, and their presence gradually adds these ions to the soil, allowing other plants to grow.

A1 It requires closely grazed chalk downland. There are now fewer sheep and rabbits to graze downland, so the vegetation grows taller and the low-growing thyme – the food plant of the caterpillars – is out-competed.

A2 Examples of its required habitat are becoming increasingly few and far between. It has not been able to adapt to live in different habitats.

A3 Left alone, short turf on chalk downland will gradually undergo succession to scrub and finally woodland. Management therefore involves cutting down bushes before they have a chance to establish. Some gorse has been left to provide shelter, as the red ants require well-drained and sheltered places to make their nests. Thyme seeds have been sown to increase the food available for caterpillars.

A4 There will be natural variation amongst thyme plants. The plants in a particular area may have slightly different collections of alleles – gene pools – than those from other areas. Sowing locally collected seeds means that the plants are more likely to have a collection of alleles that adapts them well for growing in local conditions.

Chapter 7

1 Black is dominant, because heterozygous dogs have black spots.

2a Codominance, as the heterozygote has a phenotype between that of the two homozygotes.

2b H^N could be the normal allele, and H^S the sickle cell allele.

2c $H^N H^N$ normal, $H^N H^S$ sickle cell trait, $H^S H^S$ sickle cell anaemia.

3

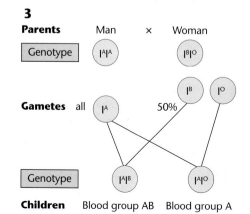

Parents	Man	×	Woman
Genotype	$I^A I^A$		$I^B I^O$

Gametes: all I^A — 50% I^B, I^O

Genotype: $I^A I^B$ $I^A I^O$

Children: Blood group AB — Blood group A

We would therefore expect half of the children to have blood group AB, and half to have blood group A.

4 Possible genotypes of the woman are $I^A I^A$ or $I^A I^O$
The child must have the genotype $I^O I^O$.
The child has therefore obtained an I^O allele from both of its parents. The man with blood group AB does not have an I^O allele, so the father must be the one with blood group A.

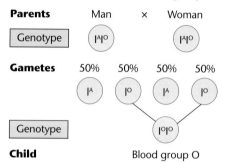

Parents	Man	×	Woman
Genotype	$I^A I^O$		$I^A I^O$

Gametes: 50% 50% 50% 50%
I^A I^O I^A I^O

Genotype: $I^O I^O$

Child: Blood group O

You could also draw a genetic diagram to show the possible genotypes of the children of the man with genotype $I^A I^B$, showing that he could definitely *not* be the father of a child with blood group O.

5 Two of Beatrice's children did not have Huntington's disease. She must, therefore, have carried the normal allele, which she passed on to these children. Since the normal allele is recessive, unaffected individuals must receive one normal allele from each parent.

6 Her children will not develop Huntington's disease.

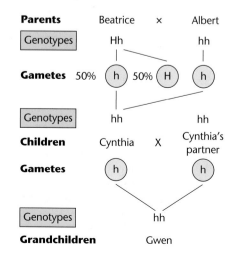

Parents	Beatrice	×	Albert
Genotypes	Hh		hh

Gametes: 50% h 50% H h

Genotypes: hh hh

Children: Cynthia X Cynthia's partner

Gametes: h h

Genotypes: hh

Grandchildren: Gwen

7 A 50% chance, since he carries the normal allele.

8 Discussion question.

9 The symptoms of the disease do not develop until relatively late in life. A person with Huntington's disease could have children before he or she knows that he or she has the faulty allele and will pass on the disease.

10a Siamese C^SC^S; Burmese C^bC^b.
10b All black coated.

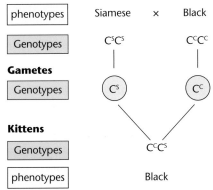

Parents

phenotypes — Siamese × Black

Genotypes — C^SC^S — C^cC^c

Gametes

Genotypes — C^S — C^c

Kittens

Genotypes — C^cC^S

phenotypes — Black

11a C^bC^S.
11b Yes, because the heterozygote has a different phenotype from either of the homozygotes.
11c The proportions of the kittens would be 50% pale brown, 25% Burmese and 25% Siamese.
11d She will have to breed a Burmese with a Siamese each time.

12a 1 : 2 : 1
12b 50%. The Manx cat must have the genotype Mm, and the tailed cat mm. All the gametes of the tailed cat will have the genotype m, whereas the gametes of the Manx cat will be 50% M and 50% m.

13a The sperm. All eggs carry one X chromosome, but the sperm can have either X (which will produce a girl) or Y (which will produce a boy).
13b The sperm could be sorted into those carrying an X chromosome and those carrying a Y chromosome. If only the X sperm are used, then all the zygotes would be XX and become cows rather than bulls.

13c Answers will vary depending on personal response.

14a 25% of children – 0% of girls, 50% of boys.
14b A mother with genotype X^CX^c or X^cX^c and a father who is X^cY could produced a red/green colour-blind girl.
14c A boy receives his X chromosome from his mother, not his father.

15 45, because there is only one X chromosome and therefore one fewer than the normal complement of 46.

16 **i** to **iv** XXY, XYY and XXXY might all be expected to have male characteristics, since all will possess the SRY gene. XXX would be female.

17 Answers will vary depending on personal response.

18a X^hY.
18b Mother X^HX^h and father X^hY.

19 $q^2 = 1/15\,000 = 0.000067$
so $q = 0.0082$
so $p = 1 - 0.0082 = 0.9918$
Proportion of carriers = $2pq$
$= 2 \times 0.0082 \times 0.0018 = 0.0163$
This means that about 1.6 in every 100 people are carriers of the PKU allele.

A1a No native Americans have blood group B.
A1b B and AB absent. Genotypes I^AI^A, I^AI^o, I^oI^o would be present.

A2a No. For example, Nigerian and Japanese people both have 23% with Group B, but their skin colour differs. On the other hand, both Nigerians and native Australians have black skin, but the percentage with group B is quite different.
A2b The black population derives from people of West African ancestry, such as Nigerians, whereas the white population is largely European in origin, where the proportion of group B is around 9%.

A3a They probably migrated to the area where they live from somewhere with a low proportion of group B in the population, such as from the population in Siberia from which some people moved east to North America.
A3b Information from DNA analysis, for example, from comparing mutations in genes for substances such as haemoglobin.

Chapter 8

1a The amount of food available, and space for example, for breeding sites.
1b The population would increase 10-fold each year. There would soon be severe competition for food.
1c Two.

2a

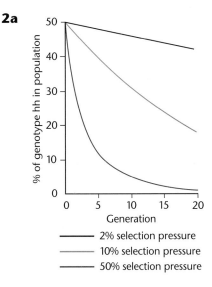

% of genotype hh in population vs Generation

— 2% selection pressure
— 10% selection pressure
— 50% selection pressure

2b At 2%, 4 200 individuals. At 10%, 1 800. At 50%, 100.
2c It would decrease.

3a **i** $q^2 = 0.02$
so $q = 0.14$ This is the frequency of the sickle cell allele in the population.
So $p = 1 - 0.14 = 0.86$
ii Proportion with sickle cell trait $= 2pq = 2 \times 0.14 \times 0.86 = 0.24$
So almost one quarter of the population (nearly 25%) have sickle cell trait.
3b We must assume that there is no significant immigration or emigration into or out of the population. All genotypes have an equal chance of survival. All genotypes have an equal chance of breeding with all other genotypes.

4

Parents	Mother	X	Father
Phenotypes	Sickle-cell trait		Sickle-cell trait
Genotypes	Hb^AHb^S		Hb^AHb^S

Gametes				
	50%	50%	50%	50%
Genotypes	Hb^A	Hb^S	Hb^A	Hb^S

Children

Genotypes	Hb^AHb^A	Hb^AHb^S	Hb^AHb^S	Hb^SHb^S
Phenotypes	Normal	Sickle-cell trait		Sickle-cell anaemic
Ratio of phenotypes	25% 1	50% 2		25% 1

The probability of their first child having sickle cell anaemia is 1 in 4 (25%).

5 The allele is relatively common because it was an advantage for people in West Africa to have sickle cell trait, as they were less likely to die from malaria. It is becoming rarer because it is no longer an advantage, as malaria is not common in the USA. The selective advantage of possessing a sickle cell allele has been lost.

6 Every individual with the dominant allele will show the unfavourable condition. Heterozygotes with the recessive allele will not show the unfavourable condition; so recessive alleles can be passed on by heterozygotes.

7 Tree trunks in industrial areas became blackened. Melanic moths were better camouflaged. Predators ate fewer of the melanic moths, so a higher proportion survived. Therefore these melanic moths would pass on their melanic allele to their offspring. The process would continue over successive generations until most moths were melanic.

8 The melanic allele is dominant; so all moths with this allele would be black.

9 A higher proportion of melanic moths were recaptured in the polluted wood, suggesting that black wing colour helped their survival. Conversely, a higher proportion of speckled moths were recaptured in the Dorset wood.

10 The proportion of the speckled form in the population is likely to increase again.

11 It is possible that there was not much variation in beak size in this species. If none of the birds had alleles for small beaks, then however strongly selection acted against those with large beaks, there would be no birds with smaller beaks to survive and breed.

A1a *A. porcatus*
A. porcatus
A. porcatus
A. porcatus

A1b The results do support the hypothesis. Each of the four species has mitochondrial DNA that is more similar to *A. porcatus* than to any of the others. This suggests that each species is most closely related to *A. porcatus*, which lives in Cuba. This would be expected if each species has evolved from *A. porcatus*. If one had evolved from another, then we would expect them to be more closely related to each other than to the Cuban species.

A2 A few lizards from Cuba may have accidentally spread to another island, for example clinging to a floating log after a storm. These lizards would have bred, producing a population. The conditions on the newly colonised island may have been different from the conditions on Cuba; so different characteristics were selected for. Over time and many generations, random genetic variation and natural selection would produce a population of lizards with a different set of alleles from the original population. Eventually, the lizards on the new island could become so different from the original population on Cuba that they could no longer interbreed successfully, even if they were brought together again. As conditions on each island are likely to be slightly different, selection pressures would also be different and so the characteristics of the populations on each island would not be the same.
Another contributory factor would be the set of alleles possessed by the original colonisers. By chance, the few lizards that initially colonised an island might not have all the different alleles in the original population on Cuba. This would affect the final mix of alleles in the population that grew on that island.

A3 The colour of the dewlap could be important in courtship, in attracting a female lizard to a male and allowing him to mate with her. So if the males in one population have a different colour of dewlap from those in another population, the females would mate only with males from their own population.

A4 They could bring together males and females of different species and see if they can interbreed to produce fertile offspring.

A5 More DNA analysis – perhaps of different regions of DNA – should be made of more individuals within the lizard populations, and also of more species of lizards in the islands of the Caribbean. This would give a fuller and more reliable picture of the relationships between the different species, which could then be used to determine the probable sequence in which one species arose from another. If it were still found that each species had DNA that was more similar to the Cuban lizards than to other species, then the hypothesis would be supported.

Chapter 9

1a If the flatworms used a taxis, they would use light receptors to compare the light intensity on each side, and move to the darker side. If they emerged in the light area they would immediately turn back.

1b If the flatworms used a kinesis, they would move in straight lines until they found themselves in a dark enough environment. They would then turn in tight circles and/or slow down their rate of movement. (This is what flatworms actually do.)

2 Positive phototropism will bring the stems out of the shade of the hedge so their leaves will receive more light for photosynthesis.

3a The roots will grow downwards into the soil, helping to stabilise the plant and to provide more opportunities for obtaining water and mineral ions.

3b Together with phototropism, this will help the stem to grow upwards into conditions of maximum light intensity.

4 Facilitated diffusion.

5 A stimulus results in the generator potential after a very short time interval. The generator potential has a shorter duration than the stimulus.

6 When in the brightly lit area, much of the rhodopsin in the rods has been broken down; time is needed to resynthesise it before the rods can become sensitive to light again. The low light intensity does not stimulate the cones, so although their pigment is resynthesised much more quickly, they are not of use in dim light.

7 Snow reflects almost all the light so the light intensity will be very high. The pigment in the cones is broken down faster than it is resynthesised. So the cones become inoperative and the person becomes temporarily blind since the rods also don't work in bright light.

8a Only the cones distinguish colours; the cones only work in bright light.

8b Several rods synapse with one bipolar cell, so sufficient transmitter substance is produced to depolarise the bipolar cell. A cone cell would not produce enough transmitter substance to depolarise its bipolar cell.

8c Each cone synapses with a single bipolar cell, so the information from it is discrete rather than being mixed with information from other cells as is the case with rod cells.

8d In bright light we can see colours right to the edge of our field of vision, not just at the centre, so there must be cones in all parts of the retina.

9a C, where there is the highest density of cones.

9b A–B and C–D because these have the highest density of rods.

9c We would not see the image because there are neither rods nor cones at B. B is the optic disc – the region where the optic nerve leaves the eye.

10 Red and green.

11 Because of the stimulation of red- and green-sensitive cones.

12 White light stimulates all three types of cone. Black objects do not reflect light, so none of the cones are stimulated.

13 These people lack red-sensitive cones, but the green-sensitive cones are stimulated by red light, so all the dots will appear green.

14

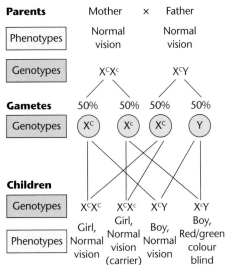

Parents	Mother	×	Father
Phenotypes	Normal vision		Normal vision
Genotypes	X^CX^c		X^CY

Gametes	50%	50%	50%	50%
Genotypes	X^c	X^c	X^c	Y

Children				
Genotypes	X^CX^C	X^CX^c	X^CY	X^cY
Phenotypes	Girl, Normal vision	Girl, Normal vision (carrier)	Boy, Normal vision	Boy, Red/green colour blind

15 Mutation of the gene for rhodopsin could have resulted in a different nucleotide sequence in the DNA.

This would add a different sequence of amino acids during protein formation. The resulting protein would have a slightly different shape and so would absorb different frequencies of light.

16 The pigment in the red cones becomes broken down faster than it is resynthesised. When you look at a white area, only the green and blue cones from this part of the retina send impulses to the brain, so the paper is seen as green-blue.

17 A greater rate of delivery of oxygen to the muscles and of removal of carbon dioxide from the muscles.

18a Rate of heart beat would reduce.

18b A lower rate of heart beat would result in a fall in blood pressure.

18c They speed up heart rate.

18d The arterioles dilate.

A1 Poor features of the design included:
Germination rate on Earth had not been tested before the experiment – in the event less than 50% of the seeds germinated in space and only 50% on Earth.
Only 3 seeds were planted in each compartment and some of these did not germinate.
There is no indication that the intensity of light reaching the seedlings was measured – only that it was different.

A2 The low germination rate and the apparently random growth meant that no valid conclusion could be reached.

A3 The response to different light intensities may be affected by lack of gravity. Growth in space was generally irregular indicating that gravity affects the translocation of growth hormones. Some seedlings grew away from directional light, the opposite to the response on Earth, indicating that both light and gravity are involved in the phototropic response. The taller growth of plants in space and the irregular growth might be explained by growth hormone remaining near its site of production at the tip rather than moving down the stem (towards the stimulus of gravity).

Chapter 10

1 Electrical insulation.

2a A neurone is an individual nerve cell; a nerve is a bundle of neurones.

2b **i** Sensory neurone – information will not pass from the receptor in the leg to the brain.
   ii Motor neurone – information will not pass from the brain to the leg.

3 30 m s^{-1}

4 $10/15 \times 100 = 66\%$ increase.

5 100 / 10 – 80 times faster

6 One advantage is that they can be grouped together in nerves without risk of impulses passing laterally between fibres. Another is that because they are much thinner, many more fibres can be gathered together in a nerve, so a nerve can carry much more information.

7 The fibre between the cell body of the sensory neurone and the synapse with the relay neurone, because an axon carries impulses away from the cell body.

8 As a flow of electrons.

9 Receptor.

10 Motor neurone.

11 Sodium, potassium and chloride.

12 Because of the impermeability of the myelin sheath, the neutral electrode would have to be positioned between the sheath and the cell membrane of the axon. This would be almost impossible to arrange.

13 Facilitated diffusion occurs along concentration gradients. The

K⁺ ATPase pumps produce a high concentration of Na^+ ions on the outside of the membrane and a high concentration of K^+ ions on the inside.

14 The Na^+ / K^+ ATPase pumps require ATP from respiration to operate. DNP inhibits respiration so there will be no ATP to operate the pumps. Without the pumps, the ions will only move through the channel proteins, and then only until concentrations on each side of the membrane are equal. At this time there will be no net difference in the number of ions on each side of the membrane, so there will be no resting potential.

15 **i** D, **ii** C, **iii** A, **iv** B.

16 110 mV (change from –70 mV to +40 mV).

17 0.5 milliseconds.

18 The internal temperature of invertebrates is largely dependent on the external temperature, and is usually lower than that of mammals. An increase in temperature will increase both the rate of diffusion and the rate of the enzyme reactions involved in active transport.

19 Receptors for the transmitter molecules are found only on the postsynaptic membrane.

20 The inhibition of acetylcholinesterase means that the neurotransmitter acetylcholine would not be broken down and removed from the synaptic cleft. The postsynaptic membrane would continue to depolarise and impulses would continually be transmitted along the postsynaptic neurone.

21 The sino-atrial node (SAN).

22 The beta-blocker molecules have a shape that will fit into the receptor molecules in the postsynaptic cells of the SAN. When noradrenaline is released by the nerve fibres, the molecules will

not be able to bind with postsynaptic receptors and so will have no effect on the SAN.

23a Because it has a complementary shape, caffeine binds to adenosine receptors at synapses and excites nerve cells.

23b Too much caffeine stimulation might cause the brain to reduce the number of adenosine receptors. A person then might feel lethargic without caffeine and have the urge to drink more.

24 Discussion question.

A1 Dopamine is released into the synapse at the presynaptic membrane. After binding with the dopamine receptor on the postsynaptic membrane, dopamine returns to the synapse where it is reabsorbed via dopamine reuptake receptors in the presynaptic membrane. The dopamine is then degraded by monoamine oxidase in the mitochondria.

A2 Dopamine will accumulate in the synapse, resulting in continuous depolarisation of the postsynaptic membrane.

A3 The change is not due simply to chance.

A4 Increase in use of cocaine; decrease in use of LSD, amphetamines, tranquilisers and cannabis. Overall decrease in drug use, bigger decrease in 16–24 group, but increase in use of Class A drugs in general population.

A5 Drug use generally higher in 16–24 group than in rest of population, in most cases at least twice as prevalent.

A6 The actual numbers were not measured but estimated from data obtained only from people who had been in contact with 'authorities'.

A7 People were asked about drug use in surveys. Their answers may not always be truthful since use of many types of drugs constitutes an offence.

Chapter 11

1 The fibre has many nuclei rather than one; the fibre is striped; the fibre is composed mainly of myofibrils; the fibre has an elongated shape.

2a This is the region where both actin and myosin filaments are found.
2b **i** Myosin filaments only.
ii Actin filaments only.

3 DOMS appears many hours after lactic acid has peaked then diminished.
Running downhill results in higher DOMS, but requires less energy and is therefore less likely to involve lactic acid production. Research indicates that damage to tissues initiates a chain of events leading to DOMS.

4 Unaccustomed muscle activity causes damage to the actin and myosin filaments. These proteins then break down causing the initiation of the inflammation that characterises DOMS.

5a No; actin and myosin filaments neither contract nor expand.
5b The actin filaments slide in between the myosin filaments so there is more overlap leading to a wider A band and a narrower area where there are just actin filaments.

6a Flow chart should follow this order.
Nerve impulse arrives at neuromuscular junction → acetylcholine released into synaptic cleft → action potential produced in plasma membrane → action potential transmitted down tubules into centre of myofibril → calcium ions released from endoplasmic reticulum → calcium ions bind to tropomyosin switch proteins binding sites on actin filaments exposed → myosin processes bind to exposed sites on actin filaments → energy from ATP used to move actin filaments.
6b To provide the abundant ATP required for muscle contraction; each swing of a cross bridge

needs the energy from one molecule of ATP.

7 Temporal summation, since impulses are arriving more rapidly from the same neurone to the muscle fibre.

8 Increasing the frequency of nerve impulses to a muscle fibre increases the strength of contraction; for a nerve fibre the number of action potentials produced would increase, but every action potential would be the same size. The nerve fibre has a refractory period but the muscle goes into tetanus if the time interval between impulses is very short.

9 Spatial – since it depends on the number of neurones feeding in impulses.

10 For example, shot, discuss, javelin, weightlifting, high jump, long jump, triple jump.

11a The liver.
11b Mainly under the skin.
11c **i** Energy can be obtained from it very quickly.
ii It is used up very quickly – there is not enough to sustain a long activity.
11d Lactic acid.
11e It dissociates into hydrogen ions and lactate ions. If hydrogen ions accumulate in the muscles they affect the tertiary structure of actin and myosin molecules, altering their shape and making contraction less efficient.
11f Slow-twitch fibres obtain most of their energy from aerobic respiration. Mitochondria are the site of aerobic reactions in respiration.

A1 Yes – endurance athletes such as long distance runners, canoeists and triathletes have highest proportion of slow-twitch fibres in relevant muscles, short event athletes such as shot putters and weight lifters have lowest proportion.

A2 Study A does not support the hypothesis, but study B does.

However, both studies involved too few athletes to give reliable results.

A3 There is insufficient evidence to support the hypothesis, but some indication that fibre conversion may take place under some circumstances.

Chapter 12

1 Its shape is complementary to that of receptors on the cell membranes.

2 Many different types of cells in the body have the same shape of receptors in their membranes and will be affected by the same hormones; whereas others have receptors with a unique shape and are affected only by one type of hormone.

3 Blood glucose concentration.

4 Enzymes are first released only when the sperm head contacts the granulosa cells, causing the acrosomal membrane to break down.

5 Polyspermy is prevented by the cortical reaction in which the zona pellucida develops a barrier to further sperm once one has entered. Enzymes from the cortical granules also destroy ZP3, so no more sperm can bind to the zona pellucida.

6 • Fewer sperm mean that the chance of a sperm reaching the egg is very much reduced.
• This reduces the quality and the amount of seminal fluid. This fluid stimulates the sperm and provides a suitable medium for them.
• Male hormone stimulates sperm production.

7 The receptor proteins on the sperm head bind to ZP3 initiating the acrosome reaction.

8 The body would produce specific antibodies against these proteins. The antibodies would lock onto their targets – the ZP3 receptors – preventing the sperm from reacting with the zona pellucida.

9a In early pregnancy progesterone maintains the uterus wall. If progesterone is destroyed by papain, pregnancy would not be possible.

9b It would be digested by protein-digesting enzymes.

10 The diploid set – 23 pairs.

11a Her reproductive life is (45 – 13) = 32 years. Her menstrual cycle was inactive for 3 years.
29 years every 28 days = 378 times

11b 45 years

12a It stimulates the endometrium to become thicker and more glandular. It stimulates the development of the corpus luteum in the ovary.

12b At low concentrations progesterone inhibits FSH; at high concentrations it inhibits LH.

13 Oestrogen.

14 Fertility drugs stimulate the production of mature eggs from the ovary, ensuring that eggs are available for fertilisation. Birth control pills work by tricking the body into thinking it is already pregnant; no mature eggs are released during pregnancy.

15 The effect of negative feedback is to bring about small changes; processes such as childbirth need massive physiological changes to bring them to their eventual conclusion.

16 Negative feedback is about fine tuning. It helps to regulate blood glucose levels within fairly narrow limits. The processes that are controlled by positive feedback are dramatic, all or nothing events that need to run their course.

17a The receptor is in the tip region.

17b The effector is in the region just behind the tip region.

18 Yes, because when the chemical messenger was prevented from reaching the effector by the mica there was no response.

19 A chemical diffused from the tip into the block. This chemical then diffused from the block into the decapitated coleoptile, stimulating growth.

20 By measuring the angle of curvature – the greater the concentration of auxin, the greater the angle.

21 To prevent lateral diffusion of the hormone.

22 Photo-destruction of the hormone on the illuminated side of the tip. Movement of hormone away from the illuminated side of the tip.

23 Gravity results in hormone moving towards the lower side of the stem. The increased concentration of auxin on the lower side stimulated growth there, so the stem curves upwards.

A1 Clomiphine prevents inhibition of FSH production by oestrogen. More FSH will be produced resulting in maturation of several eggs.

A2 Factors associated with the male.

A3a Effect of hormones varies with woman, so correct dose difficult to calculate – resulting in some failures.

A3b Screening removes 'unsuitable' embryos.

A3c Hormone level also affect 'readiness' of the uterine wall – many embryos fail to implant.

A4 Implantation. Embryo transfer rate mainly 80%+ of treatments, but successful implantation rate about 20% in most cases.

A5 Pros: higher chance that at least one embryo will survive.
Cons: Effect of failure traumatic on woman. Further treatments after failure expensive and traumatic. Possibility of multiple birth. Multiple pregnancy less likely to go to full term, so higher risk of premature birth. Difficulties for parents in raising several children of same age.

A6 Discussion question.

Chapter 13

1a Water, plasma proteins, sugars, lipids, mineral ions.

1b They could be damaged and induce clotting on their return to the body.

2 Absorption of benzodiazepines which can affect liver cells.

3 In solution in the water.

4 Facilitated diffusion.

5a New supplies from donors are required for each patient.

5b Mitosis.

5c Gene for cell division extracted from cells that normally continually divide (using endonuclease); gene inserted into plasmids using ligase; plasmids inserted into liver cells.

5d Cell division may go out of control producing cancer.

6 It will require energy since it is the opposite process to glycolysis which releases energy.

7a Production of digestive enzymes, e.g. peptidases, amylase and lipase.

7b Polysaccharide.

7c The highly branched molecule gives it a compact shape.

7d Hydrolysis.

7e Condensation.

7f The difference in concentration of the substance outside and inside the cell and the number of carrier protein molecules for that substance.

7g Glucose molecules pass through cell membranes by facilitated diffusion that involves carrier proteins. The more carrier proteins, the more 'passages' there are for glucose through the membranes.

8 Only liver cells have receptors complementary in shape to glucagon.

9a Sugary foods do not need to be digested, or they are digested rapidly by intracellular enzymes in the villi. This leads to sharp increases in blood glucose concentration. Starchy foods take longer to digest because starch molecules have longer chains.

Absorption of the sugars produced by starch digestion takes place over a longer period and does not lead to sudden rises in blood sugar.

9b Hypoglycaemia results from low blood glucose levels. These result in less than 60 mg dm⁻³ of sugar being delivered to the nerve cells in the brain. The symptoms are the result of brain malfunction caused by low blood sugar delivery.

10 A chain of amino acids held together by peptide bonds.

11 Endopeptidase, which hydrolyses peptide bonds deep within the molecule.

12 Both the types of amino acid and their sequence.

13 86 amino acids × 3 nucleotides = 258

14a Circular muscles in their walls contract.
14b The circular muscles relax; blood pressure then brings about widening.
14c

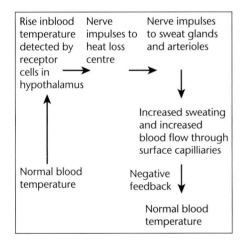

15 Sweat is soaked up by the sandwiched fibres. Because the inner layer is a good thermal conductor, it conducts heat from the body surface, cooling the body down. The conducted heat causes sweat in the fibrous layer to evaporate. The outer layer allows this evaporation. As heat is used to evaporate the sweat in this fibrous layer, more heat is conducted away from the body surface.

16 They have lots of muscle, so have a small surface area to volume ratio. This means they can lose less heat through their skin surface.

17 By the time the hypothalamus detects the fall, blood temperature will already have fallen. Blood temperature is maintained by thermal energy from respiration in all the body tissues. But if the blood entering the muscles is cold, the temperature of the tissues will fall causing a fall in the rate of respiration producing even less thermal energy.

18 Brown fat has the capacity to generate much more heat than white fat because it has far more mitochondria, the site of respiration in the cell; 80% of the energy transferred by respiration is transferred as thermal energy. It increases the rate of glucose uptake of the cells

19 Adrenaline increases the rate of glucose uptake by the cells, thus increasing the amount of oxygen for respiration.

20 As environmental temperature increases, endotherms require less oxygen for respiration to produce heat to maintain a high, constant body temperature. As the external temperature increases, enzyme activity in ectotherms will increase, including the enzymes involved in respiration.

21 During the night, the external temperature falls. Since ectotherms have little control over body temperature, the internal temperature of the crocodile falls. During the day, the temperature of the crocodile increases as the external temperature increases, up to a maximum of 30 °.

22 The 5 000 kg crocodile maintains its body temperature at the average daytime air temperature.

23 Yes, the larger the mass of the crocodile, the more stable its internal temperature.

24 The 10 000 kg crocodile has a much smaller surface area to mass ratio than the 100 kg crocodile and will therefore lose much less heat per kg body mass.

A1 Vasoconstriction – diverting blood away from the peripheries.

A2 This will divert blood away from the core and may cause rewarming shock.

A3a A drop in core body temperature caused by treatment for hypothermia.
A3b Cold blood returning to the heart may cause heart failure.

A4 Skin receptors detect increase in temperature, impulses sent to hypothalamus. Hypothalamus sends impulses to skin arterioles, diverting blood to skin. Less blood flows through muscles that are generating heat.

A5 Warm air is breathed into the lungs. This will warm the blood flowing through the capillaries that surround the alveoli. Nothing is done to the outside of the body to induce vasodilation.

Chapter 14

1 DNA molecules are longer; double-stranded instead of single; have deoxyribose instead of ribose; have thymine instead of uracil.

2 U G C U A A C A C G U G C U C

3 Both are single stranded, both contain uracil instead of thymine. mRNA is a long strand, tRNA is a clover-leaf shape; tRNA has an anticodon, mRNA does not.

4 AAC CAG CAC CUC UGC
UUG GUC GUG GAG ACG

5a 435
5b 435
5c 14

6a Lysine, arginine, serine, alanine.
6b TTC GCG AGA CGT
6c UUC GCG AGA CGU

7a The only codon would be UUU, which encodes for phenylalanine.
7b Lysine.
7c Serine and leucine. The codons would be alternately UCU and CUC.

8a Substitution, Addition, Substitution, Deletion.
8b Val, Val, Ser, Thr, Leu
Val, Ala, Ser, Thr, Leu
Val, Val, Phe, Tyr, Ser
Val, Val, Ser, Thr, Leu
Val, Tyr, Leu, Leu
8c One amino acid different.
All amino acids changed except first two.
No change (CAT and CAG code for same amino acid).
Only first amino acid the same.

9 Only mutations in developing ova in the ovaries could be passed on to children.

10 mRNA molecules transfer genetic information from a gene to a ribosome. Many are produced at each gene. Faulty mRNA molecules will only cause a small number of polypeptide molecules to be faulty in one cell.

11 p53 mutates → inactive proto-oncogene → active → uncontrolled cell division.

12a Substance Y is not made, so no substance Z can be formed from Y.
12b Enzyme A can still catalyse the production of substance X from substance W, but substance X is then not broken down.
12c Substance Y would be converted to substance Z, because enzyme B is bypassed.

13a Many plantlets from one parent; rapid; little space required; environmental conditions can be controlled for maximum rate of growth; plantlets are disease-free; sterile hybrids can be propagated.

13b All plants are susceptible to the same disease; growth in a laboratory is quite expensive, so only worthwhile for high-value plants.

14 Some people consider that human life begins at the moment of fertilisation and to use cells from the embryo infringes the embryo's 'rights'. There is also concern about the 'disposal' of unused embryos.

15 *Pros*: some successes, e.g. leukaemia treatment and growing skin for grafts; potential for using stem to study diseases and for screening drugs.
Cons: expectations of cures may be raised unrealistically; charging money for treatments where there is little chance of success.
Conclusion: much promise but cures for most diseases are a long way in the future.

16 Same shape as active part of oestrogen molecules, so able to block oestrogen receptors.

17a Single-stranded RNA that is complementary to mRNA.
17b It paired with the mRNA preventing access to tRNA molecules.

18 A control that contains all the necessary reagents except the independent variable reagent.

A1 Both block the effect of a disease agent.

A2a It blocks the genes necessary for the production of the compounds necessary to produce new virus particles.
A2b It is difficult to get the siRNA into cells where it would function.

A3a RNA silencing already occurs in liver cells.
A3b When cells die in this process, harmful chemicals are not released into surrounding tissues.
A3c Liver disease brought about by the action of the patient's own immune system.
A3d It reduced production of proteins that bring about apoptosis

A4a It protects the body's DNA from damage which results in tumours.
A4b The stem cells could replace the cells with damaged DNA.

A4c Genes that result in the production of much higher amounts of the substances they code for.

A5 There are encouraging results for experiments in vitro and using laboratory animals in the treatment of HIV, some forms of hepatitis and cancer. These successes have generated much further research, but the main difficulty in extending this treatment to humans seems to be delivering the RNAi into human cells.

Chapter 15

1a Cell membranes are composed of phospholipids. Detergents disrupt lipids.
1b Endopeptidase, or exopeptidase, or a combination.

2 Active site on enzyme; only a nucleotide sequence with the right shape will fit into this active site; therefore, the enzyme can only cut the DNA at this position.

3a A A T T
3b Only complementary bases will join together to produce a molecule with the sugar–phosphate backbones the right distance apart so that the molecule can twist into a regular double helix.

4

Bam I	C	and	CTAGG
	GGATC	and	C
Eco RII	C	and	GGACCG
	GCCTGG	and	C
Hind III	T	and	TCGAA
	AAGCT	and	T
Pst I	G	and	ACGTC
	CTGCA	and	G

5

TCC	and	AGGCCTGG – *Eco* RII
GGACCG	and	CTGCA – *Pst* I
ACGTCGGT	and	GCCAAGCT – *Hind* III
TCGAATC	and	TAG

6a 1124
6b 7 fragments: 30, 346, 1071, 622, 179, 1366, 748 bases.
6c *Bal* I and *Sna* I.

7 The bases in the mRNA correspond to specific bases in the DNA strand, i.e. A to T, C to G, G to C and U to A. This will be the order of bases on one of the strands of DNA. The other strand will be the corresponding match with DNA nucleotides.

8 51 amino acids, each with a 3 base code gives $51 \times 3 = 153$

9 To destroy the enzyme; otherwise it would cut the recombinant DNA at the junctions made by the ligase.

10a UAGCUGA
10b AUCGACU
10c Each pair of bases is complementary, so A links to U, etc.
10d The normal mRNA does not attach to the ribosomes. Therefore the enzyme polygalacturonase is not made. The pectin in the cell walls is not broken down, so the cells do not separate and make the tomatoes go soft.

11 Four, with each of the complementary bases, A, C, G and T.

12 The bacteria have enzymes that are stable at much higher temperatures than most. They are not denatured at 70 °C.

13 20

14 The enzyme only cuts the DNA at one particular sequence of bases.

15 Each copy of the DNA molecule will be identical. Each will be cut at the same positions, making fragments of the same length.

16 The backbone consists of a phosphate–sugar chain, so every nucleotide has a phosphate group.

17a 3. The smallest fragments are carried furthest through the gel.
17b 3
17c ATTGCAGTCGAC
17d UAACGUCAGCUG

18

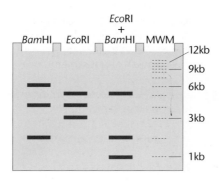

19

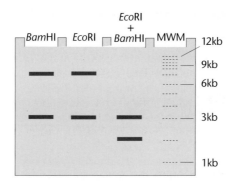

20a Substitution.
20b 394 amino acids, each with a 3 base code gives $394 \times 3 = 1182$.
20c The tertiary structure of the protein has a different shape because the substituted amino acid links up differently during folding. Therefore it does not fit into the active site of the elastase and act as an inhibitor.
20d The AAT gene is not affected, so patients continue to produce defective AAT.
20e No; someone whose emphysema is caused by smoking has the normal AAT gene already. Giving them treatment with AAT will make no difference.

21a The high concentration lowers the water potential. Water in the mucus layer tends to be reabsorbed.
21b Pancreatic enzymes cannot reach the duodenum. These include lipase, amylase and endopeptidases, so digestion of fats, starch and proteins are all affected. Fat digestion is usually particularly poor because there is no alternative source of lipase.

22 When the CFTR gene is transcribed, complementary mRNA is produced. If mRNA is present in a cell it shows that the gene is being expressed.

23 Composed with liposomes, adenoviruses are more efficient at getting the CFTR gene into the cells and may be more efficient at getting the gene expressed. However there is a possible danger that some unmodified viruses might cause infections, and patients may develop immunity to the viruses.

24 As the cells lining the airways die they are replaced by dividing cells from lower layers which do not contain the genes. Even if the genes were inserted into the cells in lower layers, they would probably not be passed on during cell division because the genes have not been incorporated into the chromosomes. Therefore the epithelial cells have to be re-sprayed with liposomes or adenoviruses at regular intervals.

25a These cells continue to divide so the healthy genes can be passed on to a good supply of cells.
25b The white cells will die due to the normal turnover of cells in the body. Most new cells will develop from untreated cells in the patient's bone marrow. Also, the genes are not incorporated into chromosomes and will probably not be effectively passed on during cell division.
25c The healthy genes would not be passed on, as only sex cells that develop in the reproductive organs are passed on.

26 Discussion question.

27 Discussion question.

28 The DNA of the inserted gene might, for example, be inserted into a section of DNA that forms a gene.

29 Gene therapy could introduce an antisense gene to viral DNA that would block its transcription and translation.

30 Genetically engineered Factor VIII is disease free, and could be cheap and widely available. It does not require donors.

31 Some haemophiliacs develop antibodies which are powerful and long lasting, making treatment more difficult.

32 Plasma contains the precursor molecules necessary for the synthesis.

33 Cells that would produce much more Factor VIII.

34a Produces a less severe immune response.
34b Changes to its molecular shape so that lymphocytes would be less likely to recognise it.

35a The time for 50% of it to be broken down is quite short.
35b Treatment would be needed less often.

36 Group discussion.

A1 Discussion question.

Chapter 16

1 Answer provided in text; 4665%

2 Answer provided in text; tree damage reduced by 9%

3 Answer provided in text; 0.6 mites/10 °C = 0.06 mites °C^{-1}

4 0.5 mites/5 °C = 0.1 mites °C^{-1}

5 8.5 / 1.4 = 6.1 MJ day^{-1}

6a The more days of snow the less the increase in population; above 12 days of snow a decrease in population.
6b Fewer birds to breed since food hidden under snow.

7 The greater the latitude of the northern limit, the lower the minimum temperature. The lower the minimum temperature, the lower the temperature below which the buds are damaged by frost.

8a The greater the area of the wood, the greater the diversity of bird species.
8b The woodpecker is a very much larger bird and will require much more food than the treecreeper. Greater competition for food between woodpeckers will mean that territories will be larger and population density therefore smaller.

9 Loss in mass: 60 g, time: 90 minutes. Rate: 0.67 g min^{-1}

10a Only A found, B does not survive.
10b Both species found; A confined to temperatures below 16 °C.

11a Light does not destroy auxin (difference in concentrations not significant).
11b Mica prevents redistribution of auxin.
11c Auxin migrates away from illuminated side.

12a 10 seeds per tray. Little competition, so no restriction to height grown.
12b In densely planted trays, conditions will vary from outside to inside – may be difficult to get a representative sample.
12c Plant 5 trays, each with 10 seeds then harvest all the plants.

13 0.62

14a The type of woodland does not affect the distribution of snails.
14b 2 categories (banded and unbanded); degrees of freedom = (number of categories – 1) = 1
14c 49
14d 27.4
14e p is < 0.0001 so the null hypothesis is rejected.

Appendix

No mark scheme supplied since there are several acceptable approaches to planning each essay.

Glossary

2,4-D
2,4-Dichlorophenoxyacetic acid; an artificial plant growth regulator, often used as a broadleaf weedkiller.

abiotic factor
Non-living environmental factor such as temperature and humidity.

absorption spectrum
Range of wavelengths absorbed by a pigment.

abundance scale
Set of letters or numbers used to represent the relative proportions of different species in a quadrat or other area being studied.

acetylcholine
Chemical that acts as transmitter substance in many synapses.

acetylcholinesterase
Substance formed from a two-carbon fragment of pyruvic acid and coenzyme A; the two-carbon compound that enters the Krebs cycle.

actin
One of the two proteins that bring about contraction of muscle fibres.

action potential
Change of potential across the membrane when a nerve cell is stimulated.

action spectrum
Efficiency of photosynthesis in a range of wavelengths of light.

acuity
Ability to use the eyes to make out detail; clarity of vision.

adaptation
Characteristic that ensures the survival of an organism.

addition
One of the three basic types of gene mutation; an extra nucleotide is inserted, so an extra base is added to the sequence.

adenosine diphosphate (ADP)
Lower energy form of ATP; consists of adenine, ribose and two phosphate groups.

adenosine triphosphate (ATP)
'Energy currency' of most cells; consists of adenine, ribose and three phosphate groups.

adenovirus
Virus that infects the membranes, for example, in the respiratory tract, intestines, urinary tract and eyes.

adrenal gland
Gland that secretes hormones to control important body functions such as heart rate and blood pressure; there are two – one just above each kidney.

adrenaline
Hormone produced by the adrenal glands; often known as the 'fight or flight' hormone.

adult stem cell
Undifferentiated cell in a tissue; multiplies by cell division to replace dead or injured cells in that tissue.

allele
Alternate form of a gene.

all-or-nothing principle
Principle which states that the size of the action potential produced by a nerve cell is independent of the size of the stimulus, once the required threshold has been reached.

alpha (α) cell
Cell in the pancreas that secretes the hormone glucagon.

ammonification
Process in which decomposers break down nitrogenous compounds into ammonia or ammonium ions.

amyloplast
Colourless organelle in plants, similar in structure to a chloroplast; stores starch.

anaerobic respiration
Respiration without oxygen; forms lactic acid in animals, and ethanol + carbon dioxide in plants and microorganisms.

antagonistic
Working against each other; for example, the actions of the sympathetic and parasympathetic fibres when SAN reduces heart rate; can also refer to 'antagonistic' pairs of muscles (such as biceps and triceps) in the upper arm.

anticodon
Sequence of three nucleotide bases in a tRNA molecule that bind to complementary bases on an mRNA molecule.

antisense RNA
Complementary copy of the base sequence on an mRNA molecule; used to deactivate action of a gene.

artificial fertiliser
Manufactured fertiliser such as ammonium sulfate.

ATP
see adenosine triphosphate.

ATPase
Enzyme that breaks down ATP, releasing energy.

atrioventricular node
Group of slow-conducting fibres between the atria and ventricles of the heart.

auxin
Plant hormone that affects processes such as growth and development.

auxin transporter protein
Protein that transports auxins across cell membranes.

axon
Long 'thread' that carries impulses from the cell body to the target cells.

barn egg
Egg produced by a hen that is kept inside but has enough room to move around and perch.

battery farming
Keeping hens, for meat or egg production, in small cages with little room to move.

belt transect
Line along which quadrats are placed to sample species distribution and abundance.

beta (β) cell
Cell in the pancreas; secretes insulin.

bioaccumulation
Increasing concentration of a pesticide or other chemical as it is passed along a food chain.

biomass
Mass of living material in an ecosystem.

biotic factor
Factor that affects a living organism and is caused by other living organisms, for example, competition for food or predation.

bipolar cell
Cell in the retina of the eye; connects a rod cell and cone cell with a ganglion cell.

blastocyst
Embryonic stage, before implantation, when the embryo consists of a hollow ball of cells.

BOD
Biological oxygen demand; rate at which oxygen is used up in a sample of water; a high BOD indicates dense populations of aerobic bacteria; this in turn indicates pollution.

broad-spectrum
Antibiotic used to kill a wide range of microorganisms; also used in terms of insecticides.

Calvin cycle
Series of reactions that take place in the stroma of a chloroplast, in which carbon dioxide is fixed and sugars are produced; the light-independent stages of photosynthesis.

cAMP
Cyclic adenosine monophosphate; chemical messenger, or compound that transmits information from cell membranes to cell proteins.

carbon cycle
Pathways by which carbon atoms pass between living organisms, and between living organisms and their environment.

carbon sink
Place in which carbon accumulates and is stored.

cardiac muscle
Muscle that makes up the wall of the heart; produces heartbeat.

cardioacceleratory centre
The part of the brain that sends impulses to speed up heartbeat.

cardioinhibitory centre
The part of the brain that sends impulses to slow down heartbeat.

carnivore
Animal that eats other animals.

carotenoid
Orange and yellow pigment found in plant leaves and other plant structures; absorbs green light.

carrying capacity
Maximum number in a population that can be sustainably supported in a particular place.

central nervous system
CNS; brain and spinal cord.

centromere
Region where the two chromatids in a chromosome are joined together.

chemical mediator
Chemical that carries information inside and between cells.

chemoreceptor
Receptor that is sensitive to different kinds of molecules.

chi-squared test
Statistical test to check if a null hypothesis is true.

chlorophyll
Green pigment that absorbs energy from sunlight; found in chloroplasts.

chloroplast
Organelle found in some plant cells; contains chlorophyll, and where photosynthesis takes place.

cholinergic synapse
Junction between two nerve cells where acetylcholine is the transmitter substance.

chromatid
Either of the two strands that make up a chromosome.

chromosome
Thread-like, gene-carrying structures found in the nucleus of a cell.

climax community
Final, relatively unchanging species that occurs in an ecosystem at the end of succession.

codominant
Alleles are said to be codominant if both have an effect on the phenotype of a heterozygote.

codon
Basic unit of the genetic code; consists of three nucleotides.

coleoptile
Protective sheath around the shoot emerging from a grass seed.

community
All the organisms, of all the different species, living in the same place at the same time.

competition
Interaction between two or more organisms that require the same resource which is in short supply.

complementary DNA (cDNA)
Form of DNA with a base sequence complementary to that of the DNA molecule from which it is produced (in a laboratory by the reaction catalysed by the reverse transcriptase enzyme).

cone
Cell in the retina of the eye; responsible for colour vision.

connective tissue
Animal tissue that connects and supports other tissue such as bone, tendons, ligaments.

constrict
Become narrower.

consumers
Organisms that eat other organisms.

coordinator
Brain and spinal cord receives impulses and forwards them to the appropriate organs.

corpus luteum
Structure formed from a follicle in the ovary after egg release.

coupled reaction
Metabolic reaction closely linked to another reaction, for example, the breakdown of ATP might release

energy that is immediately used to make another reaction take place.

cover trap
Stone or other object, raised slightly above the ground, beneath which small organisms may congregate and be captured.

cystic fibrosis
Genetic disease that results in excess mucus in the lungs and pancreatic duct.

dark adaptation
Changes in the iris and retina of the eye that allow better vision in dim light.

decomposer
Organism that breaks down the remains or waste products of other organisms.

degenerate code
DNA base sequence that codes for the same amino acid as a different DNA base sequence.

deletion
One of the three basic gene mutations; loss of part of a chromosome or part of a DNA base sequence.

demographic transition
Movement from, for example, a high birth rate to a low birth rate in a country.

dendrite
Highly branched, slender process from a nerve cell.

dendron
Process from a nerve cell that usually branches, forming thinner dendrites.

denitrifying bacterium
Bacterium that gets its energy by converting nitrates to nitrogen gas.

deoxynucleotide
Compound that contains a base, the sugar deoxyribose and a phosphate group; the building blocks of DNA.

deoxyribose
The sugar found in DNA.

depolarisation
Events at a nerve cell membrane resulting in an influx of positively charged ions, making the inside of the cell less negative.

detritivore
Organism such as an earthworm; feeds on the dead bodies or waste products of other organisms.

dicer
Enzyme used to chop mRNA strands into shorter lengths.

dilate
Widen.

diploid
Containing two sets of chromosomes.

directional selection
Type of natural selection in which organisms with characteristics away from the median value are selected for, so that the range of this characteristic changes over many generations.

disease
Impairment of a bodily function.

DNA polymerase
Enzyme that links nucleotides to form a complementary strand during DNA replication.

dominant
Property of one allele that suppresses the expression of another allele of the same gene.

dorsal root
Branch of a spinal nerve nearest to the back of the animal.

double digest
When two enzymes are used together to cut the DNA in restriction mapping.

double-stranded RNA (dsRNA)
Combination of sense and antisense RNA molecules.

dry biomass
Mass of living organisms after all water has been removed.

ecological factor
Environmental feature that affects living organisms.

ecological niche
Role of an organism in an ecosystem; what it needs to survive, and the ways in which it affects other organisms.

ecosystem
All the living organisms and non-living features in an area, that interact with one another.

ectotherm
Animal with a body temperature that varies.

edaphic
Relating to the soil.

edaphic factor
Environmental factor that relates to the soil.

effector
Structure that brings about a response, for example, a muscle or a gland.

efficiency of energy transfer
Percentage of energy that is passed on in a useful form, as opposed to that which is converted to other forms such as heat, or is lost from the system.

electron carrier molecule
Molecule situated in the inner membrane of a mitochondrion and the thylakoid membranes in a chloroplast; passes electrons from one molecule to the next and generates ATP.

electron transport chain
Chain of electron carrier molecules in the inner membrane of a mitochondrion or the thylakoid membranes in a chloroplast.

endocrine gland
Gland that secretes hormones into the blood system.

endometrium
The inner lining of the uterus (womb).

endotherm
Animal that maintains a constant internal temperature, for example, mammals and birds.

envelope
Two membranes surrounding an organelle; around a mitochondrion or a chloroplast.

ethanol
Substance produced along with carbon dioxide by anaerobic respiration in plants.

ethidium bromide.
Substance used to stain DNA fragments to make them visible.

eutrophication
Addition of nutrients to a body of water; often results in the growth and then death of algae and the growth of

large populations of aerobic bacteria, decreasing the oxygen concentration in the water.

exon
The part of a gene that is expressed.

extensive farming
Growing crops or rearing animals without large amounts of inputs such as fertilisers or artificial feeds.

fast twitch
Type of muscle fibre that reacts quickly, but is prone to fatigue.

feedback effect
Control mechanism whereby an increase or decrease in the level of a particular factor inhibits or stimulates production of that factor; important in the regulation of, for example, enzyme and hormone levels.

feedback system
see feedback effect.

fixed
Incorporation of carbon dioxide into a plant as it combines with RuBP; also the change of nitrogen into a compound such as ammonia, carried out by nitrogen-fixing bacteria.

follicle
Structure in ovary that releases eggs and produces hormones.

follicle stimulating hormone (FSH)
Hormone produced by the pituitary gland; stimulates egg maturation and oestrogen production in the ovary.

food chain
Series of organisms along which energy is transferred in the form of food.

fovea
Area of the retina with the most acute vision; consists almost entirely of cones.

frame quadrat
Area delineated by a square that is made of, for example, wood or wire, in which sampling of a community is done.

free-range egg
Egg laid by a hen that is allowed to move around outside.

fungicide
Chemical used to kill fungi.

gene
Part of a DNA molecule that codes for a polypeptide.

gene mutation
Change in the nucleotide base sequence of a gene.

gene pool
All the different genes, and alleles of those genes, present in a population.

gene therapy
Insertion of genes into a cell to replace defective genes.

genetic diagram
Conventional way of showing a cross in genetics.

genetic fingerprinting
Technique for analysing and comparing DNA samples with different origins.

genetic probe
Marker used to find the position of a gene in a DNA sequence.

genotype
Alleles possessed by an organism.

germ line
Cells that pass on from one generation to the next; gametes are part of the germ line.

glucagon
Storage carbohydrate found mainly in liver and muscle cells.

glycerate 3-phosphate (GP)
Three-carbon compound, formed in the Calvin cycle.

glycogenesis
Production of glycogen from glucose.

glycogenolysis
Breakdown of glycogen into glucose.

glycolysis
Conversion of glucose to pyruvate during respiration.

glycoprotein
Polysaccharide carbohydrate with protein attached.

gonadotrophic
Produced by the ovary or the testis.

Graafian follicle
see follicle.

greenhouse effect
Conversion of short-wave radiation to long-wave radiation at the ground

surface, and the subsequent trapping of much of this long-wave radiation by carbon dioxide and other greenhouse gases in the atmosphere.

grey matter
The part of the brain and spinal cord that contains mainly unmyelinated nerve fibres (fibres without fatty sheaths).

gross primary productivity (GPP)
Rate at which a plant produces organic materials through photosynthesis.

habitat
Place where an organism lives.

hair erection
Pulling upright of hairs by tiny muscles at their base.

half-life
Time required for half of an original unstable radioactive element to decay.

haploid
Cells that contain only one set of chromosomes.

Hardy–Weinberg equation
Equation that allows the calculation of the frequency of two different alleles in a population, if the proportion of homozygotes or heterozygotes is known.

heat gain centre
The part of the brain that sends impulses to the effectors that conserve body heat.

heat loss centre
The part of the brain that sends impulses to the effectors that lose body heat.

herbicide
Chemical used to kill plants.

herbivore
Organism that eats plants.

heterozygous
Possessing two different alleles of a gene.

histamine
Substance released by injured cells.

homeostasis
Keeping a constant internal environment.

homologous
Having the same genes in the same positions; in a diploid cell there are

homologous pairs of each type of chromosome.

homozygous
Possessing two identical alleles of a gene.

hormone
Chemical substance that is produced by a gland and travels via the blood to a target organ.

hyperglycaemia
High blood sugar level.

hypoglycaemia
Low blood sugar level.

hypothalamus
Controls body functions such as temperature control; found in the brain.

IAA (indolylacetic acid)
Plant hormone; its main function is to control plant growth.

indicator species
Species of an organism that requires particular conditions in which to live, which can therefore be used to indicate whether pollution has occurred in an area.

inner cell mass
Cluster of pluripotent cells inside the hollow sphere of a blastocyst.

insecticide
Chemical used to kill insects.

insulin
Hormone produced by the pancreas that lowers the blood sugar level.

integrated pest management (IPM)
Controlling pests by using chemical insecticides, herbicides or fungicides and also biological methods such as biological control.

intensive farming
Growing crops or rearing animals using high inputs such as fertilisers or artificial feeds.

internal stimulus
Stimulus from an organ in the body, such as the bladder when you need to use the toilet.

interrupted belt transect
Belt transect in which quadrats are placed at intervals along the transect.

interspecific competition
Competition between organisms of different species.

intraspecific competition
Competition between organisms of the same species.

intron
Non-functional part of a gene.

inversion
Reversal of the sequence of nucleotide bases on part of a chromosome.

involuntary muscle
Muscle that is not controlled by the brain.

iodopsin
Light-sensitive pigment found in cone cells.

kick sampling
Method of finding what is living in a stream by disturbing the stream bed and catching organisms in a net held slightly downstream.

kinesis
Change in direction as a response to a directional stimulus.

Krebs cycle
Series of reactions that takes place in the matrix of a mitochondrion, in which NADP is generated and carbon dioxide given off.

lactic acid
Formed in muscle cells by anaerobic respiration; contributes to muscle fatigue.

lamellae
Thin layers of bone or tissue.

leaching.
Washing out of soil.

life expectancy
Number of years one can expect to live.

ligase
Enzyme used to join together two pieces of DNA.

ligation
Surgical tying off of a vein or a duct.

light-dependent reaction
Stage of photosynthesis in which light energy is used to make reduced NADP and ATP, and to split water.

light-independent reaction
Stage of photosynthesis in which reduced NADP and ATP from the light-dependent reaction are used to convert carbon dioxide to sugars; the Calvin cycle.

limiting factor
Environmental factor that prevents a process (such as photosynthesis or population growth) from taking place any faster.

Lincoln Index
Formula used to calculate the estimated size of a population when using the mark–release–recapture technique.

line transect
Line along which living organisms are sampled to find out what lives there and the abundance of each species.

link reaction
Conversion of pyruvate to acetylcoenzyme A in respiration.

liposome
Microscopic pouch, bounded by a phospholipid membrane, used to incorporate materials into cells.

locus
Position of an allele on a chromosome.

Longworth trap
Human trap used to catch small mammals without hurting them.

luteinising hormone (LH)
Hormone produced by the pituitary gland; stimulates the release of an egg from an ovary.

marker gene
Type of gene, for example, an antibiotic-resistance gene, used to determine which bacteria have taken in modified plasmids.

mark–release–recapture
Technique used to estimate the population size of mobile animals.

mast cell
Cell that releases histamine during an allergic reaction.

mature mRNA
mRNA that has had its introns removed.

GLOSSARY

meiosis
Cell division in which the number of chromosomes is halved.

menopause
Stage in a woman's life when the menstrual cycle ceases.

menstruation
Monthly breakdown of the endometrium (lining) of the uterus during the menstrual cycle.

messenger ribonucleic acid
see mRNA.

microclimate
Particular climatic conditions in a small part of a habitat.

motor end plate
Termination of a motor neurone in a muscle.

motor neurone
Nerve cell that carries information from the central nervous system to a muscle or a gland.

mRNA
Single-stranded RNA molecule produced in the nucleus; carries information for synthesis of a polypeptide.

multiple allele
Several alleles that affect a characteristic, for example, height.

multipotent cell
Cell that is able to develop into several different types of cells.

mutagenic agent
Chemical or type of radiation that can induce a mutation.

myelin sheath
Insulating sheath around a nerve cell.

myofibril
Thin filament in a muscle; in bundles, which bring about contraction.

myometrium
Muscular wall of the uterus.

myosin
Thicker of the two types of protein filaments that bring about muscle contraction.

NAD
Nicotinamide adenine dinucleotide; a hydrogen acceptor (enzyme cofactor) involved in respiration.

NADP
Nicotinamide adenine dinucleotide phosphate; a hydrogen acceptor (enzyme cofactor) involved in photsosynthesis.

natural fertiliser
Substance such as manure; made from living material and contains nutrients which can help plant growth.

natural selection
Differential survival of organisms; having features that best adapt organisms to their environment, making them more likely to survive and reproduce.

negative feedback
A change in a factor brings about a reduction in that factor.

negative geotropism
Growth response away from the direction of the stimulus of gravity.

negative phototaxis
Movement of an animal away from the direction of a light stimulus.

negative phototropism
Growth response away from the direction of the stimulus of gravity.

net primary productivity (NPP)
Net gain of dry mass stored in a plant after it has used up some material in its own respiration.

neuromuscular junction
Junction between a nerve cell and a muscle.

neurone
Nerve cell.

nitrate
Salt of nitric acid composed of ions with the formula NO_3^-.

nitrification
Conversion of ammonia to nitrite, and of nitrite to nitrate.

nitrifying bacteria
Bacteria that convert ammonia to nitrite or nitrite to nitrate.

nitrogen fixation
Changing unreactive nitrogen gas, N_2, into a more reactive compound such as ammonia or nitrate, for use by plants.

Node of Ranvier
Small gap in the myelin sheath surrounding a neuron.

noradrenaline
Transmitter substance in some synapses; has the same effect on the body as adrenaline.

null hypothesis
A proposal stating that there is no statistically significant difference between two sets of results.

oestrogen
Hormone produced by the ovary that stimulates production of LH and stimulates the build-up of the inner lining of the uterus.

oocyte
Cell that divides by meiosis to form an ovum (egg).

opsin
Substance produced by the breakdown of the pigment rhodopsin in the rod cells of the eye.

ova
Plural of ovum (egg).

overgraze
To allow too many animals to feed on a limited amount of growing plants, so that the plant population cannot be maintained and the soil is unprotected, resulting in erosion.

ovulation
Release of an egg.

oxidation reaction
Reaction in which oxidation is gained or electrons are lost.

oxidative decarboxylation
Removal of carbon dioxide during an oxidation reaction.

oxidative phosphorylation
Production of ATP during respiration.

parasitism
Close relationship between two organisms in which one (the host) is harmed and the parasite benefits.

parasympathetic fibre
Type of fibre that slows down many body processes, for example, heartbeat.

percentage cover
Percentage of ground that is covered by a particular species.

peripheral nervous system
Parts of the nervous system, not in the brain or the spinal cord.

persistence
Long-term existence of a substance because it is not broken down, for example, an insecticide in a food chain.

pesticide
Substance used to kill unwanted organisms.

phenotype
Observable features or characteristics of an organism, produced by interaction between its environment and its genotype.

phosphate
Salt of phosphoric acid composed of ions with the formula PO_4^{3-}.

phosphocreatine
Also known as CP; reserve energy supply substance in muscle cells.

photolysis
Splitting of water, using light energy.

photophosphorylation
Production of ATP during photosynthesis.

photosystem
Group of chlorophyll and other pigment molecules found in the membranes inside a chloroplast.

pigment
Coloured substance, such as haemoglobin and chlorophyll.

pioneer species
Species which is able to colonise newly-disturbed habitats.

pitfall trap
Partly covered, small hole in the ground or sunken container, into which insects and other small animals may fall in order to be examined or sampled.

pituitary gland
Gland at the base of the brain which secretes several different hormones.

plantlet
Tiny plant produced by tissue culture.

plasmid
Small ring of DNA found in bacteria; plasmids often contain genes for antibiotic resistance.

pluripotent cell
Type of cell that can differentiate into several different kinds of cells, for example, different types of blood cells.

point quadrat
A quadrat that is so tiny that it consists of only one point.

polymerase chain reaction
Technique for producing large numbers of copies of a DNA fragment.

polynucleotide
Long chain of nucleotides.

population
All the organisms of a species that live in the same place at the same time and can breed with one another.

population pyramid
Graph that shows the number of individuals of different ages in a population at one moment in time.

positive feedback
When a change in a factor amplifies the effect of that factor.

positive geotropism
Growth response towards the direction of the stimulus of gravity.

positive phototaxis
Movement of an animal towards the direction of the stimulus of light.

positive phototropism
Growth response towards the direction of the stimulus of light.

predation
Killing and eating of an animal.

pre-mRNA
mRNA that contains both exons and introns, before the introns are removed.

pressure receptor
Usually in the skin, a receptor that is sensitive to pressure.

primary consumer
Organism that eats producers.

primary oocyte
Cell that undergoes the first meiotic division during the formation of an ovum.

primary productivity
Rate at which plants convert carbon dioxide to organic materials.

primordial follicle
Cell that begins the formation of a follicle in the ovary.

producer
Organism that takes energy from the non-living environment to synthesis organic substances from inorganic substances; plants are producers.

progesterone
Hormone produced by the corpus luteum in the ovary; maintains the endometrium of the uterus.

prostaglandin
Fatty acid molecule that acts in a similar way to a hormone.

proto-oncogene
Gene with the potential to become an oncogene, which is responsible for producing cancer.

pyramid of numbers
Graph that shows the numbers of organisms at different trophic levels in a food chain.

pyruvate
Three-carbon substance, which is the final product of glycolysis.

quadrat
Square-shaped area within which a community is sampled.

random sampling
Measuring the abundance of organisms at randomly-placed points within an area.

ratchet mechanism
In muscle contraction, a mechanism which ensures that the filaments move only in one direction.

receptor
Cell that is sensitive to a stimulus, for example, light or temperature.

receptor protein
Protein in a cell membrane which has a surface shape complementary to that of a chemical such as a hormone.

recessive
Having an effect only when no dominant allele is present.

recombinant DNA
DNA fragment, formed by cutting two DNA fragments, thereby joining them together.

redox reaction
Reaction in which oxidation and reduction take place, that is, electrons are transferred from one substance to another.

reduced NAD
NAD that has taken up hydrogen or electrons.

reduced NADP
NADP that has taken up hydrogen or electrons.

reduction reaction
Chemical reaction in which a substance gains electrons or hydrogen.

reflex action
Involuntary response to a stimulus.

reflex arc
Structure involved in a reflex action; shortest route between a receptor and an effector.

reflex escape response
Automatic response to remove an animal from danger.

refractory period
Time period after an action potential, when a neuron cannot respond to another stimulus.

replica plating
Technique in which bacteria are grown, for example, on agar containing an antibiotic, to find which bacteria have taken in modified plasmids.

repolarisation
Restoring the resting potential in a nerve cell after depolarisation.

reproductively isolated
Unable to breed with other populations, because of differences in, for example, behaviour or physiology.

resting potential
State of a nerve cell when positive ions have been pumped out, resulting in a negative charge inside.

restriction endonuclease enzyme
Enzyme used to cut DNA molecules.

restriction mapping
Using a range of restriction endonuclease enzymes to determine the sequence of bases in a DNA molecule.

retinal
Concerned with the retina of the eye.

reverse transcriptase
Enzyme involved in copying an mRNA code into a DNA molecule.

rhodopsin
Light-sensitive pigment in the rod cells in the eye.

ribose
Five-carbon sugar, found in RNA.

ribosome
Organelle concerned with protein synthesis.

ribulose bisphosphate
RuBP; the compound which combines with carbon dioxide inside chloroplasts during the Calvin cycle.

RNA interference (RNAi)
Using double stranded RNA to suppress gene action.

RNA polymerase
Enzyme involved in adding RNA nucleotides during mRNA synthesis.

RNA-induced silencing complex (RISC)
Group of proteins involved in silencing a gene.

rods
Cells in the retina used for vision in dim light.

rubisco
Enzyme that catalyses the reaction between RuBP and carbon dioxide in the Calvin cycle.

RuBP ribulose bisphosphate
Compound that combines with carbon dioxide inside chloroplasts during the Calvin cycle.

saltatory conduction
Rapid conduction of an impulse along a myelinated nerve cell; impulse jumping from one node of Ranvier to the next.

sampling technique
Method used to investigate small parts of a large area, in order to obtain estimates of the distribution and abundance of different species.

saprobiotic feeding
Feeding on an organic substrate by secreting enzymes outside the body, and then absorbing the products of digestion.

saprobiotic nutrition
see saprobiotic feeding.

sarcoplasm
Specialised cytoplasm of a muscle cell.

scavenger
Consumer that feeds on the remains of other organisms left, for example, by predators.

Schwann cells
Cells that surround myelinated nerve cells, insulating them.

SCID (Severe Combined Immunodeficiency)
A genetic disorder in which the immune system ceases to function.

secondary consumer
Organism that feeds on primary consumers.

second-messenger model
Production of a messenger substance inside a cell in response to hormone stimulation.

sense strand
DNA strand that codes for a polypeptide.

sensory neurone
Neurone that carries an impulse from a receptor to the central nervous system (CNS).

sessile
Remaining in one spot for most of the time, for example, sea anemones.

shivering
Involuntary muscle contractions to produce heat via respiration.

sinoatrial node (SAN)
Natural pacemaker of the heart, situated in the right atrium.

skeletal muscle
Muscle attached to the skeleton, which brings about movement of bones.

sliding filament hypothesis of muscle contraction
Thin actin filaments move across thicker myosin filaments to bring about muscular contraction.

slow twitch
Type of muscle fibre that contracts more slowly but is less prone to fatigue than fast twitch muscle fibres.

small interfering RNAs (siRNAs)
Short fragments of RNA used to silence genes.

snRNPs
Small nuclear ribonucleoprotein particles; help to remove introns from mRNA.

sodium dodecyl sulfate
Chemical used in electrophoresis.

spatial summation
Two or more neurones synapsing with a third neurone and producing sufficient transmitter substance to depolarise the third neurone.

species frequency
Percentage of quadrats in which a species is found.

species richness
Number of different species in an area.

stabilising selection
Natural selection which does not bring about a change in a population over time.

standard deviation
Measure of variation on a normal distribution.

sticky end
End of a DNA fragment which has been cut with a restriction enzyme, and which can be joined to another similarly-cut piece of DNA.

stimulus
Change which can be detected and which may invoke a response.

striped muscle
Muscle attached to the skeleton; striped in appearance.

stroma
'Background material' in a chloroplast.

substitution
One of the three basic types of gene mutation; one or more nucleotides in a nucleic acid are replaced by others.

substrate-level phosphorylation
Production of ATP by a reaction directly coupled to one of the steps in the Krebs cycle.

succession
Directional change in the composition of a community over time, caused by one set of species changing the environment in such a way that others are able to live there.

survival curve
Graph that shows the percentage of individuals of different ages in a population who are still alive.

sweep netting
Sampling organisms in a habitat by sweeping a net through vegetation.

sympathetic nerve fibre
Type of fibre that speeds up body processes such as the heartbeat.

synapse
Junction between two nerve cells.

synaptic cleft
Space between two adjoining neurones.

synaptic vesicle
Small sac at the synapse end of a neurone containing a transmitter substance.

systematic sampling
Measuring the abundance of organisms at regularly-placed points within an area.

systemic
Absorbed into a plant and transported in its phloem, for example, a systemic insecticide.

target cell
Cell that is receptive to a hormone.

taxis
Movement of an animal in response to a directional stimulus.

temporal summation
Production of sufficient transmitter substance to depolarise a neurone brought about by several impulses arriving at the synapse from one neurone in a short time.

tendon
Structure that connects muscle to bone.

tertiary consumer
Organism that eats secondary consumers.

thermal panting
Breathing quickly to lose heat from the body.

threshold level
Minimum level to bring about a response.

thylakoid
Membrane in a chloroplast, on which the light-dependent reactions of photosynthesis takes place.

thymine
One of the bases found in DNA.

totipotent
Able to form any kind of cell.

transcription
Copying of the DNA code to mRNA.

transcription factor
Protein that binds to DNA and stimulates transcription of the code for a particular polypeptide.

transect
Straight line across a habitat along which organisms are sampled.

transformation
Inserting DNA into a cell or a plasmid.

transgenic organism
Organism that has genes from another organism.

translation
Using the mRNA code to synthesise a polypeptide.

transmitter
Substance that carries information from one cell to another.

triose phosphate (TP)
Three-carbon sugar; the first carbohydrate to be made in photosynthesis.

triplet sequence
Sequence of three nucleotides in a nucleic acid.

tRNA
Transport RNA; an RNA molecule that picks up a specific amino acid and transports it to a ribosome for polypeptide synthesis.

trophic level
Step in a food chain at which an organisms feeds.

type I diabetes
Condition in which the pancreas does not secrete sufficient insulin.

type II diabetes
Condition in which the body cells cease to respond to insulin.

unstriped muscle
Muscle found in the internal organs, for example, the intestine.

variation
Diversity in a population.

vector
Structure, for example, a plasmid or a virus, used to transfer DNA into a cell.

ventral root
Branch of a spinal nerve nearest the underside of an animal.

voluntary muscle
Muscle that is controlled by the brain.

water trap
Container of water left on open ground to attract flying insects.

weed
Plant that grows where it is not wanted.

white matter
Parts of the brain and spinal cord that consist mainly of myelinated fibres.

zona pellucida
Glycoprotein membrane that surrounds a developing ovum.

Index

INDEX